THE RABBI'S BRAIN

Turner Publishing Company
Nashville, Tennessee
New York, New York
www.turnerpublishing.com

The Rabbi's Brain: Mystics, Moderns and the Science of Jewish Thinking

Cover design: Alex Merto
Book design: Karen Sheets de Gracia

9781683367130 Hardcover
9781683367123 Paperback
9781683367147 eBook

Library of Congress Cataloging-in-Publication Data
Names: Newberg, Andrew B., 1966– author. | Halpern, David, author.
Title: The rabbi's brain : an introduction to Jewish neurotheology / by Andrew Newberg and David Halpern.
Description: Nashville, Tennessee : Turner Publishing Company, [2018] | Includes bibliographical references and index. |
Identifiers: LCCN 2018020879 (print) | LCCN 2018021092 (ebook) | ISBN 9781683367147 (epub) | ISBN 9781683367123 (pbk. : alk. paper) | ISBN 9781683367130 (hardcover : alk. paper)
Subjects: LCSH: Judaism and science. | Neurosciences—Religious aspects—Judaism. | Brain. | Neuropsychology—Philosophy. | Psychology—Religious aspects—Judaism.
Classification: LCC BM538.S3 (ebook) | LCC BM538.S3 N49 2018 (print) | DDC 289.3/75—dc23
LC record available at https://lccn.loc.gov/2018020879

Printed in the United States of America
17 18 19 20 21 10 9 8 7 6 5 4 3 2 1

THE RABBI'S BRAIN

Mystics, Moderns and the Science of Jewish Thinking

Rabbi Andrew Newberg, M.D.
and **Rabbi David Halpern**, M.D.

Contents

Dedication

בִּלְבָבִי מִשְׁכָּן אֶבְנֶה לַהֲדַר כְּבוֹדוֹ, וּבַמִּשְׁכָּן מִזְבֵּחַ אָשִׂים לְקַרְנֵי הוֹדוֹ, וּלְנֵר תָּמִיד אֶקַּח לִי אֶת־אֵשׁ הָעֲקֵידָה, וּלְקָרְבָּן אַקְרִיב לוֹ אֶת נַפְשִׁי, אֶת נַפְשִׁי הַיְחִידָה

In my heart I will build a Tabernacle to glorify God's Honor, and in the tabernacle I will place an altar for the horns (of light) of God's splendor, and for an eternal flame I will take the fire (of passion) of the Binding (of Isaac), and for the sacrifice I will offer my soul, my unique/singular soul.

—Song by Rav Yitchak Hutner,
based on Rabbi Eliezer Azikri, Sefer Haredim

Often sung at the third meal of Shabbat, the above song stirs strong emotions of dedication and gratitude to God. It is appropriate then for us to give thanks at the beginning of this work to thank all of those that made this possible. To our families, both immediate and extended, thank you for supporting us through our journey in the creation of this new exploration. To our teachers, both in the past and present, your continued influence and lectures/discussions ignited the interest and curiosity that made this work possible. To our students, thank you for continuing to question, challenge, and bring new perspectives to everything we do on a daily basis. "Rav Chanina said: I have learned much from my teachers, more from my colleagues, and the most from my students" (Ta'anit 7a) Thank you for continuing to inspire us to study, innovate, and explore every day.

An Introduction to the Rabbi's Brain

Blessed are You, Adonai, God, Sovereign of the world, who has formed the human body with wisdom—and created in him many channels, and many openings. It is revealed and known before your throne of glory that if you opened one (of the sealed areas) or if you closed one (of the open holes) we would be unable to last and stand before you (even for a moment). Blessed are You, Adonai, healer of all flesh and wondrous actor. אֲשֶׁר יָצַר — *Asher Yatzar*

The above blessing was first described in the Talmud (*Berachot* 60b), in which the fourth-century Babylonian sage Abbaye taught that one should say these words after using the bathroom. Although this prayer was originally intended to be used privately, the prayer has become part of the morning liturgy. Psychologist Leonard Felder, who has written at length about Jewish spirituality stated, "I find it fascinating that in Jewish spirituality even the most private moment of releasing the toxins from yesterday's food is treated with mindfulness, appreciation, and deep compassion for the delicate and brilliantly constructed body we have been asked to care for by the hard-to-define Creative Source that infused us with so much life force energy."

In the Beginning . . .

Although my name is Andrew Newberg, I also have a Jewish name, which is Beryl Levy Ben Fischel HaKohen. Those who are Jewish probably understand the meaning of the last part of my name. For those who are not Jewish, the last part of my name, Kohen (plural: Kohanim), refers to a specific order of people in the Jewish population who are regarded as the rabbis or priests of the synagogue. Importantly, the Kohanim designation is passed on from father to father. Interestingly, my father used to perform several rituals as a Kohen. One of them, the *pidyon haben*, has to do with when a Jewish woman gives birth to a firstborn male by natural means. In this case, according to Jewish law, the father must "redeem" the child from a known Kohen for the sum of five silver shekels, which, for my father, was a nice amount of money.

I have always appreciated the notion that I not only am connected to the Jewish people, but specifically would be regarded as person who would seek a high level of scholarship and learning. And as I have engaged the field of neurotheology, the field of scholarship that links the brain and religion, I always have wanted to use it to embrace Judaism to see how far one might take it to address important issues and ideas in Jewish thought.

After all, the Jews represent a unique population to study. For one, they have been around for several thousand years and have maintained a tradition for as long, if not longer, than almost every group of people. They are also a small group relatively speaking, making up about 0.2 percent of the human population. This allows for more targeted investigations into how they think, how they believe, and how their brain and biology might be associated with Jewish beliefs. Jews are also unique in that they represent a cultural, ethnic, and religious group for the most part all wrapped into one. Thus, we have an opportunity to explore how these different elements of the Jewish people interact. And hopefully this information will be helpful for not only understanding the Jewish people, but for understanding the whole of humanity as well.

This latter point was reinforced powerfully for me one time in a particularly poignant sermon during a Rosh Hashanah service. The rabbi spoke about the notion that the Jews are sometimes considered, at least by themselves, to be the "chosen people." The rabbi argued that the point of this statement was not that

Jewish people were better than everyone, but quite the contrary. He argued that being chosen was supposed to show the rest of humanity that if this relatively small and lowly group of slaves can connect with God, then everyone else should be able to as well. Whether this interpretation is accurate theologically, for the purposes of this book the point is that by understanding the brain of one group of people, the Jews, we can actually understand how all of humanity, with the same basic brain functions, struggles with religion, spirituality, morals, and reality itself.

My name is David Halpern, but growing up and even now I have always gone by my Hebrew name, Dovid. Growing up within a traditional Orthodox Jewish home, I was exposed to a significant amount of rabbinic study as well as medicine. My father is a physician, and many others in my extended family, including both of my grandfathers, were rabbis. After studying abroad in Israel for two years at a yeshiva entirely dedicated to the study of Torah and Jewish texts, I became fascinated with Jewish theology and philosophy. Upon my return to Yeshiva University for undergraduate education, I became further engrossed in my studies and enrolled in the rabbinic training program there. At the same time, however, my interest in science and medicine in particular continued to grow, particularly my interest in the human brain, neuroscience, and psychology. I began to wonder just how these two worlds could remain separate for so long, when scientific research and principles about how we understand the world were so clearly highlighted in the religious topics I was studying.

My Judaism was and is a living and breathing reality for me, and it is the glasses with which I view the world. However, the world of neuroscience and neuropsychology continued to contain more and more intriguing questions for me. If there is free will, how does that pan out neurologically? How can religious identity be both an internal feeling as well as an external label when these two concepts of "religion" are so fundamentally different and are realized by using such different neurological processes? Jefferson Medical College (Now Sidney Kimmel Medical College) at the time recognized my interest in my Judaism and Jewish law to be something worthwhile and granted me a deferral to complete my rabbinic studies. Upon completion, I came back to Jefferson for medical school and engaged myself in my medical studies. Throughout the process, these questions continued to fascinate me, and when I met Dr. Newberg, we knew that we could explore this important field together with a unique approach combining a deep regard for both science and Jewish thought. This is the field of neurotheology, a field that seeks to find the link between the brain and religious phenomena.

Both of us started in very different locations for our search: Andrew came to the merger of Judaism and science from his scientific background, with a growing interest throughout the years on spirituality and helping forge the path for neurotheology. David learned and studied rabbinics, grew up in an Orthodox Jewish home, and then through his studies became more interested in the practice of science and medicine. These two converging paths led us to the thought, how did others come to their paths toward Judaism and spirituality? Part of the answer is in our biology, and part of it is in Judaism itself.

One day several years ago, while looking up new articles on the relationship between our biology and our religious or spiritual selves, we came across an article in which researchers performed a DNA analysis of Kohanim to see if they truly were more closely related at the genetic level than other Jews. The results were quite positive. There was a strong correlation on the Y chromosome, the gene passed on from father to father, in members of the Kohanim group.[1] On one hand, this supports the notion that the Kohanim go back thousands of years and have maintained some significant degree of genetic, as well as spiritual, identity. In fact, although this study relates to the Y chromosome, the more important point for our purposes is that it suggests that there may be something very specific in the genes of a Kohen (at least in the males who inherit the Y chromosome), something that makes them a natural religious figure. And if the Y chromosome is part of the source of this uniqueness, maybe there are particular genes that code for aspects of brain function that are shared by all Kohens. If this is true for the former religious ritual leaders of the Jews, could it also be true that rabbis themselves (both men and women) have genetic predispositions for choosing their line of work? Could it be that this underlying genetic similarity has led to the development of a rabbi's brain that is different from that of others? And could this have implications for the broader Jewish community in terms of various shared genetic patterns?

But to answer this question more fully, we also had to understand the mental material that arises from that brain. What are the thoughts, feelings, and experiences that Jews in general, and rabbis in particular, actually have? To that end we developed several online surveys attempting to understand the religious and spiritual beliefs and experiences that people have. And we were particularly interested to know what rabbis think, since the only thing we know in general is through sermons or writings. But what might we find out if we asked some very specific questions about the beliefs and experiences of rabbis? Our goal was to determine just how rabbis experience their Judaism and if they felt "called" in some way religiously or spiritually to make the sacrifices required to enter into a life of Jewish communal service.

More importantly, we wanted to know what they think about religion, and the big question—what do they think about God? Is God a being? A metaphor? An idea? And how do rabbis experience God in their everyday life or even during mystical experiences?

We can consider some fascinating questions such as whether there is a difference in the brain between a rabbi who is Reform, Conservative, Orthodox, or Reconstructionist. We can think about comparing the brain between a Reform rabbi and an Orthodox Jew who isn't a rabbi. We can look into other factors such as gender, socioecomonic status, or age to search for similarities and differences. Or perhaps it would be fascinating to observe how we might compare a rabbi with a priest or imam. And historically we might wonder whether Maimonides's brain was different from or similar to that of Saint Thomas Aquinas.

We might even ponder whether there is something unique about the rabbi's brain, or something that helps any individual embrace Jewish ideas and values. We would stress that a rabbi's brain or a Jewish person's brain may be different from, but not necessarily better than other brains. Calling someone's brain "better" is a notoriously difficult and subjective evaluation. But it certainly may be the case that Jewish people in general have brains that are different from those of other people. After all, there are unique aspects of the Jewish people. Of course, virtually every ethnic group shares unique qualities and abilities among its individuals. That is part of how any given group is defined. It is also the basis for many stereotypes. The focus of this book, though, is the uniqueness of the rabbi's brain in particular, and the Jewish brain in more general terms. By using neurotheology, we can consider many of the unique, ideological, and quirky aspects of Jewish people and the brain of some of their most highly religious and spiritual individuals.

Did God Create the Brain?

What is the actual beginning of the rabbi's brain? One neuroscientist, Frank Meshberger, gave a unique perspective on this question when he published the article "An Interpretation of Michelangelo's 'Creation of Adam' Based on Neuroanatomy" in the *Journal of the American Medical Association.*[2] He made the argument that the wonderful fresco in the very center of the ceiling, with God's hand stretching out to Adam, looks very much like a sagittal view of the human brain. There was the outline of the frontal lobe, the occipital lobe, and even the brain stem. Most importantly, the hand of God seems to be extending from the frontal lobe of this vague image of the brain. This had particular meaning for Meshberger, because he argued that the frontal lobes, above almost every part of the brain, are what make human beings human. The frontal lobes help us with our executive functions, help us plan our day, help us to organize our thoughts and behaviors, and help us to be compassionate to others. So perhaps the creation of human beings was not so much about the physical body as about the intricate complexities and workings of the human brain.

If it is the human brain that truly sets us apart from every other animal that has ever lived on the earth, it makes sense for us to explore how the brain works. This is where neurotheology comes in. Neurotheology explores the link between the brain and our religious and spiritual selves. Understanding this relationship helps us understand who we are, how we relate to the world, and how we relate to God, depending on your specific beliefs. From the Jewish perspective, understanding the brain helps to determine the various aspects of Jewish thought and identity. What is it that makes Jewish people unique? Ultimately, it may come down to the brain.

The topic of neurotheology has garnered increasing attention in the academic, religious, scientific, and popular worlds. Over the past twenty years, a number of books, research articles, and popular press articles have been written addressing the relationship between the brain and religious beliefs and experience. Overall, the scientific and religious communities have been very interested in obtaining more information regarding neurotheology. In fact, there is value in exploring this interdisciplinary field on many levels, from the very practical to the highly esoteric. Importantly, neurotheology is not about diminishing either science or religion, but rather about improving our understanding of both.

However, there have been no attempts at exploring how

particular religious traditions might approach neurotheology, and more specifically, how Jewish religious thought and experience may intersect with neurotheology. The purpose of this book will be to explore this fascinating area, linking the brain and religion, from a Jewish perspective. The topics to be included will be related to a neurotheological approach of the foundational beliefs that arise from the Torah and related scriptures, Jewish learning, an exploration of the different elements of Judaism (e.g., Reform, Conservative, and Orthodox), an exploration of specifically Jewish practices (e.g., *davening*, Sabbath, *kashrut*), and a review of Jewish mysticism.

Specific Jewish scholars will also be considered in terms of the relationship between their ideas/teachings and different brain functions. We can ponder how brain functions related to free will, emotions, and abstract thought have become interwoven with Jewish ideas. We will engage these topics by integrating the scientific, religious, philosophical, and theological aspects of the emerging field of neurotheology. By reviewing the concepts in a stepwise, simple, yet thorough discussion, we hope that regardless of your personal background you will be able to understand the complexities and breadth of Jewish neurotheology. More broadly, the issues we will consider include a review of the neurosciences and neuroscientific techniques; religious and spiritual experiences; theological development and analysis; liturgy and ritual; social implications; and epistemology, philosophy, and ethics, all from the Jewish perspective.

Science and Religion from the Jewish Perspective

Before we explore this Jewish neurotheology, which sits at the nexus of science and religion, it might be helpful to first answer the question: what exactly is the Jewish perspective on the science and religion debate in general? After all, neurotheology seeks to find a middle ground between science and religion, a ground that some might feel is highly useful, while others might feel to be useless. Judaism traditionally has been rather accepting of the larger scientific community compared to other religious groups. However, that does not mean that things were always assumed to be correct scientifically. The general Jewish attitudes can be divided into three main groups: a rational group, an indifferent group, and a spiritualistic group.

The rationalists, epitomized by Maimonides, the medieval Jewish scholar-physician, argue that anything that can be proved scientifically must be considered the truth. Maimonides stated in his *Guide for the Perplexed* (2:25) that even though the Bible states the world was created, he would be willing to understand the entire passage allegorically if it could be proved that the world was eternal. After all, he says, there are many passages in the Bible that describe God's physical and emotional traits, and tradition does not assume that these are to be taken literally. Therefore, whenever something known to be true conflicts with an opinion possibly presented in the Jewish tradition, Maimonides found no reason not to interpret that specific text or circumstance in a way that fit with the scientific theory of the time. However, he does add that if the scientific theory is not proved to be correct, he sees no reason why the biblical understanding must be disregarded in favor of something unproven. Modern thinkers like Dr. Gerald Schroeder use similar arguments, for example, by trying to link the days of creation to the theory of relativity. Schroeder argues that the Bible, taken from the perspective of the theory of relativity and the expanding space-time continuum, is describing the billions of years that passed since the big bang in a way that if we were to view it now, it would in fact be only six days.[3] While such approaches are highly controversial within the scientific community, they nonetheless demonstrate one way of interweaving science and religion.

Benedict Spinoza, in his work *A Theologico-Political Treatise*, takes Maimonides's conclusions one step further and states, "whatsoever is contrary to nature is also contrary to reason, and whatsoever is contrary to reason is absurd" (chap. 6, sec. 88).[4] He describes the Bible to be entirely true, but also cryptic, in order for the masses to remain inspired to serve God, through which everything that happens follows a natural course. Spinoza, while a Jew and raised among an Orthodox congregation, rejected the traditional religious doctrines of a personal God as well as the divine authorship of the Bible. Instead, he chose to embrace an entirely material world with a God and world that are predetermined, removing free choice from humans as an ephemeral perception, but lacking in reality. These attitudes ultimately had Spinoza exiled from his Jewish community. However, they are important, as they indicate the ultimate tension felt between the Jewish individual's need for rational explanation and spiritual connection. Spinoza was surely not indifferent, as he once stated:

> If anyone thinks my criticism [regarding the authorship of the Bible] is of too sweeping a nature and lacking sufficient foundation, I would ask him to undertake to show us in these narratives a definite plan such as might legitimately be imitated by historians in their chronicles. . . . If he succeeds, I shall at once admit defeat, and he will be my mighty Apollo. For I confess that all my efforts over a long period have resulted in no such discovery. Indeed, I may add that I write nothing here that is not the fruit of lengthy reflection; and although I have been educated from boyhood in the accepted beliefs concerning Scripture, I have felt bound in the end to embrace the views I here express.[5]

People in the indifferent group generally believe that the domains of science and religion do not perturb each other. This is similar to the suggestion of the famed anthropologist Stephen J. Gould that science and religion represent "non-overlapping magesteria."[6] They both say things about the world, but in completely different and ultimately noncompeting ways. An example of this approach in a recent Jewish thinker may include Rabbi Joseph B. Soloveitchik, who stated in his work *Lonely Man of Faith (pg. 10)*[7] that he has never really been bothered by issues such as evolution and creationism, or the general conflict of science and religion. To him, the issues were not theologically troubling enough to make him wrestle with the factual information in conflict. Of course, such a position might represent a naïve perspective, since it is not clear what threshold of troublingness is required to make something worthy of theological reflection. Rabbi Samson Raphael Hirsch of Germany (in his "Letter on Aggadah") discusses the question of the Jewish Talmudic sage's wisdom conflicting with known science. He states that while this might be a problem if science really proved things beyond dispute, it is well known that scientific theories in general are often changed after years of research, such as the older beliefs about the position of the earth in the solar system. As such, it is not hard to believe that as scientific beliefs change, the relationship between science and religion also might change. "If therefore, we find information in ancient works which contradict hypotheses of modern scholars, we need not hastily decide that those were false and these are true."[8]

The last position is that science cannot be trusted in general and should always be rejected and subservient to the religious perspective. The Rivash (Rabbi Isaac ben Sheshet Perfet, a Spanish medieval scholar, in his Q&A #447) states that scientists are not to be believed because they have false proofs and evidence, and it is better to rely on the words of the tradition and sages. The Midrash (*Midrash Tanchuma*) states that the righteous will live by their faith, a line that is often interpreted to mean that some faith is always needed to engage in religious practice. However, this position takes the interpretation of this passage to its extreme, claiming that it is *only* with faith and nothing else that one should ideally live. A perfect example of this position is that of the Klausenberger Rebbe (*Divrei Yatziv*, *Orach Chaim* 113), who states that if we ever think that the rabbis made a mistake in describing something at odds with science, we must believe that they had more knowledge than us and understood things we cannot comprehend. Absolute fealty to the religious tradition is kept with strict interpretation of the past teachings. This position is often assumed by the spiritualistic traditions in Judaism, especially those labeled kabbalistic and Hasidic. This may be due to their inherently antinomian way of looking at the world through a different lens or simply due to the fact that the kabbalistic explanation of the way the world functions exists independent of any scientific explanation of existence.

Consideration of the scientific and spiritualistic perspectives have permeated Judaism for centuries and continue to exert influence in various forms of Jewish practice and theology. Whether it is ritual practice, prayer and synagogue worship, communal affiliations and groups, or general theology, many different strands of Jewish thought have been woven with these two opposing ideas running together. As such, despite the fact that either the rational or spiritual understanding may be enough to explain the reasoning behind a specific practice, if one wants to study the entirety of Jewish neurotheology, one must integrate both of these approaches and recognize the murky middle ground in Jewish practice and theology in their study.

We might go even a step further and ponder what type of brain or brain processes leads one to find meaning in one of these above described perspectives on science and religion. Is the person who finds science and reason of greater value someone who uses his analytical mental processes more? Is someone who is more religious using the emotional or holistic centers of the brain more? These are the questions that neurotheology has the potential to explore.

What Is Neurotheology?

Neurotheology has been defined as the field of study that seeks to link the neurosciences with religion and theology. Neurotheology also has a broader focus than simply neuroscience on one side and religion and theology on the other. Neurotheology must take advantage of neuroscience, consciousness studies, psychology, anthropology, and the social sciences. And on the religion side, topics should include comparative religions, theology, spirituality, spiritual practices, rituals, beliefs, and faith among others. Neurotheology is a field that must preserve the integrity of science and the foundational principles of religion and theology. Thus, neurotheology should be considered a "two-way street," with information flowing both from the neurosciences to the religious perspective as well as from the theological perspective to the neurosciences. Such an approach seems consistent with Jewish thought in general, which fosters a deep spiritual belief system while fully engaging and exploring the natural world. How many great Jewish scientists also had strong religious or spiritual beliefs?

In a similar manner, research and scholarship in neurotheology should be defined broadly and can include scientific, medical, theological, sociological, anthropological, spiritual, and religious elements. At times this research might be more scientifically oriented—for example, a brain imaging study of Jewish prayer practices. Or it might be more theologically oriented—a dissertation on the implications of brain functions for understanding the nature of specific Jewish constructs such as the laws of *kashrut*.

Neurotheology asks scholars to evaluate religion and theology from a rational and scientific perspective. But neurotheology also recognizes that science might require evaluation from the religious or theological perspective. In addition, it is important to note that neurotheology is not beholden to either science or religion and hence does not specifically presume, *a priori*, that either the physical universe or God should have priority. Rather, neurotheology strives to determine the nature of that relationship and determine priority *a posteriori*.

Obviously, if someone identifies as religious or as a scientist, neurotheology does not ask for them to disregard their deeply held beliefs before engaging in its study, it simply recognizes that any study comparing and contrasting scientific and religious perspectives requires an independent foundation so that it is not to be entirely co-opted by either side.

Neurotheology is considered to have four foundational goals

for scholarship and these apply to Jewish neurotheology as well.[9] These are as follows:

1. To improve our understanding of the human mind and brain.
2. To improve our understanding of religion and theology.
3. To improve the human condition, particularly in the context of health and well-being.
4. To improve the human condition, particularly in the context of religion and spirituality.

These four goals are interrelated and also pertain in many ways to the fundamental goals of Jewish teachings. Interestingly, Judaism, like neurotheology, also spans concepts ranging from the highly esoteric or philosophical to the highly practical. And just like neurotheology, Judaism strives to find a way of integrating all of these concepts in a unified system. Thus, it might be highly useful and appropriate to work toward developing a Jewish perspective on neurotheology as well as a neurotheological perspective on Judaism.

Tapping into the Rabbi's Brain

Although we have considered the basics of neurotheology, we have to think about how we can use this approach to better understand the rabbi's brain. Many of the issues and topics we will consider bear directly on how people, particularly rabbis, think. But how do we get at the rabbi's brain? One approach is to continue to do the brain scan studies that we have done on hundreds of people doing all types of religious and spiritual practices. And we do plan to use that data to help deepen our understanding of how Judaism and the Jewish experience manifests in the Jewish people. But observing the brain is just one piece of this puzzle.

We need to take an important step toward understanding the

rabbi's brain by exploring what a rabbi actually thinks and feels. In fact, determining the phenomenology of religious experience is a fundamental requirement of any neurotheological scholarship. Without assessing what the person is thinking, we have no way of knowing how to interpret a brain scan. Just because we see the emotional centers become active on a brain scan does not mean we know what emotion the person is feeling. And even if we could relate one part of the brain to the experience of happiness, can we determine exactly *how* happy someone is? The only way to know that at all is to ask. But once you ask, the person is no longer feeling the happiness in the same way he was right before you asked. Similarly, you can't tap someone on the shoulder during prayer and ask how spiritual she feels. Once you ask, the feeling is already altered and you are only getting a post hoc evaluation by the person—that is, what she felt when she was feeling it.

In spite of these limitations, we still have to ask the questions. And as we began thinking about and writing this book, we realized that no one had ever systematically surveyed rabbis to find out what they were thinking. We just assumed that they were religious or believed in God or felt a certain way about themselves. So we decided to use a neurotheological approach and create a formal survey to pose a variety of pertinent questions to a large group of rabbis. We could then use this data to get a better understanding of their psychological and neurological status, as well as their overall beliefs about Judaism and even God.

Over the span of six months, we surveyed 160 rabbis of varied backgrounds and denominations. In our Survey of Rabbis, we received responses from 20 Orthodox, 23 Conservative, 37 Reform, and 59 Reconstructionist rabbis (21 described themselves as "other" or unaffiliated). Overall, there were equal numbers of male and female rabbis even though all of the orthodox rabbis were males. We wanted to know: Are some more optimistic, more talkative, more extrovertive than others? And are there distinctions between Reform, Conservative, Orthodox, or Reconstructionist rabbis? We also wanted to know whether rabbis always wanted to be a rabbi and what other factors may have led a person to becoming one. Most importantly, we wanted to know what they thought about God and how emotions, thoughts, and experiences contributed to their religious and spiritual beliefs. No survey is perfect, and ours, while reasonable in size and scope, still only obtained information from a limited number of rabbis. But like everything in neurotheology, it is only the beginning steps—and sometimes these early steps are the most fascinating simply because they have never been taken before. So, we will reveal the results of this Survey of Rabbis throughout this book as we explore the fascinating ways in which the brain enhances and limits a person's ability to engage the Jewish traditions, culture, and religion.

What Does the Rabbi Actually Mean?

One of Rabbi Nachman's (1772–1810) famous stories goes something like this:

Once upon a time there was a King who had a son who woke up one day, took off all of his clothing, and sat under the table clucking like a turkey. The King and his family were distraught, what had happened to their son? How was he going to rule the kingdom in such a state, and what could they do to cure him from this condition?

The King called his wise men, advisors, and got doctors, psychiatrists, and anyone he could to try to help his son, but nobody was able to touch the delusion that he was a turkey, and thus, nobody could help him. Every day he remained seated under the table, clucking like a turkey and requesting bird seed. The King announced a large reward for anyone who could help his son, but it seemed hopeless.

One day, a quiet wise man arrived at the King's court. "I can help your son," he said, "but you must promise not to question my methods until I am done, and nobody can disturb us." At this point the King had nothing left to lose, so he agreed.

The wise man immediately undressed and came to squat next to the prince under the table. "What are you doing here!?" asked the prince. "I am a turkey, just like you" replied the old man, and he began to cluck and move around the floor like the prince. Initially the prince was skeptical, but after some time, he was convinced the man was also a bird like him, and they stayed naked under the table.

After a few days, the old man asked for a shirt to be brought to him, and put it on. "What are you doing!?" exclaimed the prince. "Why I am putting on my shirt," said the wise man, "a turkey can wear a shirt if they want and still be a turkey, it doesn't change who I am." The prince thought about this for a while, and slowly nodded. After some time, he also asked for a shirt.

And so it progressed slowly over time, the wise man slowly put on more clothing, began to walk straight, speak normally, eat normally, and use utensils and objects like a person, and with each step convinced the prince that turkeys could also do such things.

Finding Definitions

Have you ever listened to the sermon of a rabbi, or any other clergyperson for that matter, and wondered, "What is he or she actually talking about?" The story above told by Rabbi Nachman of Breslov demonstrates how complicated and confusing a rabbi, and Jewish teachings, might be. The prince may have become a prince again, but he remained a turkey as well. What was the message of the story? Rabbi Nachman did not always specify, and it was left for the audience to interpret the message. In general, rabbis have been known to speak in parables and confusing examples. Maybe you interpreted the lack of clarity as your own fault in not being able to understand a knowledgeable rabbi. But one of the most common mistakes we all make when considering or discussing various issues of religious and moral importance is not defining our terms. After all, if two people are arguing over the existence of God, don't we need to know what each of them thinks God actually is? What if one person believes that God is a man with a gray beard floating in the clouds, while the other person thinks that God is a metaphor for understanding how to behave morally? In the debate as to whether God exists, these two individuals might agree or disagree, but never truly know what they are agreeing or disagreeing about! Religious and spiritual concepts are notoriously difficult to understand and define.

Part of the problem lies in where our definitions come from. One source of definitions comes from the Bible itself. Through its descriptions, stories, and dictums, the Bible helps to show what a religious person is, what a spiritual person is, and how all people should relate to God. Theologians and other scholars subsequently develop more refined definitions, in part based on the sacred texts and in part on their interpretations of those texts. In Judaism, the Midrash is an outstanding example of how various ideas and definitions regarding Jewish thought are conceived. The Midrash is essentially a genre of rabbinic literature containing some of the earliest interpretations and commentaries about the Torah, with the goal of helping to resolve various problems with respect to the ideas, beliefs, and laws arising from the Torah.

We might also compare definitions obtained in the Jewish tradition with those from other traditions such as Christianity, Islam, Buddhism, or Hinduism. We might wonder how the definition of God or religion compares between these different ideologies. Sociologists and anthropologists offer another approach toward developing definitions. For these scholars, religion and spirituality are part of a social or cultural matrix. Religiousness is not defined so much by the sacred text, but by how it is applied within the context of a society or community. Neurotheology takes definitions to another level by exploring not only what is understood from a religious or philosophical perspective, but from a scientific one as well. Thus, we might find a definition of religiousness that seeks to explore various cognitive, emotional, and experiential dimensions, particularly as they relate to the processes of the human brain. In fact, neurotheology would argue that all of these different perspectives on definitions should be considered as part of a multidisciplinary approach.

One of the primary principles of neurotheology is the importance of developing adequate definitions for a number of critical concepts such as mind, consciousness, religion, spirituality, and God.[1] This is an essential first step for opening a neurotheological discourse. While these definitions might be different depending on a person's particular perspective, we would like to consider a few key definitions in relation to Judaism and neurotheology.

Mind, Brain, and Soul

For the purposes of this book, we will try to refer to and define the "brain" as the conglomeration of neurons and support cells that exists within the human head. This also includes all of the neurotransmitters, chemicals, and blood vessels that make up and allow the brain to function. The "mind" will be defined as the thoughts, feelings, and experiences that a given individual may have. In general, these functions are "less tangible," since they cannot be measured other than by obtaining a first-person account of the thoughts, feelings, and experiences. The mind's functions are frequently associated with the structural and physiological processes of the brain, although how that relationship is understood is not always clear. Brain imaging studies and other methods for evaluating brain function can assess the physiological processes associated with our thoughts and feelings, but science has no definitive way of measuring the mind-related elements, since they are subjective. Because of this, it becomes difficult to ascertain whether the brain creates the mind, the mind creates the brain, the two are mutually intertwined, or the two are separated and only appear to be connected.

Consciousness is similarly difficult to grasp, since it too is a purely subjective experience. Like the mind, consciousness appears to be based, at least to some degree, on how the brain works. As with the mind, the crucial question is whether consciousness can be separated from the brain or is completely intertwined with the brain.

Consciousness should be described at least in comparison with "awareness." We will refer to awareness as the subjective perspective of things in the environment that are actually registered within our sphere of understanding. Awareness should be distinguished from the mere detection of things in the environment. For example, a thermometer can detect the temperature of the air around you but cannot actually "feel" cold or hot. So awareness is related to a personal experience of the environment.

Consciousness is typically defined more specifically as the awareness of the self. Thus, the individual is both the subjective experiencer as well as the thing that is experienced. It is this reflexive self-awareness that forms the basis of consciousness that we observe in ourselves. One important question for both science and religion is whether consciousness is related only to humans or might be found in other animals. There is some evidence that certain species, such as dolphins and some primates, exhibit functions that appear similar to consciousness.

From a Jewish theological perspective, it is important to recognize that there are various approaches to understanding the mind-body divide depending on the school of thought within Judaism one subscribes to. Almost all Jewish thinkers have focused on the fact that there is a distinction between our personal consciousness and the physical body. However, these two terms can also be linked together. The Talmud (*Sanhedrin* 91b) states that when a man is being judged in heaven, he will try to claim it was all his body and not his "soul." However, God places the soul on the body and judges both together, like a blind man and a lame man who team up to steal fruit off a tree are judged as one person for their crime. From this perspective, the body is seen as a component of human beings that is separate from the sense of self-consciousness but may in fact have its own consciousness or at least its own will and desires. These two components of a person may have independent drives and motivations, but neither can function properly without the other.

Thus, a man would not be a man without his mind, but he would not be a man if he was just a mind either. And the same goes for women. This understanding still allows for three general interpretations: (1) the mind can be something that is an epiphenomenon of the physical brain, independent from the "soul" of a person; (2) the mind may be another aspect of the soul, with the body simply a container; or (3) the mind is a unique emergent property of the soul interacting with the body that cannot exist without both components.

These three positions are not historically how the issue has been divided among Jewish theologians. Traditionally, the human soul and mind were analyzed within the assumption of the second position. The difference lay in the way the mind and soul interacted on the metaphysical level with God. Maimonides and Rav Saadiah Gaon, both Jewish medieval rationalist scholars, believed that the mind and soul were practically indistinguishable entities that encompassed all of our thoughts, feelings, and powers.[2] Practically, how the soul related on a metaphysical level to God was unfathomable, and therefore not discussable or arguable.

However, the mystical tradition had a slightly different take on the mind-soul relationship. Many of the great mystics in Judaism have used a conception of the mind-soul combined to describe the various levels of human consciousness and the ability to make choices on many planes of existence. Rav Chaim of Volozhin, in his famous work *The Soul of Life* (Gate 1: Chapter 15), described the classical kabbalistic understanding of the soul of a person as having many components from different spiritual and physical worlds. Much of the Hasidic literature on the soul describes the spiritual and physical as being at odds with each other, depicting each person's internal struggle to suppress our animal instinct and strive for a higher purpose. This is done with our consciousness, identifiable here as the mind as well. Ultimately, then, the mind is seen as the component of the soul that is easily viewable in the present realm of consciousness and has the job of reining in the other components of the soul. Rav Schneur Zalman of Liadi, in his book *Tanya* (*Likkutei Amarim* Chapter 9), which is studied today by Chabad Hasidim, writes that human beings have a divine soul that searches for God in the world and seeks to uplift everything we do. However, we also have a *nefesh habehemit*, an "animal soul," which is tasked with tempting us with all physical desires beyond what we should indulge in (similar to Freud's id). When we vanquish this animal soul and subdue it to serve God, we earn a higher spiritual existence within our own mind-soul.

Neurotheology suggests there might be a way to distinguish concepts such as an animal soul from a divine soul. Physical desires, particularly cravings for sex, food, or drugs, are a common area of research for present-day cognitive neuroscience. We can identify parts of the brain and different neurotransmitters that are associated with these bodily needs and desires. Thus, we might at least be capable of understanding how the animal soul relates to specific brain functions, even if we must leave the spiritual part of the soul to theology.

"Soul" itself is a fascinating term because it has come to mean many different things depending on a scholar's perspective. Plato and Aristotle both regarded the soul as the essence of the human being. Plato, in particular, considered the soul to include reason, emotions, and desires. However, it is not clear whether the soul can exist beyond the body. The present catechism of the Catholic Church defines the soul as "the innermost aspect of humans, that which is of greatest value in them, that by which they are most especially in God's image: 'soul' signifies the spiritual principle in humans." While these definitions clearly suggest that the soul is not the brain or the mind, the soul appears to have a deep relationship with the brain and mind. If the soul has reason, emotions, and desires and is the essence of who we are, then it seems apparent that the brain has an intimate relationship to the soul. Based on what we have stated previously then, Judaism would traditionally agree with the existence of a soul and with its intimate relationship with the human brain and mind. Traditional Judaism also believes in the immortality of the soul and of a spiritual life after death, though the details of such existence remain murky and inconclusive.

From a Jewish theological perspective, the rationalistic model of the soul by Maimonides follows closely to that of Aristotle. Maimonides affirms that the soul is a unique entity separate from the body and that it has total control over the body as well as the cognitive faculties of human beings. However, the singularity of the soul does not preclude it having separate definable components. In his commentary on the *Ethics of the Fathers* (Chapter 1 of "Eight Chapters" Maimonides introduction to Pirkei Avot), Maimonides identifies five components of the soul. They are the nutritive, which runs the body; the perceptive, which interprets the senses; the imaginative, which can fantasize, imagine, and plan; the emotional, which provokes responses, opinions, and desires about things; and ultimately, the rational, which integrates everything together. By these five ways of operation, the soul is able to comprehend the world and integrate knowledge toward human health and comprehension of life and God. This view of the soul is highly consistent with neurotheology, since, again, we have the ability to explore how the brain controls the body, how our emotions are associated with bodily processes, and the differences between emotions, perceptions, and thought.

This perspective is slightly different from that of the kabbalists, who see the soul as existing in many different spiritual planes. However, the *Zohar*, the mystical text of Kabbalah, states that the different parts of the soul come from five (similar to Maimonides's count) realms of existence that God created on the way to creating the physical world (See *Zohar* I:81a 205b-206a and *The Book of Reincarnation* by the Ari Chapter 29).[3] These five components are *Nefesh*, *Ruach*, *Neshama*, *Chaya*, and *Yechida*. The first level a person attains is that of *Nefesh*, from the world of *Asiya*, which means "action." This soul understands the physical world, and it is the piece that exists physically with a person's body and dies when the person dies. It may be the closest kabbalistic term to what scientifically is the "brain" or "body." The second component is *Ruach* on the plane of *Yetzirah*, or "formation," and perhaps closest to the "mind." This second level is where human emotional intelligence lies. The control of emotion as well as love and awe of God stem from this mental world and element of the soul. With greater effort, a person can earn the revelation of *Neshama*, paralleling the world of *Beriya*, "creation." This world is where intellect rules and understanding everything is key. This marks the end of the levels that people are accustomed to live in. These first three elements of the soul, *Nefesh-Ruach-Neshama*, are often used together to define the basic existence of man. However, the spiritualists go further and state that when one is able to fully submit oneself to total ego annihilation, then he or she may be able to attain the level of *Chaya*, paralleling *Atzilut*, the plane of emanation, which is entirely spiritual. Here things exist in an unrealized state, and there is pure divine emanation that one's soul is both made up of and can merge with. Lastly, the final level is that of *Yechida*—the God consciousness of the level of *Adam Kadmon*, the primordial human, that "world before worlds" as it exists outside of time and space, transcends all the worlds, and is never separated from God.

While the first three levels correspond well to the brain and mind as it is understood by the rationalist thinkers, albeit with a strong spiritual bend, the final two levels dealing with pure meditative practices and spiritual pursuits suggest the kabbalistic reference to a higher state of consciousness. This higher state is sometimes referred to in various spiritual circles as "pure awareness" or "pure consciousness."[4] In such a state, the person is believed to engage a truer experience of reality. It is the experience of awareness itself, sometimes even referred to as "God consciousness," that sees the universe as a highly integrated spiritual whole, potentially even unified with the metaphysical Godhead entirely.

Perhaps it is no surprise that human beings have come up with five parts to the soul. After all, we have these similar five basic networks for processing and regulating our basic body functions (including sex, eating, fighting), our perceptions (sight, hearing, smell, and taste), our intelligence and abstract thinking, our sense of connectedness and oneness with others or the world, and finally, our sense of self-transcendence. Each of these has specific brain structures and functions that are associated with them, which we will consider in more detail in subsequent chapters.

Thus, we might consider E. O. Wilson, who indicated that sociology recognized the belief in a soul as one of the universal human cultural elements.[5] Wilson also suggested that biologists investigate how human genes predispose people to believe in a soul. This belies the assumption that the soul is not supernatural, but a consequence of brain functions and ultimately genetics. This added analysis by which we consider various concepts such as mind, brain, or soul based on how our genetics might lead to such definitions lies at the heart of future neurotheological research. But neurotheology would also emphasize the importance of safeguarding all concepts of the soul, including those that are nonmaterial or spiritual. In fact, if research would show that certain genes, maybe even those on the Y chromosome of the Kohanim, predispose people to believe in a soul, this finding could be interpreted by a religious individual to mean that the body *requires* a soul to exist and that it is *hardwired*, not only as a means for survival, but as a natural consequence of actually having a soul that relates to God.

Religion and Spirituality

The terms "religion" and "spirituality" are essential to define when considering a Jewish neurotheology. After all, if we were to design some study to assess how the brain works in a religious person, we would clearly need to know how the term "religious" was being defined. Perhaps more important is how we might differentiate various religious individuals. For example, can a Reform Jew ever be as religious as an Orthodox Jew? Or is an Orthodox Jew by definition more religious? And while differentiating Orthodox from Reform might be easier, how might we distinguish a Reform, Conservative, and Reconstructionist Jew in terms of religiousness? The distinction between these approaches to Judaism might become even more complex if we were to ask which type of Jew was more spiritual.

Religion and spirituality are almost always defined with some degree of overlap, and the same would be true within the Jewish framework. From a Jewish perspective, "religion" or "religiousness" is often defined as the keeping of some level of the ancient legal tradition of Judaism. Many Jews will view religion as keeping all or at least some of the mitzvot, commandments believed by Orthodox Jews to have been given by God and then added to and guarded by the rabbis of subsequent generations. However, for many currently, religion also includes a general association with institutions that are related to Judaism. For example, in some circles, being a religious person means that one goes to synagogue weekly and listens to the rabbi's sermon, even if he or she keeps very little to none of the other traditional Jewish laws. Additionally, there are many levels to any group's perception of personal religiousness. In one setting it may mean coming once a year to synagogue and paying annual dues, while on the other extreme it may include going to a ritual bath every morning to prepare oneself before prayer and wearing the same garb that was worn by one's great-grandparents in Europe a hundred years ago.

"Spirituality" can be thought of as a very broad and diverse term. However, it is usually understood to mean one's personal connection to some form of God or personal otherworldly experience. Furthermore, many people consider spirituality to be a connection with the universe or creativity, especially if they are a scientist or artist. Spiritual strands within Judaism recognize spiritual experiences as important to developing one's own religious identity and, in Hasidic and kabbalistic thought, even essential to having any sort of "true" religious experience. However, the more rationalistic perspective in Judaism may not need spirituality in the sense of having an out-of-body experience or feeling that one is directly communicating with God. From the rationalist perspective, one traditionally needs to direct one's emotions and intelligence toward understanding God and worshipping God. The classical spiritual experience is almost never necessary for someone to be a Jew. One simply must follow the laws of Judaism, recognize that God commanded them, and that will suffice.

As we will do throughout this book, we must always consider how the brain handles such concepts. A person who has had a spiritual experience, which appears to be associated with a variety of neurobiological processes as well, is more likely to accept the importance of spirituality as part of Judaism. If a person has not had such an experience, a more rationalistic or practical perspective on Judaism might arise. Neurotheology would ask us to evaluate the rationalist and spiritualist perspectives both as separate ways and as integrative ways of engaging Judaism.

This leads us to consider one of the questions in our Survey of Rabbis. We asked them whether they had had any mystical experiences. We found that a little over two-thirds of the rabbis we surveyed reported having had a mystical experience. There was no differentiation in terms of denomination. Interestingly, almost 85% of male rabbis reported such experiences while 60% of female rabbis did. Although both men and women had a lot of mystical experiences, it is unclear why it was more common in men.

On one hand, the large percentage of mystical experiences in rabbis is not surprising, since rabbis are generally highly religious and/or spiritual. On the other hand, it is also interesting that about one-third of rabbis had not had a mystical experience and yet still considered themselves religious enough to be a rabbi. Thus, intense spiritual experiences are often a part of a person's religion, but by no means the only expression of religious beliefs. We will expand on the survey responses regarding these experiences in later chapters. But for now, you can see how our survey contributes to understanding religious and spiritual beliefs simply by asking the question.

"Religion" and "spirituality" may also be defined in a more scientific way. In such an approach, "spirituality" is usually reserved more for personal feelings and experiences about something sacred, while "religion" and "religiousness" typically refer to doctrinal concepts of an established tradition. There is obviously extensive overlap, however. In Judaism, like many other groups, when people are surveyed, their sense of spirituality and religious falls into a 2 x 2 box. Some people are both religious and spiritual, some consider themselves spiritual but not religious, others are religious but not spiritual, and atheists are typically neither spiritual nor religious. (Though some have argued that if one identifies as an atheist, they are by definition "religiously atheistic.") Interestingly, even in these categorizations, it is unclear what people actually mean when they respond to the questions "Are you religious?" and "Are you spiritual?" Do people even understand these terms when they think about their own beliefs and experiences?

Those people who are both religious and spiritual might be the most congruent in their overall beliefs and practices. They are spiritual in that they deeply feel their Jewish beliefs, and they hold as true the various beliefs and ideas that arise from the Torah. In most surveys, approximately two-thirds of people are in this category. The complete opposite category refers to atheists who are neither religious nor spiritual. They do not adhere to any defined religion, and they generally do not feel any kind of connection to something greater than the self, especially something supernatural. The third group of those who are spiritual but not religious has been increasing in recent years, and Jews are no exception. For these individuals, the specific doctrines, beliefs, and practices have lost their meaning, but the individual still craves for something spiritual—some feeling that connects him or her to something sacred and meaningful. These people frequently seek alternative expressions of spirituality through other traditions or even secular approaches such as meditation or yoga. Interestingly, many of these "new age" approaches are actually derived from ancient traditions such as Buddhism or Hinduism. And many Jews today seek to integrate such approaches into their lives.

Jews might have a particularly high percentage (relative to other groups) of people who fit the last, "religious but not spiritual," category, since many people consider themselves "cultural" Jews. These people feel a connection to the Jewish people, even though they do not personally ascribe to the specific elements of the Jewish faith. Within this category as well, Jews can be divided into those who believe that they are fulfilling their religious duties and maintaining religious beliefs and those Jews who continue to practice religion but have cognitively given up on its truth claims. This final group, if pushed, would not say they are truly religious, but they continue to enact "religious acts" every day due to a desire to remain within a given social and ethnic group. It is interesting to consider how a person's brain might operate to allow him or her to perform acts purportedly required by God without actually believing in God. How does a person's brain manage such inconsistencies? Certainly, there is substantial data to show how good the brain is at accepting apparently irreconcilable concepts. On the other hand, a brain scan study we performed on an atheist who meditated on the notion of God generally showed a brain that was unable to grasp that particular concept.[6] This person's brain did not activate during the meditation practice like that of other people who truly believed in what they were meditating upon. Thus, there appears to be something to be said for fully "buying into" the practices that you perform.

Historically, there are a number of scholars who have also grappled with the definitions of religion and spirituality. These might be helpful to briefly consider in the context of the Jewish experience of religiousness and spirituality. Importantly, many of these scholars tried to define religion based on concepts that can be tied to brain processes that underlie various feelings or thoughts. For example, William James defined religion as "the feelings, acts, and experiences of individual men in their solitude, so far as they apprehend themselves to stand in relation to whatever they may consider divine."[7] Friedrich Schleiermacher described the essential element of religion as an experience, a deep and transcendent sense of the divine also referred to as a feeling of "absolute dependence."[8] Rudolf Otto defined religiousness in terms of the "the Holy" (Das Heilige) and described the awe and mystery of the experience of God as the mysterium tremendum.[9] Carl Jung defined religion as

> a peculiar attitude of the mind which could be formulated in accordance with the original use of the word religio,

> which means a careful consideration and observation of certain dynamic factors that are conceived as "powers": spirits, demons, gods, laws, ideas, ideals, or whatever name man has given to such factors in his world as he has found powerful, dangerous, or helpful enough to be taken into careful consideration, or grand, beautiful, and meaningful enough to be devoutly worshiped and loved.[10]

These thinkers all line up with the spiritualists in Judaism, whose ultimate goal of religion is to allow one to experience the merging of one's soul with that of the Godhead as well as bringing God's metaphysical presence down into the physical world. These definitions also suggest the ways in which the brain is associated with this merging through the attitudes and feelings of the religious individuals who have them.

However, the sociologist Emile Durkheim defined religion as a unified set of beliefs and practices relative to sacred things, which are set apart and forbidden within a given society of culture.[11] Anthropologist Clifford Geertz defined religion as (1) a system of symbols that acts to (2) establish powerful, pervasive, and long-lasting moods and motivations in men by (3) formulating conceptions of a general order of existence and (4) clothing these conceptions with such an aura of factuality that (5) the moods and motivations seem uniquely realistic.[12] For both Durkheim and Geertz, religion was clearly constructed by human beings as a way of helping create cohesive social groups. More recently, scholars such as Scott Atran and Pascal Boyer have elaborated concepts of religion based on various evolutionary and neurobiological perspectives, usually endorsing group cohesion as an important element.[13] This is in clear distinction to the rationalistic thinkers within Judaism, who indicate that the sets of beliefs and practices have divine or at least strongly emotional significance, making their practice independent of simple group conformity. These Jewish thinkers view the goal of religion to be the perfection of the human intellect to better understand God, and in line with

Maimonides, who ascribed reasons to all of the commandments and practices in a similar vein. For these thinkers, the moods and motivations not only seem realistic, but are in fact real consequences of the actions on our emotional and intellectual life.

Another approach is not to necessarily define religion or spirituality in a specific way, but rather to consider the various dimensions that can be incorporated into religion or spirituality. From the neurotheological perspective, these dimensions or components of religion and spirituality can each be evaluated with respect to the tradition itself (i.e., theological analysis) as well as to relevant psychological and brain processes. As we review these dimensions, we can consider which brain areas might be associated with them, ranging from the emotional areas to the rational areas to the areas that support basic bodily functions.

1. **Religious belief:** This typically refers to the specific beliefs that are held as part of a particular religion and that define the religion. Traditional Judaism has been divided on how many beliefs are actually required to be considered religious. The number has ranged from one core belief in God, to Rabbi Joseph Albo's three principles (from *Book of Principles*), all the way up to Maimonides's thirteen.* Many modern branches of Judaism do not even require any of the ancient principles of faith in order to be considered a religious attendee, thus further complicating the definition with respect to Judaism. Additionally, people who are born to a Jewish mother do not have to believe anything to technically remain Jewish, thus creating another caveat within the framework of

---

* Maimonides's thirteen principles are as follows: (1) God exists; (2) God is absolutely one; (3) God has no physical body; (4) God existed before all else; (5) God should be the only one to pray to and worship; (6) there is true prophecy and prophets; (7) Moses was the greatest prophet who ever lived, and no prophet after can undo what he has said; (8) the Torah is from heaven; (9) the Torah will not be changed later on; (10) God knows the actions of people; (11) God rewards and punishes; (12) the Messiah will come; (13) the dead will be resurrected.

what constitutes necessary belief for being Jewish. Nevertheless, these beliefs still remain important as defining pieces that must be dealt with by any religiously minded individual or theologian. These beliefs can also be considered from a neuroscientific perspective, since the beliefs must be simple enough to be grasped by individuals following the religion and must make cognitive sense in the context of how the human mind perceives the world. The notion of a belief in one God requires the holistic processes of the brain to consider a God that extends and incorporates the entire universe in a divine oneness. Other beliefs such as whether the Torah is from heaven require different cognitive processes in order to assess causality and agency—that is, the Torah is caused or comes from God in heaven.

2. **Religious affiliation:** In neurotheology, affiliation first implies which general religion an individual follows, such as Judaism or Christianity. But in the more specific context of Judaism, affiliation refers to the different branches, all of which have very different understandings theologically and practically about what the Jewish religion and religious practice is about. For example, Orthodox Judaism places more emphasis on the traditional mitzvot as a means toward proper ethical development, while Reform Judaism is more about the ethical principles derived from the Bible and practical deeds that one can do to continue to live by those principles. It might also be important to determine not only what a person is today, but how that person evolved throughout life. An individual may have been raised in a Conservative household, became agnostic while in college, and then eventually became Orthodox after a trip to Israel. Our Survey of Rabbis provided evidence for such shifts; for example, most Reconstructionist rabbis were not Reconstructionist to begin with, but about half were Conservative and half were Reform. In fact, the data show that the majority of shifting appeared to come from the Conservative group. Most Orthodox remained Orthodox (73 percent), and most Reform remained Reform

(52 percent), although about 46 percent of the Reform became Reconstructionist. By contrast, only 32 percent of the rabbis in our sample who were raised Conservative stayed Conservative, with 52 percent of those raised Conservative becoming Reconstructionist rabbis, 14 percent becoming Reform, and only 2 percent becoming Orthodox. It would be fascinating to use a similar survey to evaluate everyday Jewish people to see what path they took toward becoming their present affiliation. In addition, it is important to know how an individual understands that affiliation—does she consider herself Jewish because her parents were Jewish or because she goes to synagogue every Shabbat?

3. **Organizational religiosity:** This aspect of religion relates to its doctrines, beliefs, rituals, and system of adherence. Taken together, these elements form the organizational or structural elements of a given religion. Organizational religiosity also relates to affiliation, since an Orthodox Jew is likely to have a very different perspective on the organizational elements of Judaism compared to a Reconstructionist individual. In fact, Judaism is probably one of the best examples of a significantly organized religion. Within the hierarchy of religious scholarship, there are different actions or rituals one may perform depending on the time period in which one lived, the personal level of scholarship, and one's family lineage (Kohen/Priest, Levite, or general Yisrael/Jew). The belief system that developed after the destruction of the Second Temple with the rise of the rabbinic Talmud and its series of developing levels of authority and differing positions is a prime example of how, within an organized structure, different groups of Jews can interpret the rules in a more or less stringent manner.

4. **Non-organizational religiosity:** Many individuals consider themselves Jewish without necessarily taking part in the more formal organizational apparatus of the Jewish religion. Usually this refers to the individualized aspects of Judaism rather than aspects specifically related to the synagogue. Of particular

interest would be the traditional Jewish prayer group, the minyan of ten adults. Although this group can and often does meet at the synagogue, technically whenever there are ten men together (for Orthodox, or a combination of ten men and women in the other denominations), they can make a minyan and pray wherever they choose. In addition, the daily prayers can be said by individuals without necessarily anyone else being around. This can change fundamentally the rigid association one might have with prayer as an organizational function. Whether one has a preference for the organizational or non-organizational elements of any religion might depend on a variety of neurophysiological factors, such as how important a person values his or her individuality versus social interaction, as well as how important it is for the individual to feel connected to a formal versus informal group.

5. Subjective religiosity: This is closely related to the spiritual elements of religion, since it describes how individuals experience their religion. The subjective experiences that people have can vary widely both within and across traditions. For some, the subjective aspect of Judaism might be relatively minor and manifested by brief feelings of a need to repent while in temple on the Days of Awe or by feeling happy and excited while dancing with the Torah on Simchat Torah. For others, religion might be a calling that they sense and follow throughout their life, and by simply studying the Torah they feel a spiritual connection. Still others may have powerful mystical experiences that form the basis for their religious or spiritual beliefs, as many of the kabbalists do. These subjective aspects of religiosity are an important target for Jewish neurotheological research and scholarship because they can be evaluated both through first-person reports and ultimately compared with neurophysiological and other objective measures.

6. Religious commitment/motivation: What motivates people to be religious is also an interesting issue to be considered in neurotheology. Motivation is a psychological concept that is also based in brain chemistry. People can be motivated to follow religion out of guilt, anger, fear, love, personal experience, desire of social belonging, and many other causes. Many of these religious motivators are present in Judaism, particularly the fact that Jews are often born into the tradition and raised among a group of like-practicing individuals. Being raised with a belief in ultimate reward and punishment can also be a strong motivating factor for some Jewish groups, as the desire to earn reward and avoid punishment links the spiritual endeavor to very physical realities. Additionally, the spiritual love one feels in a state of spiritual ecstasy can also be translated after such an experience into a desire for a life of service to the Godhead that one feels responsible to and interconnected with. Since rabbis appear to be particularly committed to Judaism, we tried to understand where that commitment came from. One of the questions we asked in our survey was whether they felt "called" to be a rabbi. The result is that 57 percent said they had felt called to be a rabbi. Interestingly, women were more likely to feel called (67%) compared to men (47%). It would be interesting to try to determine if there are specific aspects of the female brain that might lead to a higher prevalence of such an experience. For example, one hypothesis might be that the generally higher social functioning in women might lead to a more personal feeling of "being called". If this were the case, it would be helpful to compare men versus women in other professions that are associated with a feeling of a calling to see if there is a similar predilection in women.

Of those rabbis that felt called, for many it was a sense of wanting to teach, particularly Jewish ideas, or to serve the Jewish community. A few indicated that they had spiritual or mystical experiences that led them down the path to becoming a rabbi. On the other hand, 43 percent indicated that there was no sense of a calling, suggesting that they simply felt it matched their interests and goals in life without necessarily feeling a pull. Interestingly, none of the rabbis specifically indicated that they felt God calling them. But such an experience is not typically described in

Jewish teachings. In contrast, many Christians speak about directly interacting with Jesus or God, who brings them into the Christian faith. For the rabbi then, the sense of communal responsibility and "being a link in the chain" may have a stronger bearing on their commitment and desire to lead than any feelings of being personally chosen or spoken to by God. From the neurotheological perspective, it could be fascinating to delineate how the different feelings associated with commitment to the Jewish religion affect the brain. Are there reward circuits that make a person feel emotionally positive about being Jewish? Is it more cognitive? And what would be the difference in people motivated to follow Judaism out of guilt, anger, fear, or love?

7. Religious well-being: Well-being can refer to the physical, psychological, spiritual, or religious condition of a person. Physical well-being typically refers to an overall healthy individual without significant medical problems or symptoms. Psychological well-being usually refers to a person who is generally happy and high functioning and has no significant depression, anxiety, or other neurotic or psychotic symptoms. Religious and spiritual well-being are more difficult to define. People who are religious are not necessarily more happy or satisfied than people who are not religious. Perhaps people who are more religious feel uncomfortable with their place in life and that is why they pursued religion. People can also struggle with their religious beliefs, leading to a kind of existential crisis. Religion itself can lead individuals to very positive or very negative thoughts and behaviors. Sometimes these negative thoughts can be encouraged by a religion, such as committing violence against opposing religious groups, and sometimes the negativity can be more individualized, such as when people think that God is punishing them. Individuals can also have positive experiences that differ from the doctrines of their religion and thus cause substantial anxiety as they try to relate those experiences back to their initial beliefs. Religions can also support positive self-esteem, optimistic beliefs, and a sense of love and compassion. While Judaism does promote much in terms of religious well-being, mandating one to take care of their own body and allowing desecrations of the Sabbath for both physical and mental health, it is not without its own potential for inner psychic conflicts. Some of the biblical laws about destruction of the entire nation of Amalek or the seven Canaanite nations have continued to be sources of conflict for the Jew seeking well-being as an ultimate goal of religious identity (See Exodus 17:14–16, Deuteronomy 20:17–18 & 25:19, and Samuel I 15:3).[14] In this regard, Reform and Conservative Judaism modify the rules and regulations of the Bible, marginalizing their significance as well as their impact on the practitioner's psyche. For the Orthodox Jew, however, the struggle remains a palpable one on the forefront of religious consciousness and has been the subject of much debate and conversation regarding ethics independent of the halachic system. It is interesting to note that the name Israel given to Jacob in Genesis 32:29 is based on the Hebrew word to "struggle with God," suggesting the importance of religious struggle. The question from a theological, as well as physiological, perspective is what kind of balance should be achieved between religious struggle and well-being?

8. Religious coping: Many people turn to religion to help cope in times of crisis. People can put their suffering in perspective and can deal with that suffering in a more adaptable manner through religious beliefs and activities. A classic example of this within Judaism is the idea of the shiva when someone dies. The children, parents, and spouse of the individual sit for seven (*shiva* is the number seven in Hebrew) days in the home and are comforted by visitors from the community and tell stories about their lost loved one. This time period is often seen as a positive way for people to deal with the trauma of the loss of a loved one while continuing to move forward with life and reintegrate themselves into the community. Knowing that there are caring individuals in the religious community can also help someone recognize that there may be some spiritual being who cares for the person as well. From a brain perspective, effective coping strategies can take advantage of both cognitive and emotional processes. And the social support offered through religion can be extremely helpful for reducing stress and depression. In fact, the brain has specific areas for social interactions that support feelings of empathy and compassion while reducing negative psychological symptoms.

9. Religious knowledge: Religious knowledge provides insight into many aspects about the world based on a particular religious doctrine. In Judaism, the Torah and other religious writings and sermons can provide information and purpose regarding the physical and metaphysical world. These sacred texts, and their interpretation, thus provide knowledge regarding the workings and origin of the physical world and also explain our role within that world. Religions certainly provide knowledge regarding how human beings are to relate to God. In this way, the religion explains what human beings need to do and think in order to connect to God. The laws of prayer and religious acts are prime examples of this in Judaism. Religions can offer information on how human beings are to interact with one another. For some, this might be antagonistic, and for others, this might be compassionate. Finally, religions can establish a system of ethics, such as through the biblical Ten Commandments or rabbinic *Ethics of the Fathers*, which can help us understand how to be good or bad people.

10. Religious consequences: Religions also provide a sense of consequences that are associated with various types of thoughts and behaviors. In this way, a given religion tells us what happens if we are a good person and what happens if we are a bad person. Studies have suggested how the human conscience forms within the brain as it triggers emotions such as embarrassment or guilt when we do something wrong. Religions also provide some information about what happens to all human beings "in the end." Thus, Judaism and many religious systems, have some form of a Judgment Day that will determine the fate of a person and humanity. Interestingly, there are models that focus far more on the life after death as the ultimate fate of the person rather than something that happens in this living world. Maimonides believed that in the end of days, the world will not change except that people will understand that God exists and strive together for truth and understanding. The Talmud states a position that there is no difference between this current world and the ultimate end world except for the oppression by kingdoms (*Berachot* 34b). However, other traditions within Judaism have the end of days being filled with wonders and a massive war between two large kingdoms that consumes the world, as clearly exemplified by the discussion of the War of Gog and Magog in Ezekiel. Despite a variety of options with regard to the end of days within the tradition, Jews in general have been taught not to focus on what will happen in the end of days, and instead to focus on what God requires of them now. By recognizing that one's actions are always being judged and that there are always consequences, the religious conscience is molded and shaped from the beginning of the educational process. Ultimately, the religious individual seeks to serve without a need for punishment or reward. However, Judaism has traditionally taught some form of metaphysical consequence (good or bad) for sins (*aveirot*) and for properly following God's commands/mitzvot with the understanding that *mitoch shelo lishma ba lishma*, "from doing something not for its own sake, one can come to do the act for its own sake."

While the above described dimensions of religion and spirituality are not exhaustive, they provide a way to understand the full concept of the Jewish religion and begin to connect the various elements of Judaism to brain functions including those that support behaviors, thoughts, experiences, feelings, and emotions. This is an important starting point for Jewish neurotheology.

Belief and Faith

Webster's Dictionary defines "belief" as a "state or habit of mind in which trust or confidence is placed in some person or thing" or "a conviction of the truth of some statement or the reality of some being or phenomenon especially when based on examination of evidence." This is differentiated from "faith," which is defined as "a belief and trust in and loyalty to God or a firm belief in something for which there is no proof."[15] It is interesting to note that belief is primarily distinguished on the basis that a belief has *some* evidence, whereas faith has none. How does the brain determine what evidence is of value? This is a particularly challenging question from the perspective of the brain. Cognitive neuroscience studies have shown how the brain functions when we make decisions about things and how we use some types of evidence over others.

The issue regarding appropriate evidence is even murkier when one compares different intellectual or scholarly pursuits. For example, evidence in philosophy is different from that in theology, which in turn is different from evidence in the sciences such as biology, chemistry, and physics. Thus, the lines between belief and faith are considerably more blurred than we would like to believe.

When it comes to belief in God, since there appears to be no definitive way of *proving* God's existence, one has to have faith in God's existence without the need of proof. Faith itself derives from the Latin word *fides*, which means "trust." Thus, both belief and faith refer to trusting something or accepting something as true without definitive proof. On the other hand, the apostle Paul strongly suggested that even religious believers should strive to "prove all things" and not have complete "blind faith" (Thessalonians 5:21).

The debate of reason- and "evidence"-based approaches in religion has existed for a long time in Judaism as well. Belief, often called *emunah* in Judaism, is subject to many disputes regarding how "best" to believe. Maimonides, as well as Gersonides, a scholar after him, both agreed that faith must conform to the logic of rationalism. However, they were living in a time when the proofs and evidence used were that of Aristotelian and Platonic logic, not modern-day science. As such, they believed that Judaism could be proved rationally, and Maimonides actually goes to great lengths to prove such in his *Guide for the Perplexed* (See Maimonides Introduction to *The Guide*). Rabbi Bachya ibn Pakuda, in his *Chovot HaLevavot* (*Duties of the Heart: Gate of Unity*), also states that one who follows simply based on tradition is like a blind man being led by someone else—one wrong slip and it's all over. In the brain, there are cognitive areas that help with determining causality, logic, and comparisons, which are part of rational processing. But using such processes in the context of proving religion is obviously quite problematic. Eventually, any person striving for rational answers still comes back to a sense of faith in our experience of reality. The same may even be true of science, as we shall see in the next chapter.

Despite the strong stance on Maimonides's part, as well as many more modern Jewish scholars' beliefs that Judaism and scientific evidence work together, many traditional rabbinic authorities have consistently held that faith does not need scientific evidence or proofs purposed from apologetics. Rav Saadiah Gaon explains that in addition to the general knowledge and "evidence" the rest of the world has, Judaism has an additional source of knowledge considered as authoritative, and that is the tradition of the religious experience. According to Saadiah Gaon, the knowledge given to Jews at Mount Sinai and maintained through a tradition of faith is held as an equal if not greater source of authority and as evidence for how the world functions along with the relationship between God and humanity.

Among more recent Jewish authorities, the Chofetz Chaim[16] (Rabbi Israel Meir Kagan), Rav Eliyahu Dessler,[17] Rebbe Nachman of Breslov,[18] and Rav Moshe Chagiz[19] all believe that simple faith is the best mode of religious observance and connection to God. The Sridei Esh (Rabbi Yechiel Yaakov Weinberg) writes that the highest level of faith is that of a deep penetrating intuition into what is true (see Sridei Esh, *L'perakim* 26). This perspective maintains that faith entirely separate from logic is, in fact, a higher level of belief. The main distinction that these thinkers have with those just previously discussed is the notion that the system of logic and scientific research is not the only or even best way of approaching humankind's existence. Many of the aforementioned scholars also state that true deep faith is only found upon entrenching oneself in in-depth Torah study and analysis, and not in continued theological sophistries. It would seem that there are in fact two schools of thought regarding how faith and reason are meant to interact and that these two camps are fighting over what is the proper way for someone to have *emunah*.

Those who have opposed religion have frequently remarked, as Richard Dawkins puts it, "Faith, being belief that isn't based on evidence, is the principal vice of any religion."[20] Dawkins suggests that since religion is not based on "evidence," it cannot be valid. However, we again face a similar conundrum regarding what qualifies as evidence. Many religious individuals will argue that they have plenty of personal evidence that God exists in their lives. Judaism has a certain uniqueness about its belief system in that it additionally relies upon the claim of a mass revelation—the providing of the commandments on Mount Sinai—a fact that is often drawn upon in debates about the "provableness" of Judaism's claims. However, should we invalidate religion because it derives from evidence that is of a different type? In the New Testament, faith is defined as "the substance of things hoped for, and the evidence of things not seen" (Hebrews 11:1). Again, there is the use of the word "evidence," but this time as a way of supporting the importance of belief. Belief itself becomes the evidence that may or may not invalidate religious faith depending on your perspective.

Jewish faith more accurately translates into the Hebrew word *bitachon*, which means to rely upon God once one has the foundations of faith. Putting one's trust in God is different from simply believing in God, and that is what *bitachon* is truly about. The Chazon Ish (Rabbi Abraham Yeshaya Karelitz) in his scholarly work *Faith and Belief (Start of Chapter 2)*, states that *bitachon* is the mind-set one should have in their faith, namely that everything, good and bad, is from God and to maintain one's faith, in spite of anything that may be happening. Clearly then, the interplay of these two components—that of the basic faith in the existence and omnipotence of God and the "trust" one places in God—is key to determining how different belief systems within Judaism can interpret these perspectives and integrate them into their larger religious practices.

This problem of evidence is also why science and religion have historically been viewed as opposed to each other. What scientists and religious individuals allow for in terms of belief and evidence is quite different. On the other hand, if we consider beliefs to be related to the ways in which we organize our perceptions of the world into ideas and stories that enable us to interact adaptively with the world, then most of our higher brain functions can be related or even equated with beliefs. In a prior work, we defined a belief as "biologically and psychologically identified as any perception, cognition, emotion, or memory that a person consciously or unconsciously assumes to be true."[21] This definition might be useful on several levels. First, it is an "operationalized" definition such that we can scientifically study perceptions, cognitions, emotions, and memories. Second, we can determine how these four elements influence the formation of specific beliefs. For some, beliefs are highly dependent on our social interactions with friends, family, and teachers. In Judaism this relates to the beliefs that arise within a congregation or minyan. For others, there is extensive cognitive processing through different possibilities until a belief can be held. This relates to the kinds of beliefs that derive from individual Torah study.

This definition of beliefs also has value because it refers to beliefs as both conscious and unconscious. Research has shown that unconscious beliefs can have a substantial influence over our thoughts and behaviors. This is particularly relevant to religion, which affects not only conscious beliefs, but also unconscious beliefs so that we act and think in a moral and religiously acceptable way, even if we are not consciously trying to. One fascinating study along these lines involved posing the following situations to a group of Christian and Jewish individuals. The individuals were asked about the morality of two situations. What if person A did not like his father but treated his father well and with respect, while person B did not like his father and did not treat him well. How should these two people be viewed morally? For Christians, both individuals might be viewed as immoral, since an immoral thought is regarded as just as bad as an immoral action. However, for Jews, person A is still considered to be relatively moral, since he treated the father well. Private thoughts are considered less relevant on the moral continuum.

Neurotheology might shed important light on the concepts of belief and faith, especially for determining if there are different brain structures associated with things that people claim to believe in and those things that people claim to have faith in. We might also explore the neurophysiological effects of the social, emotional, and cognitive aspects of beliefs. These components of belief rely on different brain processes, and thus, learning how these processes work and interact with each other could teach us a great deal about our beliefs.

Jewish Theology

Theology is traditionally distinguished academically from religion or spirituality in that theology represents an analysis of a given religious doctrine or belief system. Jewish theology focuses on the Torah and its elaboration in the various texts and commentaries that have formed the foundation of traditional Jewish religion and life, such as the Talmud, as well as many oral traditions.

Theology proper thus represents intellectual deductions from the foundational Jewish doctrine as well as reasonable extrapolations upon such a doctrine. The foundational point of theology proper, at least in Judaism, is a belief in the transcendent truth of the oneness of God, both at the literal surface level and the deeper symbolic level. From a Jewish theological perspective, there are a few approaches to what the nature of God is and God's relationship to human beings. The mainstream approach in traditional Judaism is that God is in essence unknowable. However, the nature of this unknowing is different depending on the type of Judaism one subscribes to.

Maimonides, in his *Guide for the Perplexed* (See 1:58), states that God is only knowable in the negatives, meaning that when one describes God, one can truly only describe what God is not. This is interesting from a brain perspective, since the brain has areas that support binary or oppositional concepts. We can affirm or refute something, and we can think about how similar or different things are. With respect to God, we can try to claim what God is or what God is not. This is one approach the human brain can process, and Maimonides takes this negation approach. However, the Torah speaks in a language that is supposed to be more positively relatable to human beings and therefore ascribes certain attributes and knowledge to God so that people can relate to God and serve God properly. This creates a relationship that is mostly about human beings continuing to strive for ultimate knowledge of God, despite the fact that such knowledge is truly unattainable. Nevertheless, the further one goes down the road of logical postulation based on Maimonides's religious principles, the closer one can get to perfecting one's intellect and character, which he sees as humankind's ultimate purpose.

This position, however, is not and has not been the mainstream of Jewish theology. Traditional theologians recognized that God as an ultimate being had no need of human beings to begin with. Ultimately then, the purpose of creation must have been for us, humanity, due to the absolute goodness of the creator. By fulfilling God's commands, human beings can bring glory to God and God's ultimate kingdom, and thus the whole world will recognize God's glory. However, as opposed to most missionizing visions, the Jewish vision sees this occurring naturally through Jews fulfilling the Torah's commandments and rabbinic enactments. By continuing to observe the commandments, one is fulfilling God's wish and bringing glory to God in some sense, perhaps hastening the ultimate redemption. This is meant to be for the benefit of all of humanity, not for God, since God does not need humanity's efforts toward God whatsoever.

The kabbalists took this traditional understanding one step further and explained that it is not only personal spiritual reward that is generated by fulfilling God's unfathomable will, but that the metaphysical and consequently physical world is changed subtly by one's fulfilling the commandments. In this model, when one performs a mitzvah/command from God, one is actually effecting spiritual change (See Tanya Likutei Amarim Chapter 35) in the metaphysical realms of *Asiya*, *Beriya*, and *Yetzirah*, and perhaps even beyond that. From a brain perspective, there are causal processes that help us establish how our individual behaviors might cause effects in the universe. And from this spiritualist perspective, the holistic brain processes may help us understand how our actions affect the whole universe. This is opposed to Maimonides, who views the actions one takes as part of a process to continue to refine the intellect and character of individual human beings to better appreciate a true knowledge and understanding of God (See *Guide for the Perplexed* 3:27, 51).

These differing perspectives—rational, spiritualistic, and any approach in between—all can have significant consequences in how they play out within the cognitive processes of the brain.

Regardless of the approach one takes toward a theological method, it is important to observe the cognitive and emotional elements involved. For example, all theologies are based on a primary faith system. Thus, belief is first and foremost at the foundation of any theology. However, the analytical components can have an emphasis on thought, feelings, experiences, practical behaviors, or other elements that can be related eventually back to various functions of the human mind and brain. Thus, any method of theology can theoretically be evaluated from a neurotheological perspective in addition to its more traditional approach.

Defining God

Since much of theology, and consequently neurotheology, entails the examination of the relationship between human beings and God, or at least the perceived relationship, it is also necessary to carefully define the term "God." In the Judeo-Christian religions, there is the notion of a singular God. God's presence pervades and encompasses the entire universe. However, the specific attributes of God vary dramatically depending on the belief system. From a Christian perspective, God is represented as the Father, Son, and Holy Ghost. In mainstream Judaism, there are no similar distinctions. However, kabbalistic thought does view God through ten emanations, the *sefirot*, though traditionally these have been interpreted as emanations and not as separate components of God.[22]

From a Jewish perspective, God traditionally is defined as unknowable and unapproachable. However, the biblical descriptions are often used as examples of how the "Torah speaks in the language of man" (*Sifrei*, Numbers 112), making God relatable to people by describing actions in emotional and physical terms that people can relate to. As it says in 1 Kings with the prophet Elijah, "And a great and strong wind was rending the mountains and breaking in pieces the rocks before Adonai; but Adonai was not in the wind. And after the wind an earthquake, but Adonai was not in the earthquake. After the earthquake a fire, but Adonai was not in the fire; and after the fire, a small gentle sound of blowing" (1 Kings 19:11–12).

The interaction is traditionally interpreted to be that God is not physically manifested in any of God's actions in the world, but is recognized to be "present" in some way in all things, particularly those that are smallest and most humble. Additionally, this interaction and its ensuing interpretation highlights the idea that God is unknowable and at the same time relatable in some way to people. The angry God who decimated Egypt is the same God who cares for the orphans and widows and promises to redeem the Jewish people from generations of exile. The reconciliation of these different approaches and understandings of one's relationship to God in Judaism seems to be part and parcel of the religious experience. Neurotheology would also emphasize that no matter what God is, the human brain would have to have some way of utilizing its functions to help us understand, as best as possible, what God is.

An important neurotheological point is that we cannot inherently reduce concepts of God to biology nor try to equate all conceptions of God as one. In Judaism then, different doctrines and approaches to God can have similar theological underpinnings but very different phenomenological experiences, thus creating a diverse neurotheological framework even within Judaism itself. People who experience God as a personal God who cares for them and their personal mitzvot observance will have a different experience depending on how they connect to God. Two Jews who belong to the same synagogue and pray at the same minyan may in fact have very different experiences—one who feels God's presence imminently in front of him, and another who calls out to a void hoping that God will hear him and respond. Both may even have the same theological beliefs. Phenomenologically, however, their experiences could not be more different. These experiences are the cornerstone of any true religious experience and must be taken into account in any attempt by neurotheology to explain the religious individual.

In the Bible, it is interesting how God's actual attributes are construed. For example, we read in Genesis that God can speak, see, create, and rest. Throughout the Bible, God is described as having a number of humanized emotions, including anger, jealousy, love, and forgiveness. From a brain perspective, it is often helpful for us to think about things in humanized ways. We give a car a name or refer to a boat as "she." We also see emotions in objects such as the wind or the sun. But as most religious scholars would agree, these humanized descriptive terms should not be considered to be related to the actual being of God.

From a neurotheological perspective, it might be fascinating to survey members of various congregations to determine how people think about and perceive God. What if people think about God in ways that are actually different from sacred texts or that espoused by rabbis? A large survey of religious attitudes from Baylor University showed that the Americans sampled (most of whom were not Jewish) tended to embrace one of four different "personalities" of God: authoritarian, distant, critical, and benevolent.[23] Thus, some people felt God was everywhere around them and highly engaged in the world (authoritarian), while others felt God essentially created the universe and now observes it from afar (distant). And some people felt God is merciful and compassionate (benevolent), while others felt that God is generally displeased with us and allows many bad things to happen (critical). Importantly, these concepts of God lie along a continuum related to two basic brain processes. The authoritarian and distant versions of God related to the brain's spatial processing of the world around us. The brain perceives objects that are either right around us or far away. Thus, we can perceive God in the same two ways. Alternatively, we have emotional centers that help us feel positive or negative emotions, of which either can be applied to a loving or angry God. Whether Jewish people would hold parallel characterizations of God or perhaps something very different requires a similar kind of survey study and perhaps different brain processes.

We decided to ask this very question to the rabbis in our survey. We asked the rabbis how they would describe God and gave four options as well as a fifth in which they could write in what they thought about God. The results of our survey were fascinating. Of the rabbis who responded, only 5 percent agreed that

God was as described in the Torah (all Orthodox), and 15 percent described God as a universal being. Among the Orthodox rabbis, approximately one-half stated that God was as described in the Torah, with one-quarter stating God was a universal being. However, in the entire sample, 15 percent indicated that God is an abstract concept, and another 15 percent agreed with the statement that God is metaphor. These are interesting ways to define God not so much as a being but as a concept—a mental process associated with abstract thinking. The remaining 50 percent of rabbis provided their own description of God (the answers were similar across the denominations, age and gender). However, most struggled with any definition, with some literally stating that God is indescribable. This was particularly the case for the Orthodox rabbis. The following are examples from those who could find some wording to describe God:

> God is a conscious unitive; God is an energy; God is a power that enables our salvation or self-realization; God is love, growth, and transformation; God is an unnamable mystery, love, and oneness; God is both universal and intensely personal, transcendent and immanent; God is the underlying structure of the universe; God is the source of all creation; God is the potential in the universe; God is inspiration, wisdom, creativity, kindness, love; God is a dynamic force for positive change and growth.

Perhaps you like some of those definitions or perhaps you have your own. Neurotheology challenges all of us to contemplate the ways in which we think about our religious and spiritual ideas and reflect on the ways in which our brain may enable or restrict us in finding those definitions. As with the other definitions considered so far, "God" is a term for something that is essentially impossible to define. However, if we are to reflect on what God is or how people experience God, we must do our best to maintain the principle of using as clear a definition as possible so that at least we understand what the scholarly focus is at any particular moment.

The Jewish Brain

An important topic throughout this book will be to work to uncover the nature of the rabbi's brain in particular and the Jewish brain more generally. Neurotheology would approach this concept from two perspectives. On one hand, the brain of Jewish people is likely to be greatly affected by the Jewish religion and culture. On the other hand, since Jews are a relatively small population that generally marries within the religion, it is likely that there are genetic aspects of Jews that yield specific brain processes that might differentiate Jewish people from non-Jewish people. We have already considered the recent studies of the male Y chromosome in Kohanim individuals that suggest there could be similarities in the genetics that support certain brain functions. And this applies to the broader Jewish community as well. In addition, the results regarding the Y chromosome, although only found in males, has important implications for Jewish women as well. In fact, Dr. David Goldstein of Oxford University stated in Science News that "For more than 90 percent of the Cohens to share the same genetic markers after such a period of time is a testament to the devotion of the wives of the Cohens over the years. Even a low rate of infidelity would have dramatically lowered the percentage."[24] Thus, both women and men in the Jewish community formed a tight bond and likely contributed to similarities in a number of genes passed on through generations. Thus, we might keep an eye on whether certain genes might be associated with certain brain functions, whether certain brain functions are particularly prominent in the Jewish population, and whether that prominence has something to do with the overall set of beliefs and ideals of Jews.

In the end, neurotheology might be able to help determine not only the characteristics of the Jewish brain, but of the Islamic brain, Christian brain, Buddhist brain, and eventually the brain functions associated with any tradition. However, it is important to remember throughout this discussion that neurotheology must remain as unbiased as possible and thus may not determine if one pattern of brain functions is truly *better* than another, only that it is *different*.

How Science Is Performed

The mystery that remains in the sunset is the riddle of why and how a mixture of seemingly inert, unthinking atoms of carbon, hydrogen, oxygen, and several other varieties can produce humans capable of having the subjective experience we refer to as beauty, or the love that would have us kiss our kids good night. Science is no closer to answering those questions today than it was a century ago.

—Gerald Schroeder, *The Hidden Face of God: Science Reveals the Ultimate Truth*

Standing in front of the whiteboard in my study room, I began to draw out the various chemical reactions in the human body's use of glucose to generate energy. However, I did not stop there. Everything was connected. I began to link all of the various biochemical reactions we were learning together, forming a larger web of interactions that amazed me. "It is all connected." I wondered at how awesome and formatted the interactions were. If even one did not move at the appropriate pace, the body could not function. I stood back and wondered about my experience at the same time: What were my classmates thinking about these reactions? Were they scientifically or religiously inspired by the fact that everything seemed so well coordinated? I knew that some marveled at the great complexity of life, others were not, and most were just anxious about the upcoming exam. Still, the thought nagged at me: what was it about these molecules on the whiteboard and science in general that leads some down their own paths toward spiritual and religious knowledge and understanding and some down the path of skepticism and purely rationalistic inquiry?

—David Halpern

The Jewish Scientist

To understand Jewish neurotheology and how religion and science relate to each other, we must strive to understand what science actually is and how it works. This is important since any ideas we raise in terms of studying the rabbi's brain requires us to conceive of the many facets of scientific discovery. In fact, we must strongly acknowledge that neurotheology can be of great benefit to science itself. After all, trying to measure something that is immeasurable is not easy. Neurotheology challenges many preconceived ideas about science and how we measure something that is frequently a fully subjective experience or belief.

There is also another important point regarding science and the rabbi's brain. It is perhaps one of the most amazing statistics about the intersection between Judaism and science—the number of Jews who have won Nobel Prizes in science. It should not be regarded as a coincidence that approximately 25 percent of all scientific Nobel Prizes were received by Jews since the prizes have been awarded, although Jews compose less than 0.2 percent of the world's population. Clearly, either there is something about the Jewish religion and culture that is highly consistent with scientific discovery or there is something within the Jewish brain that resonates with science. Either way, there is an affinity of Jewish thinkers to actively engage scientific methods and pursuits at the highest creative levels. This might also help us to understand how the brain conceives of science and scientific experiments and how that parallels Jewish thought.

For the religious rationalists, who understand that all knowledge is useful and part of their religious growth, science will be embraced. They are traditionally fiercely intellectual and often shrug off mystical pursuits as nonsense, similar to many in the scientific community. However, these individuals often will argue that eventually science and religion will come to agree with each other in a superior intellectual setting. The spiritualist group often responds to this threat first by entirely dismissing the claims of science, as discussed earlier. However, some may actually view science as another source of God's energy in the world and believe that by understanding it we are actually slowly moving toward a state of spiritual wholeness where science and mysticism become unified. This unifying theory, strongly supported in much of kabbalistic and Hasidic thought, has been used to advocate the study of science and religion side by side in some religious schools.[1] The predominant argument in the spiritualist camp, however, is that there is no need for scientific information when it contradicts one's spiritual experience, since that experience is what is revealing the truth.

Therefore, it would appear that the implications of scientific research on religion from a Jewish perspective will be a mixed bag for a mixed group of Jews. For those who embrace scientific endeavor, the implications for religion can be either positive or negative. Some scientists view science as something that can inform religion without necessarily destroying it. Other scientists, with an atheist perspective, typically view science as eventually eliminating the need for religious or supernatural content, even if these scientists follow certain "religious" behaviors as part of their connection to the Jewish ethnic heritage. Some religious people think of science as a foreign body to be assimilated into a religious worldview, eventually integrating the two. Lastly, for religious people who think of science as a foreign entity fighting with religion, then the implications are almost always negative and result in science and religion continuing to walk down roads either parallel or moving away from each other, but never intersecting. How the brain may work in each of these types of individuals is a challenge for future neurotheological study.

It may also be interesting to consider whether the development of scientific pursuits within a given person's mind is similar or distinct from the development of religious pursuits. Such a question might prove quite important in understanding how science is performed and how the human brain, or the Jewish brain, engages scientific inquiry. Scientific training begins at a young age in most schools. Children learn about the scientific methods, about hypothesis testing and making sure that data you obtain either supports or rejects your hypothesis. As scientists continue on in their education, they learn more advanced research methods, building off of the model of there being an objective "truth" that they can find through experimentation and analysis of data. If something is unknown, it means either the data has not yet been categorized or analyzed appropriately or we need to develop better methods for acquiring the data. Once the appropriate experimentation is done, we will know what combusts at a higher temperature or what drug is the best for controlling blood pressure.

The true scientist is someone who searches for the answers and believes that those answers can always be put back to hypothesis testing and refuted should new evidence come to light. All scientists are judged by the research they put out, and so their training must be steeped in research methods. The best way for a scientist to receive that training is to work in a lab, essentially apprenticing to older researchers as they help the young scientists with their studies. These older mentors are essential to the formulation of the aspiring scientist's career aspirations and dreams and can crush or raise up a prospective researcher.

Rabbinic training can in many ways be somewhat similar. A student becomes interested in the study of Jewish law and working with the Jewish community in some communal capacity. Some study for the purpose of their own learning, similar to a solely research-focused scientist. Many others study in order to increase participation from and enhance their congregations and communities, in a similar way to a clinical-oriented scientist who does research to help increase the health of the community. Rabbinical students spend a large amount of time studying the Torah and the various rabbinic texts and discussing their religious, theological, and practical implications. Jewish law, biblical analysis, theology, and practical rabbinics are all taught at various rabbinical schools. However, rabbis in training must also be mentored by following a senior rabbi, working with them to learn about how to best utilize their skills in managing and growing their congregation both in numbers and in spiritual commitment.

These mentors will be very influential to the future synagogue or temple leader, often being the one whom the rabbi turns to in times of crisis or need.

Lastly, there is one final striking similarity between the two groups—both work from the assumption that there is a fundamental reality that can be systematically explored. The scientist's work is based on the fundamental reality of the physical world and utilizes theorems and principles of scientific research and logic as a basis for further postulation and experimentation. The scientist becomes fascinated and excited at the prospect of uncovering some new way of understanding reality. Similarly, the rabbi uses the religious foundation of the Torah as the fundamental reality and the basis for all systematic study and exploration. All new situations can be analyzed through the prism of the Torah, rabbinic, and theological framework, and new situations, in fact, engender excitement at the possibility of applying the age-old principles to new ideas that form novel ways of understanding reality.

Both scientists and rabbis seek to use their own principles in exploring the world, and both perspectives will be needed when utilizing a neurotheological approach to Judaism. We might also consider how these developmental processes affect the brain. Studies have shown that when the brain is exposed to any behavior or belief, it begins to incorporate those behaviors and beliefs into its neuronal connections. Thus, as a particular approach to solving problems is used more and more, the brain becomes engrained with using that approach. If a person uses emotions to solve problems, then future problems will also be solved more with emotions than with logic. Those people who use various cognitive functions will tend to use them to solve future problems. In future chapters, we will explore how the brain's different cognitive processes can be used to engage and resolve various problems about the world.

Those individuals encouraged to explore new questions will typically feel enriched by future explorations and will strive for maximum creativity within a given framework of understanding the world. In the case of the scientist, that framework is based on the physical laws of the universe, and for the rabbi, that framework is based on the religious laws of the Torah. Perhaps it is this parallel approach that combines substantial creative inquiry within a well-structured framework that helps Jewish thinkers create outstanding science and theology. If one is too constrained, one cannot be creative, and if one is too creative, there might be little practical application of that creativity. Again, from the brain's perspective, this might enable connections that foster creative thinking to also be constrained enough to result in paradigm shifting results.

One example of how such creativity and constraint are manifested comes from a story of Dr. Newberg's uncle Vic as a youth. It was the Sabbath, and Uncle Vic had forgotten to cut the toilet paper in his home, which was his job to do on Friday before sundown. For those not familiar with the Sabbath laws, you are not allowed to do any activity defined as "work" that was done in the Tabernacle as the Jews wandered the desert after leaving Egypt. Part of these laws are the prohibition of stitching and tearing, and that included tearing the toilet paper. Having forgotten to do this was a potentially big problem for anyone wanting to use the toilet paper. But Uncle Vic came up with a creative solution to his problem. While in the bathroom contemplating his predicament, he noticed that the sink faucet was dripping slightly. He reasoned that if he held the toilet paper under the faucet, the drips would eventually break the toilet paper naturally, solving his problem (whether such a solution would be acceptable according to actual Jewish *halacha*, is beside the point, though in fact a similar solution is recommended via the flushing mechanism of a toilet in situations where one has no pre-cut toilet paper). So a combination of science and theology helped him solve a perplexing problem. He grew up to become a pioneer heart surgeon, so clearly the challenge of being creative within a specific framework led to fostering the brain of an outstanding scientist.

The Scientific Method

Development of the scientific method that is used in the present day dates back perhaps thousands of years, since the early scientific and philosophical schools of ancient Greek, Roman, and Arabic cultures. At the time, the distinction between science and religion was much less clear. In fact, science was frequently used as a mechanism for helping people in religious or spiritual pursuits. For example, astronomy played a prominent role in early scientific pursuits due to the need for understanding the motion of the stars and planets in the sky, which were believed to coincide with various religious and spiritual events. Ancient human beings made structures such as Stonehenge and the pyramids of Egypt, displaying remarkable engineering and mathematical precision, particularly with regard to astronomy. These structures for the most part are believed to have also been linked to various religious activities that were marked by astronomical events.

Eastern traditions and ideologies arising in China, Japan, and India began their development several thousand years ago, although these took on a different perspective of nature as well as the human mind. Some of the work performed by ancient Eastern scholars demonstrates a keen interest in human psychology. This is particularly true in Buddhist thought, in which there is a strong emphasis on the mind, consciousness, and mental processes. Eastern traditions also developed a biomedical paradigm that considered the human body from the perspective of various energy systems. These energy systems are described to be coursing through the human body, affecting health and well-being, as well as interacting with the environment. These approaches led to some highly technical and complex theories and interventions, such as those associated with acupuncture, which is a complex practice involving the insertion of small needles at various points throughout the body that are associated with meridian lines through which the body's energy courses.[2] These Eastern concepts and practices are becoming more closely studied by their Western counterparts in today's field of complementary and alternative medicine. In fact, the National Institutes of Health has created the National Center for Complementary and Integrative Health in order to better organize and fund studies of such approaches.

The results and findings of these early forays into science were based primarily on observation, with little regard for methodology. Specifically, it was believed that whatever was observed was more or less accurate as far as its representation of the real world. Hence, if one observed the stars and sun moving around the Earth, then clearly the Earth was the center of the universe. Similarly, if the world appeared to be flat in its structure, then the world was believed to, in fact, be flat. The approach to making these observations in a systematic and rigorous manner was not as relevant at the time, primarily because science was in its early stages of development and there was so much knowledge that needed to be accumulated that methodology was much less significant. It should also be mentioned that due to the technical limitations of doing science with appropriate methodologies, much of the early work was based on a more theoretical approach. Therefore, we see such attempts through the works of Aristotle and Pythagoras to define various approaches to logic and mathematics. Archimedes and Aristarchus developed ideas pertaining to physics, and the work of Galen and Vesalius observed the workings of the human body.

While the medieval period in Europe demonstrated a profound lack of interest in science and scientific methodology, there were significant advances in both Arabic as well as Eastern cultures with regard to science and technology. During what is sometimes considered the "Golden Age of Islam," mathematicians such as Muhammad ibn Musa al-Khwarizmi and Omar Khayyam elucidated many complex problems of algebra. Arabic astronomers developed many concepts still used today, such as altitude and azimuth, in addition to naming many stars in the sky, such as Aldebaran, Altair, and Deneb.

However, the next major steps had to await the work of astronomers such as Copernicus and Galileo, particularly with the development of the telescope, which opened up a whole new realm for science. Science was now beginning to develop various measuring tools and technologies that allowed for a more detailed analysis of the world and the realization that the world did not always exist or function according to what was in the immediate perception of the normal human senses. When aided by appropriate instrumentations, the world revealed a whole new level of complexity. Shortly following these developments in astronomy were advances in both philosophy and scientific methodology. The work of Isaac Newton, which dramatically affected our understanding of optical lenses and light, as well as his most important contribution in the development of calculus as a mathematical tool for understanding gravity and physics, laid crucial foundations for the rapid development of science into the nineteenth and twentieth centuries. Throughout this period of time, there were many technical developments, including the invention of the microscope, vaccines, and electricity. Perhaps more important, but not completely unassociated with such scientific and technological developments, was an entire approach to experimentation. Such an approach was particularly apparent in the works of inventors such as Thomas Edison, who developed extensive experimental methods in order to test and confirm various findings.

The turn of the twentieth century demonstrated another major advance in science with an increasingly well-defined methodology in the health sciences pertaining to the development of drugs and the study of various disease states. The rise and success of modern Western medicine with its huge litany of medication and surgical procedures to help in the alleviation of health problems, as well as many technological developments in fields pertaining to imaging and observing the human body, had a tremendous impact on science. Furthermore, the health sciences developed detailed analytical approaches to "evidence-based" medicine. This included concepts such as the randomized, double-blind, controlled study as a mechanism to test various therapeutic interventions and to eliminate bias on the part of the investigator or the patient. Experimental methods in chemistry and physics also advanced, particularly with the development of ever better techniques for evaluating and measuring physical and chemical phenomena, including such instruments as the electron microscope, the Hubbell space telescope, high-energy particle colliders, and many other measuring devices currently in use.

Of course the study of the mind and brain is of primary importance to the field of neurotheology. There have been dramatic changes over the last hundred years in terms of how we understand the human mind, how it works when it works normally, and what happens when it does not work well. The cognitive sciences, which include psychology, psychiatry, neuroscience, and cognitive neuroscience, all approach the function of the brain and mind from a slightly divergent yet complementary way. Until the turn of the twentieth century, there was little knowledge about psychological disorders such as schizophrenia, mood disorders, personality disorders, and a plethora of other types of mental disorders. Most people with such problems were either forced to function as best as possible in society or committed to an insane asylum. The work of Sigmund Freud in the early 1900s changed much of how people approached psychology, with a new understanding and approach to the mind and the development of various psychological processes.[3] It was with Sigmund Freud's work that the world came to understand, for the first time, how the conscious and unconscious mind worked and the difference between them. Also, it was the first time that people had the opportunity to explore the human mind with some methodological rigor. Of course, Freud did not get much of the workings of the brain correct, which was partially due to the limitation of the technologies and methodologies of that time. However, it was clearly an important first step in the direction of understanding the human mind. As the fields of psychology and psychiatry developed over the twentieth century, a greater emphasis was put on developing extensive research methodologies for helping in the assessment of psychological problems as well as in the treatment of such problems.

Concomitant with the development of an improved knowledge of psychological problems was an improvement in how researchers understood the workings of the human brain. This included detailed research on a cellular level to understand how nerve cells function and interact with each other through various electrical and chemical processes. These studies were typically performed outside of the living body (*in vitro*). More importantly, the development of various imaging techniques allowed for the study of the brain while still in the living animal (*in vivo*). It was particularly the *in vivo* methods, which include imaging techniques such as positron emission tomography (PET), single photon emission computed tomography (SPECT), X-ray computed tomography (CT), and magnetic resonance imaging (MRI), that allowed for the first time to peer into the detailed structure and function of the working human brain. This opened up many doors for the study of various cognitive processes, including vision, sensory and motor activity, emotions, language, and abstract thinking.

As we look to the future, we would expect a progressive development of better techniques for studying various aspects of brain activity and brain function, even to the level of individual nerve cells, as well as techniques that allow us to see a broader array of different aspects of brain physiology electrical, neurochemical, and neurotransmitter activity. Methods should also continue to advance to be able to explore the subjective experiences that people have and, in particular, how these relate to religious and spiritual states. These approaches are fundamental to the field of neurotheology and can help elucidate how different brains, including the rabbi's brain, react.

Scientific Thinking and the Brain

Since neurotheology relies strongly on the neuroscientific component, it is important to understand how scientific thinking itself works. Science arose originally from natural philosophy as a way of trying to understand how the world works. The term "science" derives from the Latin word *scientia*, which means "knowledge." Thus, science refers to the methods by which we gain knowledge about the world around us. As we mentioned above, religion and science were relatively integrated in the ancient world and even up to the last century. More recently, science and religion have gone their separate ways in general. However, there are still a large number of scientists who find no inconsistency between their religious beliefs and scientific pursuits.

Of course, science is based on the scientific method, which refers to a specific systematic approach toward acquiring empirical evidence about how the world works. This includes developing hypotheses, making observations and measurements, revising theories, and developing new predictions and hypotheses. As we also mentioned above, these elements find their way into Jewish theological and rabbinical scholarship. There is constant evaluation, analysis, and reanalysis of various concepts and interpretations. As the joke goes, "How many rabbis does it take to change a light bulb? It has not yet been determined, since they are still searching for a Talmudic reference to a light bulb." The point is that all problems start with the fundamental reality of the Torah but also require extensive analysis from that point forward.

The elements of scientific thought allow for an exploration of the natural world in great detail and with a high degree of precision. However, we must always remember that science is only as good as its last study or theory. With each new study, our scientific knowledge continues to adapt and change. Although knowledge may change, the essential assumptions about the scientific method generally do not change, even though the ways of going about observing and measuring the natural world clearly change.

Science is based upon several essential *a priori* assumptions that direct and guide how science works and how scientists think. Perhaps the most important assumption made as part of scientific discovery is that the world, as we perceive it, is measurable. Furthermore, there is the assumption that the world is generally a stable place, so that an experiment performed today in the United States should yield the same results if performed next week in Germany. Science assumes that over vast times and spaces, the universe is relatively stable. Again, in a similar way, the rabbi views the Torah, and God in particular, as being present in some way everywhere and a stable foundation as the base for theological and legal development.

Another fundamental assumption of science is that simplicity tends to be better than complexity. While there are entire fields of study based on complexity and chaos, much of science still resorts to the notion that a simple description is better than a complex one. This is the principle of parsimony, also known as Occam's razor. More recently, the work of Karl Popper and Richard Swinburne similarly argue that "other things being equal—the simplest hypothesis proposed as an explanation of phenomena is more likely to be the true one than is any other available hypothesis, that its predictions are more likely to be true than those of any other available hypothesis, and that it is an ultimate *a priori* epistemic principle that simplicity is evidence for truth."[4] Scientists have frequently utilized Occam's razor and the notion of simplicity as a way of dismissing the argument for the existence of God. Theists, on the other hand, have often argued that problems associated with scientific arguments without God are equally complex compared to arguments that include God. Many scientists, however, reject this argument, claiming that information and investigation of the natural world require the scientific method with its attempts to find the simplest answers to describe the universe.

It is also interesting to note that the fundamental concept of Judaism—that there is only one God rather than many—may be akin to the principle of parsimony. Jewish thought teaches that God is a unity that binds the universe. In that sense, God is simple. There is only God, and everything derives from God. God is the reality that we can address through either a scientific or religious approach, but the wholeness/oneness of God is the basic concept that lies at the heart of Judaism.

The materialistic approach of science can sometimes lead to the position of "scientism," which is the belief that science will ultimately be capable of explaining everything about the universe. The essential aspect of this stance is that the universe is purely material in nature and that scientific method will uncover any and all facts about the universe. In this way, someone believing in scientism will reject any perspectives that appear irrational or supernatural. Further, it is considered that the natural sciences would have authority over all other interpretations of reality including those that are sociological, psychological, religious, or spiritual. While this particular stance would likely be too limiting from a neurotheological perspective, as with all belief systems, it must be properly evaluated and accommodated within any overarching theoretical framework regarding the nature of the universe.

One rebuttal against scientism, and even Occam's razor, is what we previously suggested as neurotheology's razor. Specifically, the actual Occam's razor is stated, "Plurality should not be posited without necessity." This means that scientists should not postulate more than what is necessary to describe any given phenomenon. This rule was designed to prevent people from suggesting hypotheses for which there was no clear way of measuring and that were not actually necessary in order to explain something. For example, one might suggest that electricity works by having millions of massless, energyless, little green men pushing electrons around a wire. There would be no way to prove that the little green men did or did not exist based on any current scientific methods. However, the point in some sense is moot since it is not a necessary requirement in order to explain how electricity works.

But religion and religious beliefs present a different set of problems for science because in some ways, the notion of God is faintly familiar to the little green man hypothesis. Science has no way of measuring God or God's influence in the world. This is why scientists often evoke Occam's razor as a way of disproving God's existence. Neurotheology, in its goal of seeking a balanced understanding of science and religion, would suggest a different razor as the following: "Necessity should not be posited without plurality."[5] The point of this statement is that until we have done sufficient science, we don't know exactly what is necessary to describe a phenomenon. How do we truly know that little green men are not required to explain electricity? While we might be able to make the argument that electricity really is explained fully by Maxwell's equations and Benjamin Franklin's kite, the issue about God is a bit more problematic.

The problem with God is that while electricity is a phenomenon *within* the universe, God has a standing that goes *beyond* the universe. Although religions like Judaism don't have much problem with this, science struggles with measuring things that are at the limits of what exists. Thus, we should be careful about how quickly we draw conclusions about neurotheological research on topics such as prayer, meditation, mystical experience, or beliefs, since these are notoriously difficult to understand fully. Thinking that science can explain these phenomena in a simple way may simply not work very well.

Scientifically Measuring the Brain

To measure how the brain functions, neuroscientists have used several different methods. By measuring electrical activity in the brain, they have been able to locate areas of the brain that are associated with certain functions. For example, a person may be asked to repeatedly tap his left hand. While the person is doing this, electrodes surrounding the brain pick up electrical changes that can be isolated to a general area of the brain (this is called electroencephalography, or EEG, which literally means "graphing the electrical activity of the brain"). Computers have been used to generate three-dimensional images of this electrical activity so that better localization of brain function can be obtained. However, the ability to pinpoint the electrical activity is only good to a resolution of about one inch, which is often larger than many of the brain regions being studied.

Newer brain imaging methods have been developed in the past three decades and have improved the resolution and accuracy of identifying structures in the brain. Computed tomography (often called CT or CAT) scans use X-rays to generate a three-dimensional image of the brain's structure. CT scans have been used in medicine to study a large variety of problems including brain tumors, strokes, and head injuries. But CT scans can only look at structure and cannot tell you what it is doing. Magnetic resonance imaging (MRI) uses magnetic fields and computers to generate images of the human brain. MRI has the advantage over most other imaging modalities of an excellent spatial resolution of two to three millimeters (about the thickness of two pennies stacked together). Thus, MRI can be used to study parts of the brain that are only a few millimeters across. MRI has been very useful in the study of strokes, tumors, and other neurological disorders. Clinically, MRI is still used mostly to assess structure.

In the 1990s, the proclaimed "Decade of the Brain," MRI was developed to observe functional changes in the brain, and this ushered in the field of cognitive neuroscience. Functional MRI (fMRI) now has several different ways of looking at the brain's function. One approach is called BOLD imaging (which stands for blood oxygen level dependent imaging), which uses electromagnetic signals to measure changes in the oxygen level in the blood going to the brain. These changes reflect alterations in blood flow, which is an excellent marker for brain activity. Another type of fMRI scan is called ASL (which stands for arterial spin labeling) and similarly allows for the blood flow of the brain to be observed, but ASL is a more direct and quantitative measure.[6]

The brain is much like the engine in your car. When you want to go faster, you step on the gas pedal and more gas goes to the engine. More gas means more speed. The brain is similar in that when a certain part wants to go faster to help solve a problem or respond to something in the environment, more blood flow goes to that part of the brain. The blood brings sugar and oxygen, which is the gas for the brain. Many studies, including our own of various spiritual practices, have utilized this fMRI technique. The main downside is that the subject has to be in the scanner while doing whatever task you want to study. So if you want to study prayer, you have to have the person inside the scanner while he or she is doing the prayer. This approach can be very effective but has one major drawback: it may not be easy doing the practice or task in the scanner. And even if you can do the practice in the scanner, is it the same as doing the practice in synagogue? So fMRI scans of meditation and prayer can be very useful but have to be taken with a grain of salt.

MRI has also been used for several newer purposes, all getting at different ways of measuring the brain's function. One method is called resting BOLD, which uses the same BOLD imaging mentioned above but only acquires data while the person is resting comfortably. The purpose is to determine how different parts of the brain interact with each other in a network. This works by observing structures whose activity goes up and down in the same way. This is referred to as "functional connectivity" and enables scientists to see whether certain parts of the brain are connected to each other in certain ways and whether that might change as the result of some type of intervention or practice.

We recently published an article on the changes in functional connectivity associated with undergoing a seven-day silent spiritual retreat based on the exercises of St. Igatius.[7] We found that after the retreat, a number of brain areas were connected differently than before the retreat. The areas involved included emotional centers and cognitive centers that help us perceive and experience reality. Thus, the retreat changed the way the participant's brain helped them to view reality.

MRI has also been used for something called diffusion tensor imaging, or DTI. This looks at how water diffuses through the brain and helps to show structural connecting fibers and the directions in which they are connected. If you have ever seen the scans showing the thousands of fiber tendrils running in all sorts of colorful directions, you are usually looking at DTI. This can be used to observe how the different parts of the brain are structurally connected in disease states, but perhaps also in response to being religious or doing some type of spiritual practice. The final MRI technique we will mention is called magnetic resonance spectroscopy, or MRS. MRS allows scientists to take a chemical picture of the brain. Using the MRI scanner, we can observe the concentrations of various chemicals and molecules. We can observe whether there is more serotonin or dopamine in certain areas of the brain, or we can observe how the molecules of energy are used. An interesting limitation is that MRS only shows how much of a particular molecule there is in the brain, but it doesn't tell you what it is doing there. So there may be a lot of serotonin around, but none of it is binding to its receptors. To test that, you need a different kind of imaging technique.

The field of nuclear medicine has two imaging tools that have been extensively used for studying the function of the brain. These two techniques are called SPECT and PET. SPECT requires the injection of a small amount of a tracer containing a radioactive isotope that emits a single photon of light (hence the name). The photons (packets of light energy) are detected by a special camera and then converted into a three-dimensional representation of the brain by a computer. PET is similar except that the tracer's radioactive isotope emits a positron (an antimatter particle that is technically an antielectron). When this antielectron meets an electron in the body, the two annihilate each other (due to the reaction between matter and antimatter, just like in science fiction shows), and this creates two photons that travel in opposite directions. These emitted photons can then be detected by a camera and converted into an image. Of course the science behind these techniques is quite complicated, but the important thing to remember is that SPECT and PET yield functional images. They are considered functional because the tracers that are injected work like natural substances that normally occur in the brain. For example, some tracers act like water (which follows the path of blood to the brain), some act like glucose (which is used by the brain for energy), and some act like neurotransmitters (the chemicals that allow nerve cells, or neurons, to communicate with each other).

One of the particularly relevant advantages of PET and SPECT is that in addition to general brain function as measured by cerebral blood flow and metabolism, these imaging techniques offer the opportunity to explore a wide variety of neurotransmitter systems within the brain. In fact, a large number of radiopharmaceuticals have been developed over the past thirty years that may be of use for studying the effects of religious and spiritual practices and experiences. Neurotransmitter analogues have been developed for almost every neurotransmitter system, including the dopamine, benzodiazepine, opiate, and cholinergic receptor systems.[8]

PET imaging and SPECT imaging both involve exposure to small amounts of radioactivity and have a significant advantage of not requiring the subject to be in the scanner for all studies. This latter point is crucial. As we mentioned above, using MRI to study religious and spiritual practices raises certain problems, since the person needs to be lying down and completely still and will hear very loud knocking sounds that are made by the MRI magnet (this sound is as loud as 110 decibels, which is as loud as an airplane during takeoff). Obviously, such issues may present major obstacles for the study of spiritual experiences if one is trying to create a relatively natural setting.

The two most common techniques, using fluorodeoxyglucose (FDG) to measure glucose metabolism with PET and tracers that measure blood flow in the brain using SPECT, can potentially be performed with the person in a more natural environment. The subject can be in any posture and can move, and there can be limited distractions in the room. In fact, the person simply needs to be connected to an intravenous catheter to inject the radioactive tracer, and this catheter can be inserted long before the injection so that the person is not even aware of when the injection is performed. Shortly after the injection is given, these tracers become fixed in the brain so that any additional activities, thoughts, or feelings that the subject might have become irrelevant to the scanning. In other words, if a person is injected during a prayer session and continues to pray for approximately five to ten minutes after the injection and then engages in talking, walking, eating, laughing, or any other activity for the next half hour, none of those other activities will interfere with the scan, which can be taken up to one hour later. The scan at that time will still reflect the brain activity levels at the time of injection—in this case, during the prayer session.

There are some PET tracers that do require a person to be injected and imaged while in the scanner because of the very short period of time that the tracer remains in the person (some lasting only a few minutes). If a study were to utilize these short-lived tracers, then the subject would have to be in the scanner, which could present similar problems to those described above for MRI with the exception of the noise, since the PET scanner is typically fairly quiet. While PET imaging provides better spatial resolution than SPECT, if the goal is to make the environment relatively distraction free to maximize the chances of having as strong an experience as possible, it is sometimes beneficial to perform these studies off-hours (especially if there is a busy clinical or research service). This may complicate the use of PET because the radiopharmaceuticals such as fluorodeoxyglucose may not be readily available. Thus, while PET and fMRI offer certain technical advantages, SPECT also provides a potentially important option for the study of spiritual practices.[9]

Measuring the Subjective

Ultimately, we must always ponder how good any imaging study is at uncovering what happens in the brain during religious or spiritual phenomena. Finding brain changes only tells us what is happening in the brain. These scans cannot prove or disprove whether religion is right or wrong or whether God exists or not. These scans merely tell us how the human brain approaches these issues. It is also crucial to not only observe brain changes, but to do our best to evaluate the subjective nature of the religious or spiritual experience. If we assume that saying the *Shema* prayer automatically leads to a religious experience, then any brain scan findings during that process would be of interest. If, however, the *Shema* must be said with concentration and devotion to be a religiously significant event, such a scan may be completely useless if the person is merely saying the *Shema* without truly feeling it. We need to understand as well as possible the phenomenology of the religious and spiritual experiences and beliefs.

There are two basic ways of assessing the subjective, and both require the person being studied to report what he or she felt. One way is through standardized questionnaires. Hundreds of different types of questionnaires have been developed over the years to assess all aspects of religious and spiritual experiences.[10] Some of these questionnaires ask about the various feelings that a person has. Some questionnaires assess specific elements of these experiences, such as fear or hope. Other questionnaires explore the person's beliefs, such as whether they believe in God, and if so, how they understand what God is. Orthogonal questionnaires explore other aspects of the human psyche as they may relate to religious or spiritual phenomena. We might wonder whether religious people are more or less depressed than nonreligious people, and this requires an understanding of how to define "religious" and how to define "depressed." Questionnaires exist for both of these measures.

The value of standardized questionnaires is that they are standardized. They typically have been validated on large populations and also have an ability to compare people both within traditions and across traditions. They are frequently brief and can be used easily in the context of a research study. The downside of such questionnaires is that they are standardized. In other words, they ask specific questions and in specific ways. This can become a problem when applying them to various religious populations. For example, if a questionnaire asks whether a person believes in God and that person does not understand what kind of God is meant, the person may answer in the negative even though he or she has a strong belief in God. This is particularly the case with many questionnaires that have a Christian-based perspective. Questions that may specifically relate to Christianity (e.g. Do you believe that Jesus is the Savior?) would all be answered in the negative by a Jewish population. The result would appear that the Jewish population is not religious, when that is not true. Finding questionnaires that are more general may be useful across traditions, but sometimes they might be so general that they fail to measure specific elements of a given tradition. A questionnaire may ask whether someone participates in religious holidays, but that creates confusion for a Jewish person who participates in the Sabbath and is not sure whether that should be counted as a holiday. In the end, it is most likely important to utilize standardized questionnaires carefully so that the data is useful but does not represent the full understanding of the person being assessed. It is particularly helpful to combine standardized questionnaires with more open-ended assessments of people's beliefs. This leads us to the second basic way of assessing the subjective aspect of religious phenomena—asking the person directly what his or her experience felt like.

To ask a person directly is relatively easy, but one has to be careful as to how much or how little people may say about their religious beliefs or spiritual experiences. It is also different if you ask the person face-to-face to get a verbal answer compared to letting the person write it down. In our Survey of Rabbis, we had several places in which the respondents could provide a description of their beliefs or experiences. This allows for a thorough understanding of the person's religious and spiritual views. We can then review the narratives both individually and together using a method called content analysis. In content analysis, the various words a person uses are evaluated for their frequency and how they are connected to others.

An initial survey that we did prior to the Survey of Rabbis received almost two thousand descriptions of spiritual experiences across all denominations. Some people described naturally occurring religious experiences, near-death experiences, drug-induced experiences, and a variety of others. However, all of them were considered "spiritual." Some of the descriptions from people who were Jewish include the following:

> During my psychedelic experiences, the *kelipot*/"shells" were shattered, and I was able to "pierce the veil" and see the infinite interconnectedness underlying my experience of normative reality.

> Whereas I would normally perceive of me being physically separate from others, I saw myself as being intertwined with them in an infinite dance of energy that was thinly veiled by physical existence.

> It went on and on, hour after hour as powerful waves of surging energy, ever intensifying. I was totally immersed in intense ecstasy. Then, with another volcanic eruption suddenly my center of total awareness burst forth from the top of my head as a tiny point in a massive torrent of energy and my awareness emerged into a vast yet intimate space where I felt connected to all things in the cosmos. I wondered if I had died. I had no fear. I was merely a point of consciousness with no body at all. I was overcome with immense gratitude for the privilege of having been alive, and for my life with its joys as well as its sorrows and the total improbability of just being. I thanked God.

> A blinding light appeared above me and grew in size until it totally enveloped me. Though the light had no form, I

> immediately recognized it and would have known it from anywhere. In a soft yet powerful voice I heard, "You are the creator of your reality." And with that instant it was over.

These experiences are classic intensely "spiritual" experiences described by some who have used drugs to induce them and others who simply had the experience happen to them suddenly and naturally. But these descriptions give us an essential window into understanding the nature of spiritual experiences. Furthermore, we can use these descriptions to tie various elements, such as the sensation of light, the hearing of a voice, the sense of energy, and the sense of unity and connectedness, to various parts of the brain and their function.

What would these experiences show us on a PET or SPECT scan? We can only speculate without the actual results. However, one thing is certain: the field of neurotheology currently requires further elucidation of these phenomena on a theological and teleological level as well if the scans are to be understood well. Can one with no religious background who has a psychedelic experience on mushrooms really be said to have the same experience as someone who, having meditated for years and led a religiously devoted life, has a spiritual epiphany and revelation one day? On the surface the scans might even show great similarities, but the nuances of the theological differences in their thoughts and perceptions of reality might very well be what shape those experiences in the mind. Overall, without a basic understanding of both religion and science, we will never be able to fully comprehend the neural activity of the religious experience, and bridging these two fields is exactly what neurotheology is about.

Relating brain processes to these extraordinary experiences does not negate the experience in any way, but provides a unique, neurotheological perspective, which may forever change our understanding of religion and spirituality. Before we continue to explore such experiences in more detail, let us first review basic brain processes that may be attributable to religious and spiritual phenomena.

Neuroscience and the Religious Brain

I soon learned that it did not require a great brain to do original research. One must be highly motivated, exercise good judgment, have intelligence, imagination, determination and a little luck. . . . One of the most important qualities in doing research, I found, was to ask the right questions at the right time. I learned that it takes the same effort to work on an important problem as on a pedestrian or trivial one. When opportunities came, I made the right choices.

—Julius Axelrod, Jewish neuroscientist who discovered important information about how neurotransmitters work, winner of the Nobel Prize in Medicine in 1970

L'fum tzara agra, **according to the effort/pain is the reward.**

—Ben Hei Hei, *Ethics of the Fathers* 5:26

Historical Perspectives

Before furthering our investigation of the rabbi's brain and neurotheology, we must develop a foundation in both the neurosciences as well as religious and theological thought. As Julius Axelrod states above, doing such research requires many important brain functions, such as those related to intelligence, creativity, and determination. Combining these elements has led many great neuroscientists, both Jewish and non-Jewish, to uncover the incredible mysteries of the human brain. This chapter will describe various aspects of how the brain functions, and in later chapters, we will use this foundation to explore the neurophysiological basis of Jewish myth, ritual, and theology. In order to help those with a less scientific background, we have highlighted in ***bold italics*** the key points regarding the parts of the brain and their functions that might directly relate to understanding neurotheology and the Jewish mind.

Neuroscience has an extensive history dating back several thousand years. In fact, the first report of anything related to the neurosciences appears to have been recorded around 4000 BCE in Sumerian records regarding the euphoric effects of the poppy plant. Around 2700 BCE in China, Shen Nung developed the concepts associated with acupuncture. While this conception of the body and nervous system is somewhat distinct from Western medical science, it nonetheless describes in great detail the workings of the human body and how various points along the skin surface appear to be connected to internal organs and how these interrelationships can be manipulated to cause various responses.

The Edwin Smith Surgical Papyrus,[1] named after Edwin Smith, an American Egyptologist, was written around the year 1700 BCE but is based on texts that go back to approximately 3000 BCE. This document may actually be the first medical document in the history of humankind and may have been written by the great Egyptian physician Imhotep. The papyrus is a description of forty-eight cases, some of which are directly related to neuroscience, because they discuss the brain, meninges (coverings of the brain), spinal cord, and cerebrospinal fluid (the fluid surrounding the brain) for the first time in recorded history. In fact, there is even a case in which a specific head injury appears to have resulted in a loss of language in the patient (called aphasia).

In approximately 500 BCE, Alcmaion of Crotona actually dissected sensory nerves from cadavers. One hundred years later, the famous Greek physician Hippocrates discussed epilepsy as a disturbance of the brain and stated that the brain is involved with sensation and is the seat of intelligence. Philosophers such as Plato and Aristotle also commented on brain function, with Plato teaching that the brain is the seat of mental processes. Aristotle, on the other hand, suggested that the heart is the seat of mental processes. In 177 CE, another Greek physician, Galen, gave a lecture entitled "On the Brain."

As with many scientific endeavors, little happened in terms of neuroscience in Europe during the medieval period. However, the fifteenth and sixteenth centuries observed a dramatic increase in knowledge about the human body. In 1410, an institution for the mentally ill was established in Valencia, Spain. In the 1500s, Leonardo da Vinci produced the first molds of the internal structures of the human brain, helping to provide a "map" of the brain in three dimensions. Around the same time, Belgian physician Andreas Vesalius made a number of important contributions including a description of several brain structures and disorders, such as when there is too much fluid within the brain. In 1649, René Descartes described the pineal gland (which sits at the base of the brain) as the control center of the body and mind. While this concept was later proved incorrect, the notion that the brain was able to control the body and mind was an important conceptual step. In the latter part of the 1600s, several texts of the brain's anatomy were published. Thomas Willis coined the term "neurology" and also published *Cerebri Anatome*, which dealt not only with the structure but the function of the brain. In the 1700s, Antony van Leeuwenhoek, using the microscope, first described neuronal fibers, the connections between different cells in the brain and body. Around the same time, Franz Anton Mesmer first introduced the concept of "animal magnetism," which was later referred to as hypnosis.

Franz Joseph Gall, in the early 1800s, first published his idea about the location of various processes within the brain, a concept called phrenology.[2] Phrenology was in many ways the first comprehensive attempt at assigning various cognitive and emotional functions to different parts of the brain. Gall considered some functions as pertaining to both animals and human beings, such as the instinct for reproduction, affection, and eating. His list also included functions pertaining only to human beings, such as music, a sense of metaphysics, and religiousness. It is interesting to think about how the brain of humans might relate to that of other animals and whether this has an impact on the relationship between the brain, rituals, and religion. Abstract concepts such as metaphysics, a sense of purpose, and religion appear to arise in the human brain. But how might these be related to the other, more "primitive" brain processes? While most, if not all of these "organs" or structures in the brain have been disproved scientifically, it was a major step in understanding that certain parts of the brain appear to be more specifically responsible for particular functions. Detailed understanding of these functions had to wait another two hundred years before scientific methods and techniques were advanced enough to be able to discern how specific areas of the brain relate to various mental processes.

One area of neuroscience that was more thoroughly described by a number of scholars in the 1800s is the study of pathological processes and their association with various parts of the brain. This technique provided important information regarding how certain brain structures function. Typically, a patient who suffered a brain tumor or stroke would be observed to lose one or more cognitive functions. Upon the death of the patient or during a surgical procedure, the areas affected by the pathology would be uncovered and would then be associated with that function. Take as a simple example a woman who had a stroke and lost her ability to comprehend language. Upon her death, if the stroke was found to be located in the left temporal lobe, then that area would be associated with the function of comprehending language. Historical examples include Jean-Baptiste Bouillaud's cases of loss of speech after frontal lobe lesions and Marc Dax's paper on left hemisphere damage effects on speech. In 1861, the famous neuroscientist Paul Broca described brain localization particularly with regard to language and later described the functions of the limbic system, the primary area of emotional response.[3] Also, in the mid-1800s, William Benjamin Carpenter proposed that the thalamus, a central brain structure, is the seat of consciousness, and Bartolomeo Panizza showed that the occipital lobe (in the back of the brain) is essential for vision.

Toward the later part of the 1800s, in his work *Hereditary Genius*, Francis Galton wrote that intelligence is inherited. Such a notion has perpetuated into modern times about how much influence our genetics and our environment play. In our consideration of Jewish scientists, we can ask a similar question. Are Jews inherently good at science, even down to something in their genes, or does it have to do with their environment that encourages questioning and skepticism? And does the Jewish culture foster skepticism because it is part of some inherent biological makeup?

In the late 1800s, Emil Kraepelin coined the terms "neuroses" and "psychoses" to describe various disorders that later would be related to psychological disorders such as anxiety, depression, and schizophrenia. Some of these concepts were later made famous by Woody Allen in his comedy and movies documenting the humor in human neuroses, such as playing a game on the Neurotics baseball team in which he would steal second base, feel guilty, and go back. While these seem to be part of virtually every person and culture, there also appear to be characteristic elements that pervade specific groups that help define them in relation to other groups.

In 1891, Wilhelm von Waldeyer-Hartz coined the term "neuron" as the primary functional cell of the brain, and several years later, Charles Scott Sherrington described the term "synapse" as the space between neurons through which various neurotransmitters have their function. The 1900s began with a tremendous shift in the understanding of the function of the brain and mind with the work of Sigmund Freud (also Jewish), whose publication *Interpretation of Dreams* described the complex relationship between the subconscious and conscious mind.[4] Borrowing somewhat from biblical concepts of good and evil, Freud also described the relationship between the id, ego, and superego. The story of Adam and Eve might represent the very first expression of these intrapsychic forces, with Adam representing the ego, the apple or snake representing the id, and God representing the superego. Of course, there are probably many ways to interpret these stories and ideas, but it is important primarily to realize how the mind is constructed and functions. The brain appears to house structures that also support these concepts. There is a classic balance between the cognitive and emotional centers much like Freud's ego (cognition driven) and id (emotion driven). Freud's work clearly had a tremendous impact on the fields of neurology and psychiatry in terms of how to understand and manage mental processes and disorders. From the perspective of religious experiences, Freud also had a significant impact, since religion was considered to be a "universal obsessional neurosis."[5] The notion of religion as a "disorder" lay the foundation for a significant dichotomy in medicine during most of the twentieth century.

In 1928, Walter R. Hess reported emotional type responses when the hypothalamus, one of the most central brain structures, was stimulated. Following up on this work was that of James Papez, who published work on the details of the limbic system circuitry—the structures that ultimately would be considered the primary emotional centers of the brain. In 1957, Wilder Penfield and Theodore Rasmussen devised a motor and sensory homunculus that delineated which areas of the brain were responsible for which motor and sensory functions throughout the body. This research was based on neurosurgical analysis of these areas by electrically stimulating parts of the brain during neurosurgery procedures for epilepsy while subjects were actually awake (since the brain has no pain receptors) and could therefore relate whatever experiences they were having. Penfield found that when certain parts of the brain were stimulated, they might trigger a motor response, but more importantly, certain parts of brain stimulated a visual memory, perhaps even from childhood. Each time that particular area was stimulated, the person reported the same experience. This enabled him to develop a "map" of the brain and its functions with great detail. We will see how brain stimulation research has evolved, particularly in terms of inducing various spiritual states. In 1970, Julius Axelrod, Bernard Katz, and Ulf Svante von Euler won the Nobel Prize for their work on neurotransmitters, the chemicals that neurons utilize to communicate signals throughout the brain. In 1981, Roger Wolcott Sperry was awarded the Nobel Prize for his description of the function of the brain hemispheres in work on split-brain patients demonstrating that each hemisphere has related, but distinct functions.

The 1990s, the "Decade of the Brain," produced an explosion of research in brain function. One of the most exciting areas has been neuroimaging, which has utilized techniques such as positron emission tomography, single photon emission computed tomography, and functional magnetic resonance imaging that each can study which areas or chemicals in the brain are associated with certain disease states, thoughts, language processing, quantitation, emotions, and behaviors. All of these studies, and the entire history of neuroscience, provide a backdrop from which to begin to explore neurotheology.

Several points are important to consider prior to reviewing the basics of brain function. On one hand, the nervous system is incredibly complex, and it has taken thousands of years to reach the current state of neuroscience, which is still far short of understanding the complex structures and functions of the human brain. Thousands of studies have explored many aspects of how the nervous system works, from the very microscopic to the macroscopic level. Studies have focused on individual cells in the brain, various brain chemicals, and even individual atoms. Studies have also pursued complex cognitive processes involving many different parts of the brain at the same time. It is certainly not possible even for neuroscientists to comprehend the breadth of knowledge that science has provided about the human brain.

Regarding the various structures of the brain and their respective functions, it is important to note that while a number of functions might ultimately be related to specific areas of the brain, this should not necessarily be held to with great rigidity. All of the structures of the brain ultimately work in concert to provide human beings a consistent understanding of the world and a coherent behavioral response to that understanding. In other words, while one particular area might be highly associated with language or vision, it does not operate in isolation. All of the different parts of the brain rely on each other in one way or another. Additionally, the emerging research on neuroplasticity suggests that areas of the brain can "take over" for other areas that have ceased to function normally for various reasons. However, there are certain functions that appear to be reasonably localized to certain structures, and this review will describe the most relevant ones for neurotheology, even though many others may eventually be found to play an important role.

The Neuron

The basic functional unit of the brain is the nerve cell, or neuron. The word "neuron" is derived from Latin and Greek words meaning "sinew," which may have been appropriate since the nerves that run throughout the body appear like sinews. The neuron is a cell that consists of a cell body with the nucleus and a long axon that extends from the body toward other neurons (see figure 4.1). Each neuron also has small extensions called dendrites located at the cell body that contain receptors for various neurotransmitters.

Within the nerve cell is a tremendous machinery that is geared primarily to receiving signals from other nerve cells and relaying those signals to other neurons. This is accomplished by a number of cellular processes that include the metabolism of glucose to useful energy for the cell. This energy is primarily responsible for maintaining electrolyte balances throughout the length of the axon, which can be as long as several feet. When the nerve cell must conduct a signal down its axon, this electrolyte gradient changes rapidly so that the electrical charge spreads down the length of the axon in milliseconds until the signal reaches the axon terminal. There, the neuron produces various neurotransmitters that are released across the synapse (the gap between the axon end and the next nerve cell). The neurotransmitters migrate across this gap and land on specific receptors on the next nerve cell, which then signal to that cell to begin propagating its own signal. This is obviously an oversimplified version of what

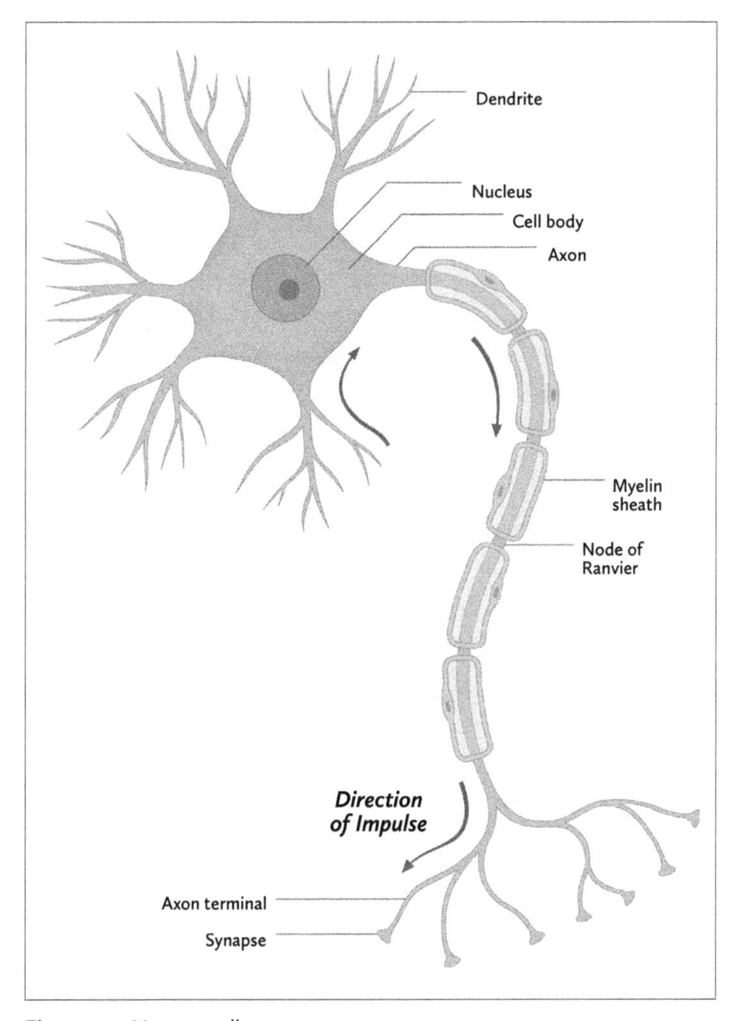

Figure 4.1 Neuron cell

actually happens in a nerve cell, but this gives some idea of the basic elements of nerve cell function.

Several specific points are worth mentioning with regard to nerve cell function. One is that nerve cells can be interconnected with hundreds or thousands of other nerve cells in terms of both receiving and sending signals. *Incoming signals are referred to as afferent, and outgoing signals are referred to as efferent.* This terminology will be discussed later when considering various spiritual states. There are many different neurotransmitters that have a variety of specific effects in the brain, which will be considered in more detail later in this chapter. However, it is important to realize that some neurotransmitters cause excitation of the nerve cell so that the signal is sent on to the next nerve cell, while other neurotransmitters cause an inhibition of the nerve cell. *The ability of different neurotransmitters to excite or inhibit different neurons is of critical importance in brain function.* Obvious examples of when these functions go awry are seizures, in which cells cannot be inhibited from firing, and depression, in which cells cannot be excited to fire. And for religious and spiritual practices and experiences, we might expect some areas of the brain to be turned on and others to be turned off.

In addition to having multiple ways of connecting and communicating, nerve cells also develop very specific functions. Some nerve cells have simplistic functions responding only to one or two other cells or types of input, while others receive information from hundreds of other nerve cells and integrate information into complex thought or language. Some nerve cells respond to specific sensory stimuli, such as a color, shape, or motion. Other nerve cells contribute to memory or higher cognitive processes. Some nerve cells contribute to moving different body parts and are connected to specific muscle groups to move the legs or arms. When there is damage to a group of these nerve cells, then the function that they subserve is altered. This can be observed in patients with a stroke in which they lose the ability to speak or to move one side of their body.

Unfortunately, nerve cells do not easily grow back or develop new connections once damaged, and therefore many brain injuries result in permanent changes. Great effort is under way to discover how to regrow damaged nerve cells, a process called "neuroplasticity," but research in this field has progressed relatively slowly. On the other hand, healthy nerve cells can develop new connections throughout the life span of the individual, which allows for learning new concepts and behaviors. Human beings can continually adapt and change even into old age provided that the nerve cells remain healthy. There is growing evidence that practices such as meditation or prayer can help nerve cells stay healthy and enable to them to grow and connect more with other neurons. Of course, the most dramatic ability to develop new connections occurs in childhood, when basic knowledge is acquired regarding mathematics, language, problem solving, practical knowledge, and social skills. The brain of the child is constantly forming new connections between nerve cells so that the child can learn all of the new things that are required to grow and adapt.

While nerve cells are the workhorse of the brain, human beings would not have a single thought if it were not for the many supporting cells that also compose the nervous system. Perhaps the most important type of cell other than the nerve cells are the myelin cells.

These myelin cells line the nerve cells and form a kind of insulation to ensure that the electrical signals that pass down the axon actually get to the axon terminal. In fact, without the myelin cells, most nerve impulses would not make it to the next nerve cell, resulting in severe if not altogether dysfunction of the brain. There are several neurological disorders that affect the myelin cells, such as multiple sclerosis, which can have devastating consequences for the afflicted individual. Other nerve cells help to support the structure of the brain by providing a type of scaffolding. While the other cells in the nervous system are not as likely to be principally involved in religion and spirituality, they are nonetheless necessary for adequate functioning of the brain as a whole.

The Autonomic Nervous System

The autonomic nervous system is one of the most basic parts of the nervous system and helps connect the brain to the body. The autonomic nervous system is responsible, with input from the rest of the brain and central nervous system, for maintaining baseline bodily functions. It also allows the body to respond to various environmental stimuli. The autonomic nervous system also plays a crucial role in the overall activity of the brain as well as in the expression of fundamental emotions, such as fear, joy, happiness, sadness, and shame. And the autonomic nervous system likely plays a critical role in the religious and spiritual feelings we have and how they are experienced throughout the body.

The autonomic nervous system is traditionally understood to be composed of two subsystems, the sympathetic system and the parasympathetic system.[6] Both of these systems innervate almost every body organ and generally have a "push-pull" effect, so that if one side increases the heart rate, the other side decreases the heart rate; if one side dilates the eye, the other side contracts the eye. The sympathetic system supports the so-called fight-or-flight response and is the physiological basis of our adaptive strategies either to noxious stimuli or to highly desirable stimuli in the environment.[7] *In short, the sympathetic system causes a sense of arousal—an arousal system.*

The second part of the autonomic nervous system is the parasympathetic system. This system is essentially the antithesis of the sympathetic system. It might be thought of as a calming or quiescent system. The parasympathetic system is typically understood to be responsible for maintaining homeostasis (the overall balance of body function) and conserving the body's resources and energy. *In short, the parasympathetic system causes a sense of calming or quiescence—a quiescent system.*

Each arm of the autonomic nervous system innervates and regulates many of the body's organs such as the heart, lungs, digestive system, and eyes. These two arms of the autonomic nervous system have often been described as "antagonistic" or "inhibitory" to each other. Normally, increased activity of one tends to produce a decreased activity in the other. Thus, each system is physically designed to inhibit the functioning of the other under most circumstances. This helps prevent an excess of the activity of either system. Some researchers have argued that the specific balance between these two systems at baseline represents the emotional set point we bring to the world—whether we are "uptight" or "laid-back." (See figure 4.2)

As additional studies have been performed, it has been realized that the interaction between the sympathetic and parasympathetic systems is much more complex. There is some evidence to suggest either that there can be a rapidly alternating activation of both the sympathetic and parasympathetic systems or that both may even be activated simultaneously.[8] These more unusual types of interactions may occur when one system is excited to extreme amounts. For example, continued physical exercise such as marathon running may increase the sympathetic system to very high levels. Deep stages of meditation may result in an increase in the parasympathetic system to very high levels. It may be at these high levels of activity that there are reciprocal interactions between both sides of the autonomic nervous system.

A number of investigators of spiritual experiences have considered the activity in the autonomic nervous system very relevant to these experiences. Thus, experiences that result in great exaltation and a powerful sense of energy overwhelming the body might be related to activity in the sympathetic nervous system. And experiences that include an intense sense of calmness or blissfulness might be related to activity in the parasympathetic system.

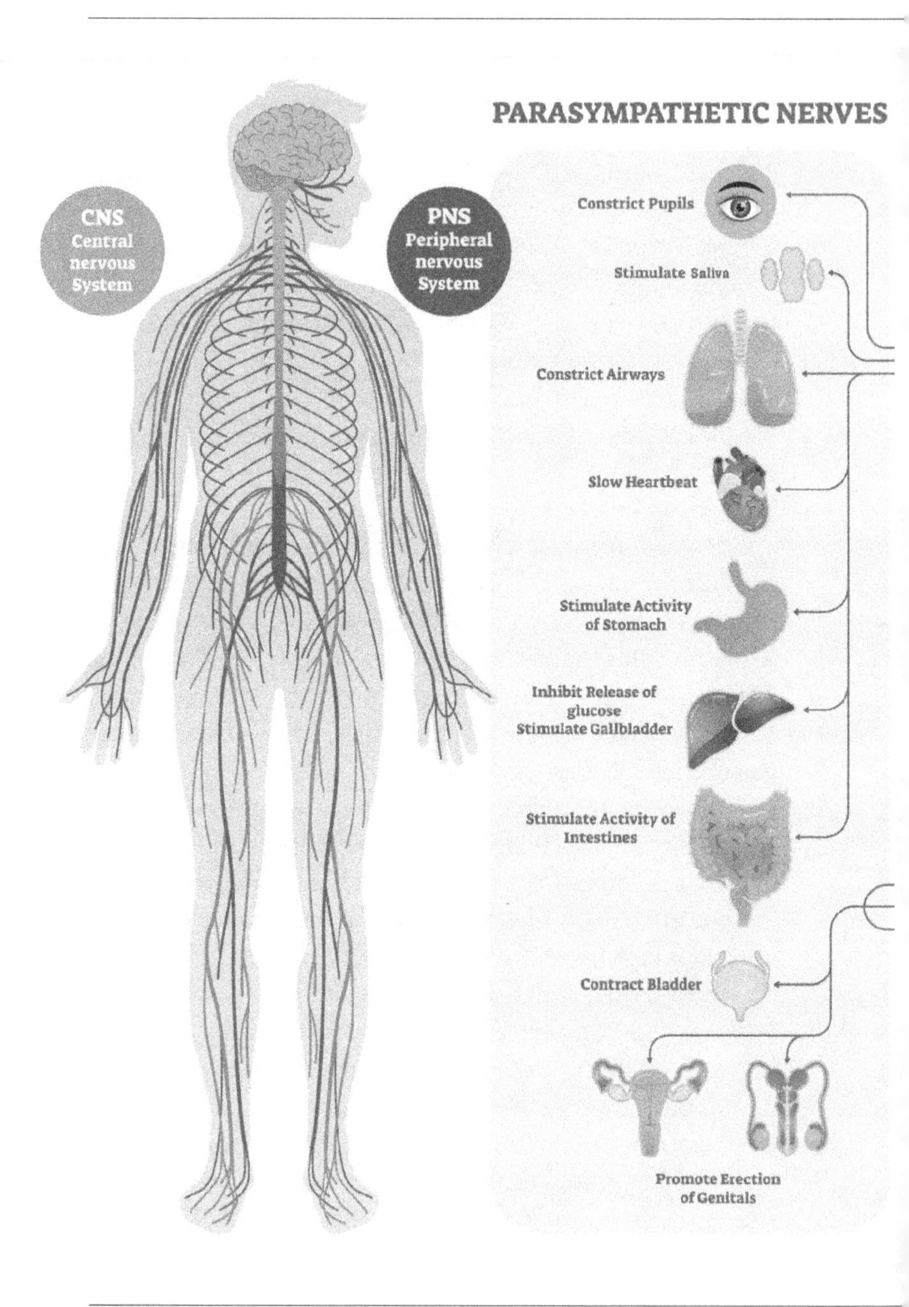

Figure 4.2 Autonomic nervous system

SYMPATHETIC NERVES

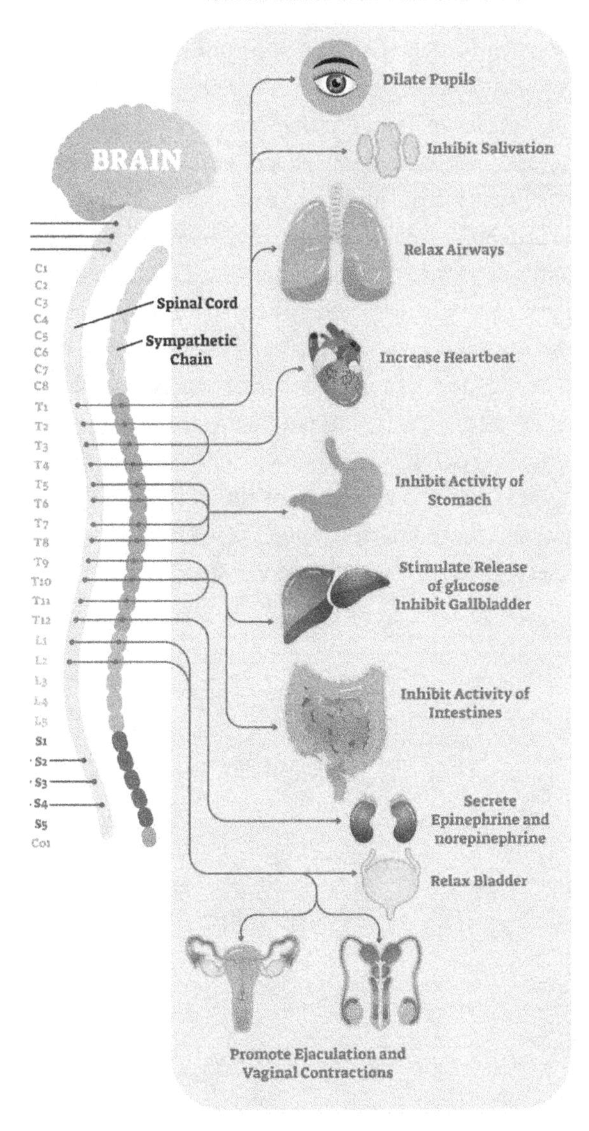

The Structure of the Brain

Neuroscientists have divided the brain into numerous subdivisions in order to make locating parts of the brain easier. *The first subdivision divides the brain in half so that there is a left and right hemisphere.* Anatomically, both hemispheres look almost identical except that they are mirror images of each other. Each hemisphere contains the cerebral cortex, which is generally considered to be the seat of higher-level cognitive functions. The vast majority of the cerebral cortex is also called the neocortex (meaning "new cortex"), because this is the most recently developed part of the brain from an evolutionary perspective. It is this part of the brain that is believed to separate human beings from other animals, since it is the seat of our most distinguishing characteristic, our intelligence. It is also the evolution of the neocortex that is typically attributed to the major developments in language, myth, art, culture, society, and, ultimately, religion. (See figure 4.3)

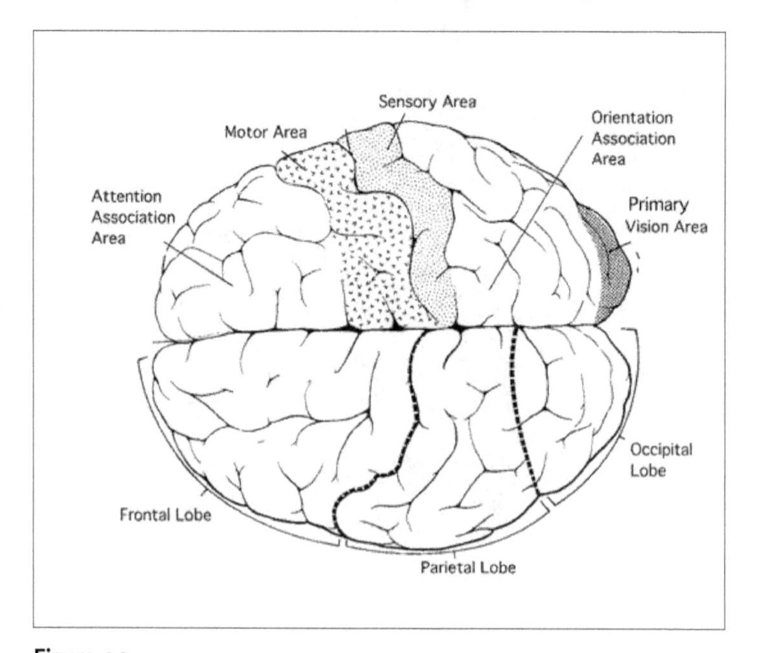

Figure 4.3

The hemispheres also contain subcortical structures (i.e., below the cortex), such as the thalamus, the hypothalamus, and various midbrain structures. These subcortical structures are involved in basic life support, hormone regulation, and primal emotions. Thus, these subcortical structures are also critical for connecting the brain with the rest of the body. The subcortical structures are also involved in sensory processing and relay higher-order information from the cortex to other parts of the brain. In addition, to the neocortex and the subcortex, there are a group of structures that are collectively referred to as the limbic system. (See figure 4.4)

The limbic system is associated with complex aspects of emotional expression and is involved with assigning emotional value and content to various objects and experiences and directing these emotions to the external world via our behavior. The limbic system is also intimately connected to the autonomic nervous system to help connect various emotional responses to feelings in the body. For example, when we cognitively become aware of our loved

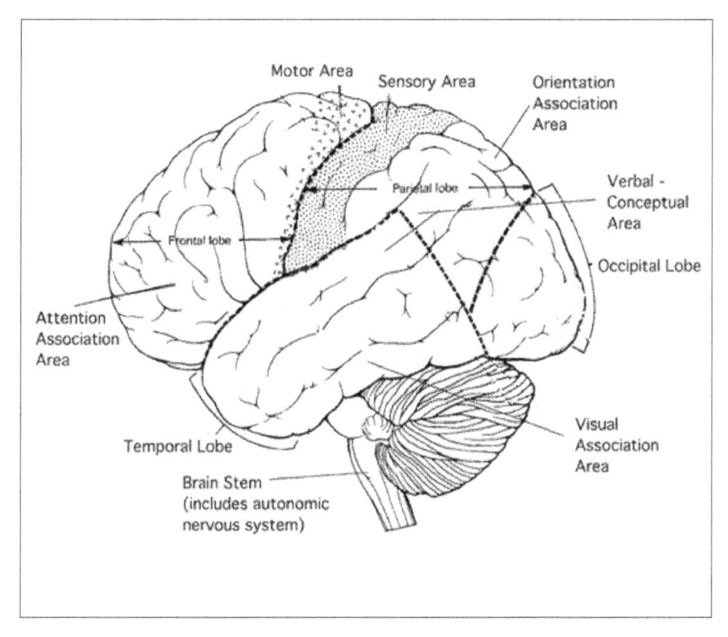

Figure 4.4

one entering the room, the limbic system relates the feeling of love and joy, causing a release of adrenaline via the sympathetic system. This reaction causes us to feel the heart rate increase and the overall level of alertness to increase—we become excited to see that person. In this way, we not only become aware of our positive emotions, but we experience the whole body's reaction to them as well.

There is a great deal of similarity between the two cerebral hemispheres. For example, the left cerebral cortex receives and analyzes sensation from the right side of the body, and the right cerebral cortex receives and analyzes sensation from the left. The left cerebral cortex generates movement in the right side of the body, and conversely, the right hemisphere generates movement in the left side of the body. Despite the similarities, there are also many differences between the two hemispheres. The classic teaching is that the left hemisphere is more involved with the analytical and mathematical processes as well as the time-sequential and rhythmical aspects of consciousness (for example, the time-sequential aspect allows us to perceive what we sense as the passage of time). The left hemisphere is also the usual site of the language center, which is that part of the brain that understands and produces written and oral language. It is because of its language capability that the left hemisphere has often been referred to as the dominant hemisphere, while the right has been termed the nondominant hemisphere. The right hemisphere is usually more involved with abstract thought distinct from language, nonverbal awareness of the environment, visual-spatial perception, and the perception, expression, and modulation of most aspects of emotionality.[9] This perspective on the left and right hemispheres has been substantially revised and clarified. It turns out in the end that both hemispheres are typically involved in most mental processes. For example, by having the left and right hemispheres working together, we can read the words in the Torah, feel the emotional effects of prayer and group religious activities, and understand the rational concepts of theodicy, free will, and omnipotence.

In order for the left and right hemispheres to work together, there must be a connection between them so that communication can occur. The two hemispheres are able to interact with each other via three structures, called the corpus callosum, the anterior commissure, and the posterior commissure. These connecting structures contain nerve fibers that link the left and right hemispheres. The true nature of the communication between the hemispheres is still under investigation, although it appears that these connecting structures do not allow all information to be transferred from one hemisphere to the other. It appears that only certain types of limited information can cross between the hemispheres. Complex thoughts or perceptions in one hemisphere cannot be transferred to the other in an exact manner. Only simplistic representations or nuances of a thought or perception can be transferred.[10] However, the general activity of one hemisphere can be transferred to the other hemisphere via these connecting structures.

There are some fascinating things that occur, though, when the hemispheres are viewed separately. In fact, they appear to be able to maintain two separate consciousnesses.[11] Furthermore, these consciousnesses can function independently of each other. Roger Sperry did some of this work with "split brain" in patients who had the two hemispheres surgically isolated from the other by cutting the fiber tracts between them (performed in patients with seizures that do not respond to medical therapy). If a picture of a hammer is presented so that the image is shown to only the left hemisphere visual system (which is possible because each hemisphere receives visual information from the other side of the body), then the person will be able to write the name of the object—"hammer." If the picture is presented so that the image is shown so that only the right hemisphere visual system experiences it, then the person will not be able to write the name of the object but will be able to draw a hammer.

In order to truly understand how the brain works, we must consider the cerebral hemispheres as they work both separately and together. Therefore, the left and right hemispheres work together so that the left hemisphere can understand what is being said, and the right hemisphere can understand how it is being said in terms of emotional nuances.[12] In the context of neurotheology, such an ability is crucial for understanding how various individuals interpret sacred texts. Do we take the story of Adam and Eve literally or figuratively and with which types of emotions? And might we argue that a rabbi who interprets the Torah more emotionally is dominated more by the emotional centers and right hemisphere compared to a rabbi who is more analytical? Could one interpret the two narratives of the creation of Adam in Genesis to be referring to the two "Adams" within our own minds? Certainly splitting Adam into a "scientific" task-oriented Adam and a "creative" experiential Adam coexisting within the same body has been postulated before (see Rabbi Joseph B. Soloveitchik's work *The Lonely Man of Faith* for further elucidation of Adam 1 and Adam 2). Perhaps it is even possible that such differences pertain to the more "male" an "female" oriented brain processes. Although on the surface, the brain of men and women appears the same, we all are aware of the potential differences between how men and women process cognitive, emotional, and experiential information. It has even been argued that originally the male and female aspects of humanity were joined in one being and separated to create Adam and Eve.[13] In some sense, the brain maintains both aspects but also keeps them separate.

Another example of these two opposing neural stances may also form some of the basis around the interesting interpretation of the Song of Songs (*Shir HaShirim*). The story, which is placed in the biblical canon, is on the surface a love story, where the woman longs for her lover who has left her, ultimately missing her chance to have him return to her, and then continues to search for him. The Mishnah (*Yadayim* 3:5) states that the story itself was considered potentially too inappropriate to include in the canon. Rabbi Akiva, however, argues that *Shir HaShirim* is the "holy of holies" and most certainly should be included, and so it was. However, for what reason was *Shir HaShirim* included? Many of the traditional commentators explain that the verses are not to be taken literally at all and instead are an allegory to the relationship of God and Israel, with God being the *dod*, or the lover, and Israel being the woman. The story is seen as a depiction of Israel failing to adhere to its relationship with God, but there is still hope for the future. The traditional commentators seek to explain the emotional content of the love story not as simply referring to a man and women, but of a much larger emotional-religious connection: that of human beings and God. The interpretations range, however, from those that are more grounded in relating everything in the allegory to the text to those that suggest that the parable is the only piece that should be looked at. It is possible that of all the rabbis of the Talmudic period, Rabbi Akiva may have had a significantly larger impact from his right "emotional" hemisphere, especially considering he started formal education at age forty. Given his unique understanding and place in the tradition, it made sense for him to be the one to argue for the inclusion of *Shir HaShirim* in the canon, as its emotional draw was palpable and spiritually satisfying, even though to those of a less emotionally inclined nature, the story might be deemed too vulgar for canonization. Ultimately, the fact that the story was included may reveal aspects of the nature of how the rabbis felt human beings should attempt to interact and connect with God as well.

Each cerebral hemisphere is also divided into four major regions or lobes (for a total of eight): the two frontal lobes, which are located at the front of the brain; the parietal lobes, which are located near the back-top of the brain; the temporal lobes, which are located along the side aspect of the brain: and the occipital lobes, which are located at the very back of the brain. Within each lobe exist smaller regions, which either have names, like the hippocampus or amygdala, or are categorized according to a cerebral map that was first developed by Korbinian Brodmann (for example, Brodmann's area 17).[14] Over the course of time, neuroscientists have tried to determine which brain functions are associated with what parts of the brain.

The Association Areas

The brain processes information in a somewhat stepwise manner. *Thus, the brain has primary processing areas in which basic information is either received or initiated.* For example, the primary visual cortex in the occipital lobes does the initial processing of the lines, shapes, and colors coming in from the external world. Neurons in this area respond to specific shapes and colors. *Input from the primary processing areas is ultimately sent to association areas that put vast amounts of information together.* The visual association area then helps put together a vivid, three-dimensional visual representation of what we see. There are also association areas that combine visual, auditory, and other sensory input to give us our full experience of the world around us.

We will focus on association areas that appear to have been implicated in one form or another with regard to overall higher brain functions and may be particularly relevant to certain aspects of spiritual, religious, or theological phenomena and concepts. Once again, it is important to realize that this review is not meant to be exhaustive, but to provide enough information for those not as familiar with neuroscience. We can also consider how these might specifically relate to Jewish ideology.

The superior parietal lobe is located toward the top of the parietal lobe. This area receives touch and body position information from the body but also receives input from the visual receiving areas, from motor and non-motor areas, auditory areas, and the areas of the brain involved in language and cognitive processing. *Thus, the superior parietal lobe is heavily involved in the analysis and integration of higher-order visual, auditory, and body sensory information.* In fact, a single neuron in this area can monitor activities occurring in many different parts of the body simultaneously. This area is able to inform us of the position and movement of the arms, trunk, and legs at the same time.[15]

Combining auditory and visual input, the superior parietal lobe is able to help create a three-dimensional image of the body in space.[16] Further, this region helps determine our position in space and can compare our internally felt location to the coordinate system of the outside world. This ability to generate a body image might allow for phenomena such as out-of-body experiences. This region of the brain is also critical for fixating on visual objects, the determination of the three-dimensional position of various objects in our visual field, and the identification of the relationship of these objects to the body and to other objects in the external world.

Overall, the superior parietal lobe appears to help us establish a spatial sense of our self. However, there may be some differences in function between the superior parietal lobe on the right and the superior parietal lobe on the left. Studies have found that patients with strokes or tumors in the right superior parietal lobe have deficiencies involving depth perception and the ability to determine location, distance, spatial orientation, and object size.[17] While the right parietal lobe appears to play an important role in creating a sense of spatial coordinates and body location, it has been proposed that the distinction between self and world may ultimately arise from the left superior parietal lobe.[18] Furthermore, the distinction of self and world probably involves other parts of the brain as well, including other parts of the parietal lobe. So the actual role of the superior parietal lobe, while important, must be understood in the context of the functioning of the rest of the brain. However, it may be that the "self-other" or the "self-world" distinction that philosophers and theologians have discussed throughout the ages, while a function of the brain in general, may be associated more specifically with the function of the left superior parietal lobe.

In Judaism, the idea of the self and other, famously described by Martin Buber's "I-Thou" relationship, is present from the first biblical story of creation. The way Rabbi Joseph B. Soloveitchik describes Adam's cognitive development is one that clearly implicates the superior parietal lobe to enable him to gather greater understanding of the world around him and his place in it. Initially Adam is asked to label and categorize all of the animals, which enables him to recognize that there is an "other" in front of him that he must confront. Additionally, some Jews have a practice when praying to visualize themselves in front of the Temple Mount or to visualize God's four-letter name in front of them when they pray. This experience requires orienting the mind toward that object and establishing a sense of how the self relates to that sacred object.

The prefrontal cortex has many critical functions including the ability to focus attention, to control emotional responses, and to initiate muscle movement and other behaviors. It is situated in the most forward aspect of the brain and is intimately and richly interconnected with the limbic system.[19] Likewise, this area is profusely interconnected with all the association areas.[20] The prefrontal cortex receives fibers from all sensory modalities (vision, hearing, touch, taste, and smell) as well as from other association areas.[21] The prefrontal cortex of each hemisphere is connected to the prefrontal cortex of the other by fibers running across the corpus callosum. Thus, it seems that part of the function of the prefrontal area is as a multimodal association area. This means that the prefrontal cortex can help integrate a wide variety of sensory and motor information. Further, this area functions to help give us a sense of "egocentric spatial organization," or how things are spatially oriented to ourselves.[22] However, we have already considered how another area, the superior parietal lobe, is more likely involved with our sense of space itself. As one might presume, there are many interconnections between the prefrontal cortex and the superior parietal lobe. These areas are both active whenever a person attends to a visual-spatial problem. And our research has shown a relationship in function between the prefrontal cortex and the parietal lobe during practices such as prayer or meditation.[23]

It should be noted that the prefrontal cortex is situated in the larger structure of the frontal lobe, which also contains the part of the brain that is responsible for movement—the motor cortex. Similar to the general coordination of the sensory system, motor function has multiple processing steps. The primary motor cortex (located toward the back of the frontal lobe) is responsible for the basic movement of every part of the body. Slightly in front of this area is the premotor cortex, which is the secondary motor association area. The premotor cortex is involved with the control of more complex, integrated movements. These movements are often associated with goal-oriented behavior such as reaching for a desired object. The prefrontal cortex acts as the motor association area with regard to coordinating highly complex movements. These areas are of great importance for various movement-related rituals. *Davening* likely begins via activity in the prefrontal cortex, which initiates and coordinates the body's movements so that they are enough to allow for repeated, rhythmic bowing, while also preventing us from tipping over. However, the functions of the prefrontal cortex are highly complex and go beyond mere motor function and include our ability to concentrate and modulate emotions.

Brain imaging studies of concentration tasks and working memory tasks implicate the prefrontal cortex as critical for helping with concentration. In humans, the loss of the ability to concentrate is a characteristic feature of prefrontal disorders, as is the loss of the ability to plan and to orient oneself to future behavior. Several reports have suggested that the prefrontal cortex, especially on the right, may function to filter out extraneous sensory input in order to allow for an adequate focusing of attention.[24] This function would be essential to all attention-related functions of human consciousness because there is otherwise an excess of sensory input. This abundance of sensory input would likely distract a person from any task that is being concentrated on. For example, this is what allows you to read a book when there are other people around or while the television is on. The prefrontal cortex helps in pushing all of the sensory stimulation into the background as redundant input and allows you to continue to concentrate on the book. Thus, patients with injuries in the prefrontal cortex not only lose the ability to plan and orient themselves to future activity, but also suffer a severe deficit in carrying out complex perceptual and conceptual tasks that require concentration or attention. They find it difficult to complete a lengthy sequence of activity or to organize complex behavior, to integrate sensory input, and to anticipate novelty.

The prefrontal cortex is also richly interconnected with the emotional centers of the limbic system (described below) and therefore plays a significant role in regulating emotions. Patients with lesions in the prefrontal cortex exhibit flatness of emotion and apathy and tend to have difficulty controlling emotion. In his earlier work, Joaquin Fuster noted that patients with injury to the prefrontal cortex seem to display a "profound indifference" to events and objects in the environment.[25] These patients cannot alter their patterns of response once established, and thus they are usually incapable of accomplishing anything beyond their initial pattern of behavior. On the other hand, sometimes patients with frontal lobe damage may not be able to control outbursts and may become enraged quickly and act on that rage. In fact, many violent criminals have abnormal functioning of the prefrontal areas, which may predispose them toward violent behaviors.[26]

From a religious perspective, the prefrontal cortex seems to be one of the most significant areas of our brains. Emotional modulation and integration of the various sensory and motor systems to modulate and focus attention is of paramount importance in all aspects of religious observance. Sitting down to study text requires a disciplined focus that requires significant modulation from the prefrontal cortex. Any religious action or physical behavior that one does is often assumed to be done for some sort of emotional or educational purpose, as we have explained previously. However, without focusing on why one is kissing the mezuzah on the door or saying the words one is reciting while praying, it is simply an action without thought behind it. The emotional regulation must be maintained in order to experience the feelings associated with the Exodus from Egypt at the Passover seder. To relive an experience one has never before had, one must be emotionally present in the moment, drawing on memories and textual stories and creating an image in one's mind one may not have ever seen. This would be the job of the prefrontal cortex.

The inferior parietal lobe, in conjunction with the superior temporal lobe, may be the area of the greatest integration of sensory input in the brain and appears to house a number of higher cognitive processes such as abstract thought, causal reasoning, logic, and language. In a sense, it is an association area of association areas and maintains rich interconnections with the vision, hearing, and touch association areas. This area also has extensive interconnections with the other association areas throughout the neocortex.[27] The inferior parietal lobe is also responsible for the generation of abstract concepts and relating them to words. It does this through rich interconnections with the language center, which is primarily located in the left hemisphere and incorporates much of the temporal lobe and parts of the frontal and parietal lobes. The inferior parietal lobe is also involved in conceptual comparisons, the ordering of opposites, the naming of objects and categories of objects, and higher-order grammatical and logical operations. As we shall see, this region might be very important in the development of consciousness and the expression of that consciousness through language.

In Judaism, the inferior parietal lobe would be the brain region engaged when performing traditional textual analysis and analytics in Talmud and Jewish law. The relation of the abstract to concrete might also lend the inferior parietal lobe to the formulation of the kabbalistic concepts and spiritual phenomena into the current theology after one has experienced what at first might be considered an indescribable spiritual experience.

The Limbic System

The limbic system is that part of the brain that generates and modulates most feelings and emotional responses. The limbic system is responsible for the broad array of emotions in human beings, ranging from simple feelings such as fear, happiness, sadness, aggressiveness, and love to subtle feelings such as envy, embarrassment, and melancholy. Since many of these responses, even if they are on a more primitive level, are absolutely necessary to life, it seems appropriate that all animals should require a limbic system in order to interact with the external world. After all, animals must be aggressive to find food (especially predators). They must have some form of affiliative behavior (i.e., primitive love) for there to be offspring and, in higher animals, so that they care for their offspring until they can survive on their own. Animals also must be able to fear predators and to avoid life-threatening situations.

In human beings, the limbic system has a similar function in terms of modulating and generating emotions.[28] In order for human beings to have a full range of emotions and the ability to apply them to various experiences and thoughts, the limbic system must be intimately interconnected with the entire brain. This allows human beings not only to have complex thoughts, but to assign emotional value to these thoughts. Thus, human beings may contemplate God, arrive at some conclusion, and back that conclusion with a strong emotional response. The four most important areas of the limbic system are the hypothalamus, the thalamus, the amygdala, and the hippocampus.

The hypothalamus is a central structure in the brain and is responsible for maintaining many of the body's functions, including the heart rate, respiratory rate, blood pressure, as well as most of the hormonal systems. The hormone systems that are controlled by the hypothalamus include the sex hormones, thyroid function, and stress hormones such as cortisol. These are actually regulated by the hypothalamus in conjuction with a small projection from the brain called the pituitary gland. The part of the hypothalamus that is closest to the midline represents an extension of the parasympathetic system into the brain, thus connecting the parasympathetic system with the rest of the brain. By contrast, the outer edge of the hypothalamus seems to be an extension of the sympathetic system into the brain.[29] The hypothalamus is involved with fight-or-flight responses, with sensations of terror or rage, and with the sensation of positive emotions ranging from moderate pleasure to bliss. An important aspect of the emotions generated by the hypothalamus, whether positive or negative, is that they tend to be stimulus bound. This means that they respond to a specific stimulus and then die off very quickly when the stimulus is removed.

The amygdala, located in the middle part of the temporal lobe, is more recently developed evolutionarily than the hypothalamus and is preeminent in the control of higher-order emotional and motivational functions – it is the watch dog of the brain. It has extensive interconnections with various regions of the neocortex and with subcortical structures, through which it is able to monitor and determine which sensory stimuli are of motivational significance to the organism. In addition to emotional and motivational functioning, the amygdala is involved in attention, learning, and memory. After all, it makes sense that the brain should help us remember things that are emotionally important. Like the hypothalamus, the amygdala is divided into a middle part (toward the midline of the brain) and an outer part (away from midline). Although the function of the amygdala is complex, it is becoming clear that the amygdala has primarily an arousal or alerting function. However, the amygdala does have some parasympathetic functions. Further evidence of the amygdala's sympathetic function is the fact that it performs environmental surveillance and maintains attention if something of interest or importance appears in the environment.[30] Further, the amygdala modifies the function of the hypothalamus and can subserve the emotions of fear, rage, joy, and compassion, important emotions for religious and spiritual experiences.

The amygdala might be considered one of the primary mechanisms for how religious and spiritual relationships with God are experienced. The emotional responses of love, anger, or awe as well as an array of positive or negative feelings can be directed toward religion in general, God, or even the rabbi or spiritual leader. While only theoretical, this conception does allow for an explanation for just how it is that "deeply religious" individuals can see events in wildly different ways from others on an emotional level. We have previously argued that the amygdala is crucial for identifying the importance of various religious ideas and objects. Sacred things are of great emotional and cognitive significance, and the amygdala is likely the part of the brain that helps identify that which is sacred.

The hippocampus, which is shaped like a telephone receiver, is located slightly behind the amygdala in the temporal lobe. The hippocampus plays a major role in information processing, memory, new learning, cognitive mapping of the environment, attention, and some orienting reactions.[31] The hippocampus is greatly influenced by the amygdala, which in turn monitors and responds to hippocampal activity. Thus, the hippocampus and amygdala complement each other and interact with regard to the focusing of attention and the generation of emotions linked to images, as well as learning and memory. Thus, the amygdala may identify the sacred, and the hippocampus helps us remember it.

The hippocampus also regulates the activity in another structure that connects the autonomic nervous system to the neocortex, called the thalamus.[32] Since the thalamus is the major sensory relay system in the brain, the hippocampus often appears to be able to block information input to various brain areas. While the amygdala may enhance information transfer between brain regions, the hippocampus usually tends to do the reverse. Thus, these two structures create a push-pull effect in the transfer of information between brain regions as well as help balance emotional responses.

Several other structures require brief mention, including the thalamus, insula, and cerebellum. The thalamus is a central structure in the brain that functions as a key relay between our sensory organs and the higher cortical areas of perception as well as between interacting parts of the brain.[33] The thalamus allows sensory stimuli, particularly vision and hearing, to enter into the brain for additional processing. The thalamus also connects structures such as the frontal and parietal lobes to enable them to communicate more effectively. In fact, the thalamus has so many connections with different parts of the brain that some have suggested that it is the seat of consciousness.[34] When anesthetics are used, activity in the thalamus is particularly decreased. It would certainly make sense that the thalamus plays a vital role in complex rituals including meditation and prayer, as many different parts of the brain become involved with these practices and associated experiences. And we have found brain imaging evidence that the thalamus is particularly affected during meditation and prayer practices.

The insula is a structure that sits between the limbic system and the cortex and appears to play a critical role in the comprehension and evaluation of our emotions. The emotions appear to arise primarily in the limbic system but are understood in a more cognitive way through the activity in the insula and its connections to the cortex.[35] In this way, our amygdala may fire as we feel angry, but our insula helps us to understand why we feel angry and directs that information to the cortex to determine the best behavioral response. Practices such as meditation and prayer that might actually affect our emotional state may do so in part via the insula. In addition, sacred texts that consider various emotions such as love, joy, or awe, are comprehended in large part via the functions of the insula.

The cerebellum sits beneath the main cortex of the brain. Located more toward the back of the brain, the cerebellum has traditionally been understood to be involved in the coordination of motor function.[36] People with damage to the cerebellum typically have great difficulty with balance or with moving their hands and feet in coordinated ways. More recently, cognitive neuroscientists have been exploring other functions of the cerebellum, particularly with respect to emotions. It appears that the cerebellum coordinates not only motor activity but emotional activity as well. Several studies have demonstrated the importance of the cerebellum in regulating emotions as well as the response of the body to those emotions.[37] Although little is known about the relationship between the cerebellum and religious and spiritual experiences, it would seem to be an area of future investigation that might be quite important for neurotheology.

Activation and Inhibition in the Brain

Another aspect that is crucial to understanding how the brain works is that nerve cells in the brain can be either activated or deactivated, and information traveling between nerve cells can similarly be enhanced or blocked. It is through these various functions that the brain actually works. If there was no inhibitory mechanism, then the brain activity would build and build until it became out of control. This is exactly what happens during a seizure. On the other hand, if areas of the brain could not be activated, then no thoughts or behaviors could be elicited. In some sense, this is what happens in a coma or when a person is unconscious because of various sedating medications. How the brain turns different parts on or off is also relevant. For example, a number of studies have shown that more intelligent people actually use less brain energy to solve problems than less intelligent people. In this case, gifted people use their brain more efficiently. The brain itself is always "on" in the sense that the nerve cells always require a certain baseline amount of metabolic activity. Therefore, the overall activity can actually be turned up or down rather than on or off. If an individual is concentrating on a particular problem, the areas focusing on the problem will likely be activated and the areas that are not required turned down. This enhances the concentration and efficiency of solving the particular problem.

It is also possible for certain parts of the brain to regulate the nerve information flow between different parts of the brain. If a person is trying to solve a mathematical problem, then the prefrontal cortex is activated, and its stimulatory output is sent to the areas that perform quantitative operations. At the same time, the auditory input of people doing work in the room next door is blocked from reaching the attention areas so that the focus of attention can be maintained. ***This ability to block information is called deafferentation. Deafferentation means that incoming information (or afferents) into a brain structure are "cut off." This cutting off is an actual physiological process, which may be partial or total.*** Deafferentation can be caused either by physical interruption, such as by a destructive tumor or surgery, or by "functional" deafferentation. Perhaps the most investigated form of physical deafferentation involves patients with epilepsy. As noted above, in epilepsy, one part of the brain starts to get overexcited. This excitation can spread throughout the entire hemisphere and then can cross over to excite the other hemisphere. The result is a generalized seizure with heightened activity throughout the brain, which has traditionally been the hallmark of epilepsy. Sometimes, in severe cases that cannot be helped by medication, patients undergo surgery to cut the connector tracts between the two cerebral hemispheres. This procedure, called a commissurotomy, prevents communication between the hemispheres so that seizure activity cannot extend to both hemispheres. This deafferents the hemispheres from each other so that neither hemisphere is aware of what the other hemisphere is doing.

Functional deafferentation occurs when inhibitory fibers from a brain structure block transmission of information between two other structures. Thus, impulses from inhibitory fibers can block input into a neural structure. For example, one hemisphere can be prevented from knowing what is occurring in the opposite hemisphere via the inhibitory actions of the frontal lobes.[38] There is similar evidence that information transmission in specific areas of the brain can be partially or totally blocked by impulses originating in the prefrontal cortex. This is accomplished via inhibitory nerve fibers of the hippocampus.[39]

A particularly interesting aspect of deafferentation is what happens when a certain structure is totally deafferented. Normally, all parts of the brain are affected by numerous other parts. Therefore, the function of any structure is determined not only by itself, but by its interaction with many other areas of the brain. If deafferentation of a structure occurs to a significant degree, the neurons within that structure no longer are under the influence of any other parts of the brain, and they begin to fire on their own. These deafferented neurons either fire randomly or, more likely, function according to their own "internal logic."[40] This internal logic derives from the function for which a given structure evolved.

Let us consider an example of how deafferentation may result in generating unusual experiences that may be associated with altered phases of consciousness. Remember that the superior parietal lobe receives information from the sensory areas and generates a sense of space and time. It does this by integrating our senses of touch, sight, and hearing and creates an overall concept of space. If this structure is totally deafferented so that it receives no input from the outside world, then it cannot form a sense of space and time abstracted from sensory input. However, it is still trying to generate an orientation in space and time. It is still working by its internal logic. It continues to attempt to generate a sense of space and time even without input from the external world to work on. The result is a sense of "no space" and "no time," or conversely, it might be described as infinite space and infinite time. The person also loses the sense of self or feels as if he or she has become one with the universe of God. Such experiences are a fundamental part of many mystical experiences, including those arising from kabbalistic practices. We will discuss such experiences in detail in later chapters.

Neurotransmitters: The Chemical Communicators in the Brain

All of the above described functions are based on either clinical studies of patients with various neurological disorders, studies of the brain after someone has died, or brain imaging studies. These general functions can be reasonably attributed to the brain structures described above. However, it is also critical to understand how the nerve cells communicate information to each other so that the nerve cells and the structures know how to function together. After all, the brain really only works when all of the parts function as an integrated whole. In order for the different nerve cells to communicate with each other, they send signals via a large variety of chemicals called neurotransmitters. These neurotransmitters allow for the information to be transmitted from one nerve cell to the next. Each different neurotransmitter has distinct functional domains. While the field of neurotransmitter research cannot be described here in detail, it is important to at least have an understanding that these chemical messengers are responsible for relaying information throughout the brain and that some neurotransmitters in particular have been implicated in religious and spiritual experiences. Therefore, in order to briefly consider the neurotransmitters of the brain, only those that seem directly related to religious and spiritual experiences will be described, and these will also be considered in the specific context of studies that have measured their involvement in such experiences.

Several studies have demonstrated an increase in gamma-aminobutyric acid (GABA) in the blood serum of individuals during meditation.[41] And one study of meditation using magnetic resonance spectroscopy showed an increase in GABA concentrations inside the brain. *GABA is the principal inhibitory neurotransmitter in the brain.* In other words, when certain parts of the brain are turned down or when overexcited areas need to be decreased, GABA appears to be the principal neurotransmitter in this regard. This might also relate to how different areas of the brain might be deafferented during practices such as prayer or meditation. *The major excitatory neurotransmitter in the brain is glutamate, which turns various parts of the brain on.* During meditation or prayer, the increased activity in brain structures that support the act of performing practice are likely turned up with glutamate.

Serotonin is an important neurotransmitter involved in emotional processing, as well as influencing the visual association areas.[42] The cells of a central structure called the dorsal raphe produce and distribute serotonin.[43] Moderately increased levels of serotonin also correlate with a positive mood, while low levels often signify depression.[44] This has clearly been demonstrated with regard to the effects of the selective serotonin reuptake inhibitor medications such as Prozac or Zoloft, which have been widely used for the treatment of depression by causing an increase in serotonin levels in the brain. When the serotonin receptors in the temporal lobes are significantly activated, the stimulation can result in a powerful visual hallucinogenic effect. Psychedelics such as psilocybin and LSD seem to take advantage of this mechanism to produce their extraordinary visual associations.[45] Interestingly after meditation, breakdown products of serotonin in the urine have been found to significantly increase, suggesting an overall elevation of serotonin during meditation practices.[46]

Norepinephrine is a neuromodulator produced by the brain stem and is a stress-related hormone.[47] In the body, norepinephrine can produce significant increases in heart rate and blood pressure and ready the body for the fight-or-flight response. Not surprisingly, norepinephrine has generally been found to be reduced in the urine and plasma during meditation, because of the overall relaxation response.[48] Cortisol is a hormone, regulated by the hypothalamus and pituitary gland, that is also associated with the stress responses, and most studies have found that urine and blood plasma cortisol levels are decreased during meditation.[49]

Another neurotransmitter that is also very critical to normal brain function is dopamine, which is released as part of the "reward" system in the brain that leads to positive emotions and even euphoria. Dopamine is substantially involved in motor function but also has a role in schizophrenia, mood, attention, and higher cognitive processing. Brain cells that rely on dopamine are damaged in disorders such as Parkinson's disease, in which patients have problems with initiating movement and controlling balance. Patients with schizophrenia have classically been treated with medications that specifically alter the dopamine system. Acetylcholine is another common neurotransmitter in the brain, which allows for adequate nerve cell communication and is therefore involved in much higher order cognitive processing and memory. Patients with Alzheimer's disease tend to have decreased acetylcholine and sometimes respond to medications designed to enhance acetylcholine levels in the brain. The endorphins are the natural opiate system of the brain and have been implicated in religious and spiritual experiences primarily because of the pleasant and euphoric feelings that they help produce.

In all, the brain—its cells and its neurotransmitters—is incredibly complex. Although we can discuss various individual structures and functions, the brain works as an integrated whole. Hence, if there is a spiritual part of the human person, it is most likely the entire brain that helps support our full repertoire of emotions, experiences, thoughts, and behaviors. We will see throughout the rest of this book how all of these processes relate to the Jewish brain, Jewish practices, and Jewish beliefs.

Searching for Beliefs in the Jewish Brain

Standing at the Western Wall, I looked up at the ancient structure, once a retainer wall for the Second Temple, now one of the holiest sites in Judaism, as it is close to the Temple Mount itself. The experience was awe-inspiring. Staring at the stones that Jews have prayed at and placed letters in for millennia elicited a strong spiritual and emotional response in me, as well as in many of the others who were standing beside me at the wall. Here there was something special, something that bound all Jews together. However, as my time in Jerusalem continued, I began to realize that there was as much that divided Jews as united them. Even the Kotel itself became a source of contention between different groups of Jews with varying religious beliefs. So then what is a Jew? And what is it about Judaism that attracts people to that wall next to the Temple Mount just so much?

—David Halpern and Andrew Newberg

General Approaches to Jewish Beliefs

When one is talking about Judaism, one first needs to define what beliefs, practices, and theology make up the religion known as Judaism and then focus on the mental processes associated with those processes. However, a more fundamental question needs to be asked first about Judaism in general: Does one need a basic belief to begin with in order to be Jewish?

Some modern Jewish thinkers have argued that this is not the case. Simply having a distinct culture with emotional and moral/ethical principles informed by Jewish values are enough for some to label a person as Jewish. For this camp, neurotheology might be helpful in discerning the biological underpinnings of cultural practices and behaviors that are subsequently enough to explain the general ethical impulses and religious experiences of Judaism. This perspective seems in line with the often quoted idea that Judaism is a religion of deed and not creed. Others use the older matrilineal descent to label someone as Jewish. In other words, if your mother was Jewish, then you are Jewish, no beliefs necessary. While this is generally true as well, there does exist the possibility that someone might lose their identity as a Jew if they were knowledgeable enough to actively walk away from the beliefs that make up traditional Judaism (See Rabbi Aharon Lichtenstein's article in "Leaves of Faith: The World of Jewish Living." Chapter 3: Brother Daniel and the Jewish Fraternity, pages 57-84).

How about for rabbinic training? Is there a certain belief that one must have in order to be a rabbi? To be eligible for training as a rabbi? Depending on one's denomination, this response will vary. All denominations of Judaism will agree that in order to be a rabbi one must have the belief in the positive value of Judaism and the belief that there will be a Jewish future that one must cultivate. Beyond that, however, on a more spiritual plane, the different denominations may have very different required beliefs.

And what about the rabbis themselves? Well, we asked the rabbis in our survey what we thought would be the most fundamentally important religious question: Do you believe in God? Only 42 percent of all participants stated that they "fully believed" in God. Of those surveyed, 52 percent said that they believed in God but had significant questions about God's existence and what God was. Five percent stated that they were agnostic, and less than 1 percent said that they did not believe in God's existence at all. The proportions did vary by denomination somewhat. Of the Orthodox rabbis, 65 percent said they believed in God fully, but 35 percent had questions. Of the Conservative rabbis, about 52 percent believed in God fully, and 43 percent had questions.

For the Reform rabbis, the values were generally reversed, with 59 percent stating they had questions and 38 percent believing fully. The Reconstructionist rabbis answered similarly, with 56 percent having significant questions and 34 percent believing fully. The Reconstructionist rabbis also had the most agnostics. This brings us to the main point, which is to determine what beliefs are necessary to be considered a rabbi or even a Jew.

But before further engaging in an analysis of what rabbis believe, we must understand where they come from. How does one receive rabbinic training? Where does the accreditation come from? In the past, rabbinic ordination, or *semicha*, as it is commonly referred to among those in the Orthodox Jewish community, was given from one teacher to his students when they had reached a level of training and proficiency in study that they could go off and be arbiters of law on their own. Over time with the creation of the large centers of Jewish learning in Europe during the medieval period up until World War II, *semicha* was given through largely oral examinations covering a variety of topics on Talmudic and Jewish legal practice, where one undergoing the exam often did not know beforehand what would be asked. The exams were therefore meant to reveal one's breadth of Torah knowledge and acumen.

There were differing levels to the exams as well, with *Yorah Yorah* given to one who was proficient minimally in the basic Jewish legal questions congregants or laypeople would come to ask their rabbi. Another level, called *Yadin Yadin*, was and is required in order for a rabbi to sit on a rabbinical council known as a *beit din* to judge financial and other legal disputes. However, with time, many different models of *semicha* emerged. Now, in order to get rabbinic ordination from Yeshiva University's Rabbi Isaac Elchanan Theological Seminary, students must enroll in a four-year program, completing specific exams and topics relevant to a rabbi, as well as intern under a supervisor in order to gain practical knowledge of the world of rabbinics that they will be entering. Much of this change has to do with how the rabbi has come to be perceived and how the role has changed substantially throughout history.

Many rabbis now are the primary counselors and first-line responders to congregants' needs when a crisis or any sort of family event occurs. The rabbi is not necessarily viewed as a repository of Torah knowledge, though he or she should be minimally able to decide on many of the most common Jewish legal questions per the training. Instead, the rabbi is often sought after as a communal leader and liaison between parties in dispute as well as someone on call to respond to the needs of congregants. However, this is not always the case. Many rabbis spend their days studying Jewish legal texts, the Torah, and the related commentaries and texts, in order to establish an atmosphere of Torah study in their communities. Others run nonprofit organizations or outreach programs, and some simply have a rabbinic degree and continue on to other professions (such as being a physician). Whatever the case, it is also important to recognize that many rabbis do not have formal rabbinic training or recognition from a specific program. Some can simply be called "rabbi" due to the job they have, and others can be ordained from a variety of programs via distant correspondence and simply sitting for one exam. The level of training is quite disparate and varies with the programs and personal rabbinic teachers. Therefore, most rabbis today are judged not on their background and training, but on their recommendations and personal interactions with congregants, as well as their level of Torah knowledge and collection of other social and political skills, depending on the job they are seeking.

What is most interesting about this rabbinic training and confirmation is that there seems to be, in many ways, an external validation of the religious process. With any religious certification, the most important component, the internal commitment, is never entirely discernible to the outside examiner. As such, those strands of the tradition that have valued study and analysis of the text as the main arbiter of the right to call oneself "rabbi" will often have a more rigorous vetting process for the giving of *semicha*. One could posit that the intellectual tradition of Maimonides and the subsequent rationalist tradition that bases itself off of his theological writings in part would subscribe to this position, while those of a more "spiritualist" perspective might be less inclined to attribute much significance to a specific rabbinic training program. The spiritualists instead base their assessment on the persona and personal perception of the individual's religious commitment. The ability to function as a kind of psychologist and interpersonal mediator is of prime importance. One might wonder what brain processes support those rabbis who make these interpersonal interactions the primary component of their practice, compared to others who focus on Torah study. It is well known that the various social areas of the brain can help people communicate more effectively and be more compassionate and empathic toward others. These brain areas are distinct from those associated with abstract reasoning and logic, which might be used to evaluate the Torah or Jewish law.

As is often the case with modern Judaism, the reality is more complex than this bifurcated model would allow. As we discussed above, the giving of rabbinic ordination takes place in many circumstances across many different groups of Jews, some of whom do not recognize the ordination of the others. Using this analysis, one might say that the concept of the rabbi itself has changed significantly over the centuries, as well as who gets to be a rabbi and how they implement and understand their own personal religious convictions. Rabbis may be seen as arbiters of Jewish law, as connectors who help facilitate communication between people, as therapists for those in search of spiritual and psychological guidance, and sometimes as spiritualists who have access to places and understandings that are beyond the basic initiate.

In the Hasidic movement of Europe, the rebbe was seen as a dynamic spiritual *tzadik*, or righteous person, who was fundamentally spiritually different. Rav Schneur Zalman, in his work the *Tanya (See Likutei Amarim Chapter 27)*, states that the *tzadik* fundamentally does not have the same struggle as the average person. He does not push away evil inclinations all day, engaged in a constant struggle, but rather has a direct conduit to the spiritual oneness of God and can use this as a way to help and assist the masses who need it. This belief in a dynamic spiritual anchor has been one that has influenced Jews for generations since. Even though the particular Hasidic movement does not constitute the majority of Jews by any means, their ideas have penetrated deep into the minds of the Jewish communities around the world. Many in search of a rabbi will often believe that the rabbi can help them in some way, that he or she has a connection that they cannot or do not have. Many rabbis of the more rationalistic school will counsel people against such perspectives, often recognizing that they may be psychologically pairing the rabbi with God, and thus blaming the rabbi or praising the rabbi for things beyond his or her control.

From a neurotheological perspective, the brain of a rabbi will fundamentally have the same building blocks as any other individual. However, the components that are emphasized in rabbis' religious experiences will be different. For example, we surveyed rabbis with Orthodox, Conservative, Reform, and Reconstructionist affiliations and found that there were some telling trends of differences between the groups. Many more of the rabbis in the Conservative, Reform, and Reconstructionist branches of Judaism had experienced some form of a mystical experience. In addition, the Orthodox rabbis as a group were far less likely to think of there being multiple pathways to God and religion and were far more likely to characterize God as described by the Torah. That is not to say that many of the Orthodox rabbis were 100 percent sure of their characterizations. Some did express doubts, but that doubt was not enough to stop them from recognizing their beliefs as true to them, to the exclusion of other possible religious belief systems.

When one analyzes these perspectives, it would appear that the more classic spiritual experiences we have been describing until now would more easily fit in a group whose framework is less rigidly confined to intellectual stimulation and engagement and more focused on the spiritual experience itself. One Reform rabbi from our survey stated that during his experience he felt "connected to a sacred dimension all around me and holiness within me." This description implies that those segments within Judaism that ascribe less significance to the literal rendition of the Torah and the laws learned from it would have more similar attitudes to those of the generally considered "Eastern" religions and meditative processes. We might think that such individuals have had one or more times in which their brain functions contributed to powerful spiritual experiences of oneness and selflessness, an experience that may be associated with changes of activity in the self-orienting part of the brain called the superior parietal lobe (located in the back-top part of the brain).

One might even suggest that those more oriented toward scholarly study, without the spiritual epiphany, as being more prone to focus on the inferior parietal lobe, located more toward the midlevel of the brain, which allows for abstract thinking, language processing, and the categorization of objects. These individuals are not looking for the intense experience of the kabbalists or other spiritual seekers, but a grounded slow creation of a worldview built on study and intellectual reasoning. This perspective allows for a more concrete worldview, at the expense of limiting the scope of the religious experience to those who can engage textually and intellectually with the material. This is obviously not 100 percent the case even for Orthodox rabbis, who quite clearly, based on personal experience as well as answers in the survey, also have significant emotional and spiritual experiences to draw from. However, the additional emphasis on scholarly research and the existence of an objective intellectual truth often places the Orthodox rabbi in a fundamentally different category.

For Conservative rabbis, the answer may be somewhere in the middle of the continuum, depending on their specific training and life experiences. One may be more spiritually inclined but still feel drawn to the study of text as a way to approach God. On the other end of the spectrum, many Reform and Reconstructionist rabbis surveyed felt that their spiritual experiences are what led them to become a rabbi. They had felt a calling and continued to do so, though the specifics about how to be a Jew and what to tell their congregants in terms of practice were less clear-cut. These rabbis themselves all had different standards of *kashrut* observance, and many were far less strict outside of their own homes. How any of these particular modes of thought would correlate to the layperson is up for debate as well. It would seem likely that for the most part, the differences would remain the same throughout depending on one's affiliation.

For many Orthodox and Conservative Jews, belief in specific theological positions is fundamental to their religious experiences. Maimonides decidedly agreed with the importance of Jewish beliefs. He wrote in his *Mishneh Torah* (Laws of Prohibited Relations 14:2) that a convert must take on all of the founding principles of Judaism, which he enumerates elsewhere to be thirteen. Not all medieval Jewish commentators agreed with Maimonides on the number of required beliefs. However, they did for the most part assume that some belief was required. Ibn Ezra, a medieval commentator and poet, writes (commenting on Deuteronomy 5:16) that thoughts can be considered sins and that since the purpose of religious commands are to "straighten out one's inner essence," we must involve our minds in the process of religious service as well. In this perspective, religion is not one aspect of a person's life, but something that defines his or her entire existence.

Biologically, this might have a lot to do with how our brain establishes our sense of self. Our self has both a spatial representation (i.e., where we are in the world) and also a memory representation that takes all of our autobiographical data and helps us to remember who we are. In this way, a thirty-four-year-old Jewish male recognizes his own history growing up in Philadelphia, playing Little League baseball and hitting the winning home run, becoming a bar mitzvah, being rejected by a girl in college, and getting into graduate school. All of these, and many others, make up the memories that define each one of us. In addition, our religious and spiritual beliefs and memories also help us define who we are—an Orthodox or Reform Jew, or a Jew versus an atheist. These beliefs and ideas are held in the vast memory stores of the brain and can be elicited almost instantly when recalled.

Is it possible that there are specific distinctions in the brain of people who hold the perspective that beliefs are more important compared to people who feel that culture and deed are more important? Several interesting studies have explored how the brains of religious individuals see the world differently compared to nonreligious individuals. One study showed people very blurry pictures and asked people what they saw.[1] Those people who were religious were often able to see what was actually in the picture, but sometimes they saw things that were not really in the picture. Nonreligious individual never saw things that were not in the picture but sometimes missed the things that were in the picture. Both groups made mistakes about their perceptions of reality, but in different directions. Interestingly, when the nonreligious people were given a drug that stimulated dopamine in the brain, their answers became more in line with those of religious people. The suggestion is that more of the neurotransmitter dopamine may lean one's brain toward spirituality and inclusion of less material perspectives, while less dopamine may lean one's brain more toward science and reason. As we have always maintained, neither of these views of reality is necessarily better or worse, just different.

This study also corresponds with a genetics study by Dr. Dean Hamer, who published his results in *The God Gene*.[2] As a behavioral geneticist at the National Institutes of Health, Dr. Hamer had a long career in studying how genetics related to specific behaviors. He found that a gene that coded for a particular receptor in the brain was significantly associated with people's feeling of self-transcendence. Self-transcendence is the feeling of getting beyond yourself and connecting to something greater, such as God. The gene that was correlated with such a feeling helps to regulate dopamine levels in the brain. So we see that certain brain processes such as dopamine activity might be important for helping people to be more or less religious.

Is there a particular genetic predisposition toward religion that comes from being genetically Jewish? The genes on the Y chromosome related to the Kohanim don't typically have much direct influence on brain functions but probably influence the degree to which the brain is "masculinized" by various hormonal influences.[3] Is it possible that the brain of Kohanim is affected to be more or less masculinized, which influences how they question or struggle with various religious concepts? And what about Jews in general? Do Jewish people have more or less dopamine in their brain compared to other ethnic groups? Does a Jew who chooses to become a rabbi have a larger genetic component or physiological tendency toward the spiritual or rational? These are interesting and yet potentially dangerous questions, which neurotheology could start to address.

Such issues also raise larger questions about the nature of human beliefs. In the book *Why We Believe What We Believe*, the foundations of human beliefs are outlined as coming from the confluence of our perceptions, thoughts, feelings, experiences, and, importantly, social interactions. Each of these elements of belief can also be related to our upbringing and the religious and cultural milieu in which we live. And all of these elements bear directly on how our brain works. Most specifically, the more we use a particular part of the brain, the more the nerve fibers and connections grow in that area. And the more we use all of the different parts of our brain, the more complex are its connections and functions. If people around us encourage a questioning attitude, then our brain holds a belief system that questions things. And if we are raised in an environment that limits free thought, then our brain ends up with very few neuronal connections and an inability to explore and think broadly about the world.

This may in fact be similar to a statement of Rabbi Eliyahu Dessler about free will. Rabbi Dessler states in his work *Michtav*

M'Eliyahu[4] that human beings have something called a "*Bechira* point," where free will operates. We have natural limitations based on our experiences and upbringing that limit what we would logically choose in a given situation. Someone raised by bandits will not think twice about stealing but may still have a choice regarding killing. The area that a person operates in is limited, but the choices made in that area push the boundaries of what a person may choose. By continuing to choose good, one allows for even better choices in life. However, if one continues to choose the wrong path, the options to do good begin to become so foreign as to no longer be a choice at all. This is consistent with how we understand the brain to work, because the more specific neuronal connections are used, generally the stronger they become. As we repeat 1 + 2 = 3 in first grade, the neuronal connection that supports that concept becomes stronger, and the connections that support 1 + 2 = 4 fade away. Similarly, the more we focus on love and compassion, the stronger those circuits become in the brain, and the more we focus on hate and anger, the stronger those circuits become. Additionally, a recent fMRI study showed that the more people lied, the less their amygdala reacted to the lying. It seems that they became immune to the negative emotions associated with lying. These studies help explain the great difficulty in resolving some of the long-standing conflicts in the Middle East and other places in which hatred has been fostered for so long. The brain can have great difficulty moving away from negative beliefs and behaviors. Consciously refocusing the mind and brain onto more positive emotions may be the best chance at resolving such difficult problems, but this is not an easy proposition.

The notion that the brain develops based on the overall environment and processes that one starts with has sometimes been suggested as a cause for the Jewish brain. The Jewish approach typically encourages a lot of questioning and analysis of various ideas, even from an early age, such as with Uncle Vic. Whether as an adult or a child, questioning leads to a brain that does more questioning. And perhaps this contributes to a brain that helps to seek answers either through the Torah, through science, or through both.

Key Jewish Beliefs

Rav Saadiah Gaon, in his *Book of Beliefs and Opinions* 5:4, says there are three types of people who abandon faith: those who do so to worship multiple other gods, those who abandon God altogether, and those who have doubts about their faith in their thoughts. Rav Nachman of Breslov, in his *Book of Ethics and Faith* (*Bitachon* Part 1:29), says that with faith there is no fear and one will always speak the truth, but without faith in God one will naturally speak lies and become a hypocrite. For these theologians and many religious Jews, faith is seen as a fundamental necessity for all aspects of a religious life.

However, this does not mean that simply because it is required in principle, it therefore happens in practice. Many Jews regularly or infrequently attend religious services simply due to other pressures (work-related, familial, communal, and emotional) and do not actually believe or feel any of the aforementioned religious principles. Therefore, when we study how the neural connections might parallel religious practices and beliefs, it is important to remember that not every person has the same system of belief behind what they are doing. As opposed to a meditative state that one chooses to enter into, the act of prayer in a communal setting (see more below in specifics on prayer) has many more complex factors, such as the type of prayer, the particular type of service (i.e., Sabbath or High Holy Days), with or without music, and the degree to which others around you are engaged. And of course, whether a person is truly religiously inspired may have the greatest impact on the prayer experience.

That being said, it is still important to delineate what exactly is a basic concept in Judaism that one would certainly need to enter into the usual realm of religious practice. The most obvious and yet amorphous idea is that of monotheism, the concept of there being one God. Ordinarily this is assumed by most adherents of Judaism. The notion of a single God likely has a place within the brain as well. In much of our earlier work, we identified the parietal lobe as the part of the brain that is involved in our holistic notions about the universe. The parietal lobe typically takes our sensory information and helps us to construct our sense of self and the sense of space around us. We have found evidence that as this part of the brain shuts down, we concomitantly lose our sense of self and the sense of boundaries between objects. When taken to extreme states, we experience a profound sense of oneness, which may be the basis of the concept and experience of one God. It might be fascinating to determine if a Jewish person's brain reacts differently to the concept of God compared to that of a Hindu, who perceives many expressions of God. The God-experience itself can also be secondary to following God's laws, which are assumed to be a continuous imperative from the initial revelation at Sinai. The revelation is continued every day one sits and studies God words, the Torah, and prays to God. As such, those who understand the tradition this way will be more inclined toward rationalist thinking and have far less classically spiritual tendencies and experiences. From a brain perspective, we might consider how these various cognitive and meditative practices affect the brain and vice versa.

However, there are many who subscribe to an alternative religious narrative. The Midrash (*Shemot Rabbah* 28:6, see also Talmud *Shavuot* 39a) says that not only were the Jews who were at Sinai there, but every eventual Jew through birth or conversion was there and had that revelatory experience, accepting the commandments. This is a common narrative among Orthodox Jews, and they recognize prayer and studying the Torah as ways to reconnect to that experience and forge new relationships with God. However, God, as always, is seen as uniquely one that is never able to be fully comprehended. This perspective is clearly expressed by Maimonides in his work the *Mishneh Torah* (Laws of Prayer 1:1), which is his corpus of all of Jewish law. There he discusses the emotional closeness one must feel in connecting to God.

Emotions are more closely associated with the limbic system and the insula in the brain. The limbic system helps us to feel our emotions, while the insula helps us interpret them. Thus, this view of a more emotional perspective might tip the balance of the brain's processes more toward the limbic system. However, Maimonides is also clearly a staunch defender of understanding our relationship to God as one that is rationally based and given to philosophical discourse. This might derive more from a frontal and temporal lobe function. These structures not only help us to hold abstract ideas, but also balance activity in the limbic system. The frontal lobes and limbic system operate much like a seesaw in which increased activity in one suppresses activity in the other. The more emotional we get, the less rational we become and vice versa.

Given this interrelationship between emotions and rationality, we thought it would be interesting to see how rabbis utilize these two aspects of themselves in terms of guiding their religious beliefs. Overall, 57 percent of rabbis said that emotions guided their beliefs and practices a moderate or significant amount, 26 percent a mild amount, and 17 percent little or not at all. In terms of rational thought processes, 90 percent of the rabbis said that such cognitive processes guided their beliefs and practices a moderate to significant amount. Only 10 percent said that thoughts affected their beliefs only a little bit or not at all. The relationship between emotions and thoughts might be more specifically different across gender, but according to our survey, there was no substantial difference between men and women rabbis. Both seemed to place similar emphasis on emotions, thoughts, and experiences in terms of how they approach Judaism. Thus, all rabbis appear to feel that rational thoughts are more important than emotional responses when it comes to religious beliefs. It would be fascinating to study the results of a similar set of questions from other religious traditions to determine whether emotions or thoughts are more important in swaying beliefs.

A breakdown by denomination reveals some additional interesting information. For Reform, Conservative, and Reconstructionist rabbis, over 90 percent stated that their thoughts influence their beliefs a moderate or significant amount. But 70 percent of Reform, 59 percent of Reconstructionist, and 39 percent of Conservative rabbis stated that emotions had a moderate or significant influence on their religious beliefs. What is interesting about this distribution is that all of them follow the concept that rational thoughts are essential for religious beliefs. However, the importance of emotions in religious beliefs seems to differ markedly between the denominations. Conservative rabbis are the least influenced by emotions, and Reform rabbis are the most influenced. Interestingly, the Orthodox rabbis were the closest to being evenly balanced, with 80 percent stating their thoughts influence their beliefs moderately or significantly and 50 percent stating emotions influence their beliefs moderately or significantly. Thus, the Orthodox rabbis report having the most balanced relationship between emotions and rationality. Whether one perspective is better or worse is up for neuropsychological and theological debate, but the results certainly suggest that areas of the brain involved in rational thought lead the way when it comes to Jewish beliefs. It would be of further interest to observe how the general congregants in each of these denominations follow the same patterns of influence between emotions and cognitions.

In terms of practice beliefs, Judaism emphasizes several important ones, including prayer and charity. Neurotheology takes an analytical approach to prayer right from the start, recognizing that there are many different types and purposes of prayer. And this is true for every religion and certainly for Judaism. The different types and purposes of prayer are also associated with different brain processes. For example, one fMRI study showed that people praying in a conversational manner—simply talking to God—activate the same brain areas used when talking to another person. This is an important finding because it shows that the brain uses its own inherent processes to engage our religious selves. Other types of prayer that elicit more emotional responses typically activate the emotional centers, or limbic system, of the brain.

The religious purpose of prayer has been debated since its inception among Jewish theologians and scholars. Most assume that prayer can effect some sort of change in the physical world via God's intervention. The Talmud (*Berachot* 54b) states that if one does pray without the assumption that the prayer will be answered, then prayers can effectively save someone or make a change. The *Sefer HaChinuch* (Book of Education #433) states that God wants and commands people to pray to God with their thoughts and desires, no matter how small. Interestingly, Maimonides also writes (*Mishneh Torah*, Laws of Prayer 1:1–2) that in prayer one should pray every day, first praising God and then asking for one's needs. Prayer in this vein is seen as something that can produce actual results in the physical world but is also not necessarily about a close spiritual connection with God. Connecting to God is not as important as recognizing God's absolute authority and one's dependence on God.

Another stream of thought highlights the existence of God and pushes requests to the wayside. *The House of God* (written by a kabbalist, Rabbi Avraham Cohen Irira; "Prayer," chapter 2) states that the purpose of prayer is not to receive anything at all, but to acknowledge God as the ultimate existence and to rely on God. The *Zohar* (215a) says that sometimes God will answer the prayers of real *tzadikim*, people who have achieved a high level of unity and understanding of God. However, Rabbi Chaim of Volozhin (rabbi and student of ancient Kabbalah, who was opposed to its spread via the Hasidic movement and sought to reemphasize the importance of Talmud study as the main means of achieving closeness with God) writes in his work "Soul of Life," *Nefesh HaChaim* (2:11), that it is not correct for people to pray for their own needs, but rather for the larger *tikkun*, "fixing," of the entire world and higher worlds. These perspectives stem from a strong notion of spiritual communion with God being the ultimate goal as well as diminishing the importance of one's current physical standing in the world. Only by working to achieve *devekut*, divine communion with God, can one truly be answered. Prayer itself is not enough.

A similar concept underlies the custom of "*davening b'avodah*" often popularized by Chabad Hasidim. The idea is that true prayer requires real internal perspective-taking, and to turn into one's own self and think deeply about the experience that one is having. Prayer is often called *"avoda sh'balev"* in the traditional sources, which means literatlly "a work of the heart." It is to this concept of "work" that the tradition speaks to: one cannot begin to unify with God simply by having an intense emotional outpouring, one must work with careful meditation and contemplation with the words of the *Tefilot* prior to engaging with God through *davening*. The *tefilah* themselves are seen as mechanisms to do this work, as well as pre and post- *tefilah* meditation. As it says in the Talmud (*Berachot* 30b), the "original *Hasidim* would wait for an hour (in meditation) prior to *davening* in order to have proper intention towards their Father in Heaven (i.e. God)."

This concept is most clearly seen by those who spend time preparing before prayers thinking about God's greatness and how they can even begin to approach God in Tfilah. The idea is that through mental and emotional preparation, one can come to recognize that the *tfilah* service is not about having an intense emotional experience, or even about getting the things asked of God, but rather it is about having the relationship in the first place. From a brain perspective, this notion is consistent with other approaches to powerful spiritual experiences in which an individual augments specific patterns of brain activity by working immersively on a spiritual practice, thus enabling an intense spiritual experience to occur. Our recent study of people participating in a spiritual retreat program has shown that the brain is altered through this process. These alterations appear to make it more likely that a person will have an intense spiritual experience. In addition, our research has provided evidence that the more engaged one is in the practice of prayer or meditation, the more extensively the brain changes. Presumably, such changes lead to powerful spiritual feelings and a deep sense of a connection with God.

Tfilah is about the encounter with God, and that unity with God, *devekut*, is something that one must prepare for earnestly according to the Chabad literature. The uniqueness of Chabad Hasidus is not that this is seen as an important part of prayer, it always has been technically recognized as an important perspective. They believed that any person at any level of spiritual development, given the right amount of preparation, can do *tefilah b'avoda* and have that kind of experience for themselves. Again, the effort required for connecting with God is both intellectual as well as emotional, thus blending the entirety of a person's existence into their his or her encounter with the Godhead. And as we have mentioned, through these processes, it is likely that areas of the brain, such as the parietal lobe, become deeply affected to help enable such an intense unity experience.

Charity is another essential element in Jewish belief. On one hand, charity is seen as a supernatural way of removing the evil decree placed on someone for the year to come. As the High Holy Day liturgy states, "Repentance, prayer, and charity remove the evil decree." Rav Schneur Zalman of Liadi (in *Tanya*) states that charity is one of the greatest commandments. By giving charity, one is taking something that one earned oneself and sacrificing it up to fulfill God's will and help another. The emphasis is on the transformative process that charity has on the individual who is giving, not on the help the other is receiving.

On the other end of the spectrum are the communal requirements for charity. The Bible states (Exodus 23:11) that the poor are allowed to go to any field and harvest the leftover wheat missed on the first pass. Charity is not only seen as something that is spiritually uplifting for the individual, but something that is required from people once they earn enough and for the community to distribute and oversee as needed. There is an obligation for all people to give one-tenth of their earnings to charity, and the Hebrew word for charity, *tzedakah*, comes from the root *t-d-k*, which means "righteousness." When people give charity, they are doing what is considered a righteous act independent of their own spiritual transformation. Maimonides describes eight levels of charity, and the highest level is helping someone become independent and no longer need support. Clearly, supporting others who need assistance is a major component of why charity makes sense. However, the second highest level for Maimonides is to give charity without knowing who will be receiving it. Why is this important? It shows that even for Maimonides, the rational reasons behind charity giving extend to the individual person who gives charity as well. If one does not know to whom the charity is given, it is seen as a greater act of kindness and will positively affect the giver as well. This makes sense because for Maimonides, imitating attributes of God is one of the high points of intellectual achievement in Judaism (See *Mishneh Torah*: Laws of Human Dispositions/Deot 6:10). As the Talmud (*Shabbat* 133b) commenting on the verse "This is my God and I shall glorify Him" (Exodus 15:2) states "*mah hu rachum v'chanun, af atah rachum v'chanun*" which translates to "Just as He (God) has compassion and mercy, so too you should have compassion and mercy." The act of charity in Deuteronomy 10:17–18 is even directly used as a description of God caring for the orphan and widow.

Studies on charity are few, but one study in which people had to decide to donate or not showed that donation, especially when there was a social benefit, activated the reward system in the brain (which would involve the neurotransmitter dopamine). In this way, the person feels an internal reward for an external cost. For the person, the act of charity is therefore beneficial, at least on a neuronal level. Such a finding might help explain how charity becomes viewed in religions such as Judaism. However, more studies would be required to differentiate some of the concepts mentioned above, such as the difference between giving charity to someone you know compared to giving to an unknown person.

These rituals, prayers, and other community involvements that exist within Judaism only serve to highlight the fact that there are always at least two opposing elements in much of Jewish thought and practice. Individual practitioners may subscribe to one, neither, or both, but the identifications remain important to any study of the neurotheology of Judaism. Both the intellectual and spiritual elements of religious acts can and do make lasting impacts on many members of the Jewish community, but their interactions and specific changes can only be understood once these practices are mapped onto a larger neurocognitive map.

The Kabbalah

Mystical experiences are something completely different and are associated with an individual's personal encounter with God. A smaller, yet well-known segment of Jews, the kabbalists, understand this mystical relationship with God similarly to that of mystics from other traditions. Medieval kabbalists felt that by meditating on Hebrew letters (much like the Buddhist meditating on a mantra) and kabbalistic secret understandings of the world, one could connect directly to God in an intimate way. The mystic's theology stems from the idea that somehow God's "self-light" constricted, thereby creating a finite space that allowed for this world to be created. The entire world is therefore at some level an extension of the Godhead's light, and people can reabsorb themselves in their own light origins by intense meditation. The Kabbalah also has a complex understanding of the metaphysical universe that God's light travels through to create the world.

The Ari (Rabbi Isaac [ben Solomon] Luria Ashkenazi of Safed) was an original thinker in kabbalistic thought and explained that the ten *sefirot*, spheres that the light of God traveled through, had been broken in the initial attempt to create the world and that the job of the Jew and the world is to bring the broken "sparks" out of the darkness back to holiness.[5] These ten *sefirot* are *Keter* (God's "crown"), *Chochmah* (Wisdom), *Binah* (Understanding), *Chesed* (Kindness), *Gevurah* (Strength) *Tiferet* (Glory, often integrating *Chesed* and *Gevurah*), *Netzach* (Victory), *Hod* (Splendor), *Yesod* (Foundation, often integrating *Netzach* and *Hod*), and *Malchut* (Kingship). When one is able to unify the ten *sefirot* that have been broken with the current ten *sefirot* of the world, then there will be a great *tikkun*, healing of the world, and God will be revealed. From a brain perspective, this is a common and yet powerful experience of merging a multitude of objects, in this case ten *sefirot* with the oneness of the universe and God. As an aside, these numbers also speak to the importance of the quantitative processes of the brain that help apply meaning to specific numbers such as ten, eighteen, or forty. We will consider the importance of the quantitative processes of the brain with respect to Jewish thought in more detail in chapter 11.

The kabbalistic perspective is shared by many of those who subscribe to Hasidic theology and is a common theme in Jewish meditation. The vessels that the light travels through are known as *sefirot*, and by traveling through these ten levels, God created the world. However, there is much debate as to whether the ten vessels and the light from God are considered extensions of God's self or are merely like the light shining out of the sun. Many scholars believe that early kabbalists viewed the *sefirot* as actual projections of the Godhead into the world, creating a much more pantheistic perspective of Judaism. However, the traditional religious understanding has run a more conservative approach, culminating in the explanation of Rav Schneur Zalman of Liadi, who states (in his work *Tanya: Likutei Amarim Chapter 52*) that it could never be God's self in the light or the *sefirot*, as they are projections into this world, just like all physical material. God is parallel to the sun, and the *sefirot* and our world are simply the light, not taking away anything from the sun but still deriving from it somehow. This explanation allows for the spiritualist to still be absorbed as close to the sun as possible without believing that he or she is somehow partaking of a physical union with God.

Therefore, when analyzing these positions, it pays to look at how various general beliefs play out in their depiction of God. Rabbi Judah HaLevi, a medieval Spanish Jewish philosopher, in his *Kuzari* (1:67), writes that the Bible will never contradict the reality of logic. Gersonides, medieval Talmudist and astrologist from France, in *War of God* (in Hebrew, *Milchemet HaShem* 3:6), states that if there is a conflict between a simple reading of the Bible and philosophy, one should not rely on the simple understanding of the Bible, but reinterpret it in accordance with the correct philosophical underpinning. Maimonides was of a similar opinion as long as the proof was provided to move away from the simple biblical understanding with certainty. As Maimonides scholar Menachem Kellner states, "According to Maimonides, God, as it were, wrote two books, one called Torah and one called Nature—therefore they cannot contradict each other, and if one wants to learn about God, one must study both."[6] The *Sefer HaChinuch* (Book of Education #25, authorship undecided from the medieval period) states that the command to believe in God is of paramount importance. One God brought everything into existence and brought the Jews out of Egypt, thus making them indebted to God forever to help serve God. This last explanation of the relationship God has to the Jews bespeaks of an entirely different relationship from the individual ones we have been exploring. In this perspective, which is also mentioned by many Jewish theologians and seen prominently in many of the biblical narratives, the relationship with God is not an individual one, but a national one. The nation of Israel has a unique relationship with God as a whole and is treated as a whole. The distinction of this perspective might imply a different set of brain processes involved with each of these viewpoints. The importance of the individual relationship with God may be based more on brain areas involved in our sense of self and how that self relates to the world or God. The importance of the group or nation relationship with God may be based more on brain areas involved in social interactions and a sense of group cohesiveness.

All of these perspectives see the idea of God and the story of the Jews in the Bible as central to the story of the Jewish religion, and all agree that the religious principles must be upheld rigorously. However, they also have a strong grounded religious perspective based on philosophical and scientific knowledge. The student of the rationalist tradition may not appear any different on the street from someone who is not religious in the slightest (with the exception of those who wear required religious garb such as the *kippah* and *tzitzit*). They are looking not for an ecstatic experience, but for a long-term consistent commitment to God based on their fulfilling commands they believe to be rationally correct. If there is no rationale for a command given, they will investigate and believe with faith that they will eventually find a reason. In much the same way, a scientist believes that there is an answer for understanding any phenomenon. And as for the brain, the ability to seek out causality and abstract logical reasoning as a way of understanding both the Torah and the natural world speaks to the synergy between the Jewish brain and the scientific brain.

However, there is another dimension to the religious Jewish experience entirely. The Alshich (Moshe Alshich, student of Rabbi Joseph Caro) commenting on *Devarim* 5:4, states that God did not want people to outright reject religion as crazy or to easily buy into it, so instead of making it obvious, God made sure that there could never be a proof that would 100 percent secure religion and instead made people rely on their own experiences as proof that God existed. Such a conception appears supported by our survey results in which 88 percent of all of the rabbis stated that their own personal experiences either moderately or significantly influenced their religious beliefs. Thus, along with rational thinking, rabbis place a strong emphasis on personal experience to guide the development of Jewish beliefs.

This spiritual understanding based on the kabbalistic and Hasidic doctrines discussed above assumes that there is a truth beyond grasp by rational means but that can be perceived with proper training in meditation and cultivation of a religious mindset. Rav Schneur Zalman of Liadi explains, "The purpose of creation is that there shall be a place (for God) in the lower worlds. That there shall be an existence, and the destruction of the existence. . . . That there shall be a destruction of the existence to nothingness" (*Likutei Torah*, pt. 5, 28). In other words, the metaphysical withdrawal of God's immediate presence allows for the world's creation, but in spite of this withdrawal, human beings have never fully left God. It is a matter of trying to engage God through spiritual practices and means.

By climbing back up to the Godhead, human beings can once again become "one" in a mystical sense and potentially lose their personal sense of free will by joining in the greater will of God. In the most extreme version of this perspective, Rav Tzadok HaKohen of Lublin explains that the Hebrew word for repentance, *teshuvah*, means "return," which he interprets literally as returning to God as the ultimate cause of all things, thus removing humankind's freedom to choose, and sin, with it: "Repentance [lit. "return"] means to return the matter to God, which means to say that one realizes that everything is from God's action. . . . And in this way, after repenting completely one turns one's willful sins into merits (*Tzidkat HaTzadik*, sec. 100). The belief that man is not free and is in fact just a component of God is not that of mainstream Jewish thought. However, the influence of these ideas on the meditational strands of Judaism cannot be overstated. In addition, there are some potentially interesting brain-related correlates to the notion of oneness and connectedness to God along with our perceived sense of personal free will. Where free will may be in the brain is a question that we will address later in this book. But for now, we can at least ponder how the brain and our will might intersect or dissolve depending on how one experiences the relationship of the self with God.

From a larger perspective, we can consider how Jewish thought and theology might inform the sciences and our approach toward science. Judaism believes in the human ability to conquer and understand the world. The Bible states in the beginning of the story of Adam in the Garden of Eden, "And God took the man and placed him in the Garden of Eden to cultivate it and to guard it" (Genesis 2:15). By using the scientific tools at humankind's disposal, Judaism can help create a framework of understanding more complex religious thought processes and help determine which brain functions are part of the religious process and which are not. The fact that there are multiple aspects to an individual's religious experience means that scientific research and analysis must be multifaceted. By analyzing the different experiences one can have in Judaism and their neurological correlates, one might be able to create a larger neurotheological framework of meaning and understanding behind religious acts.

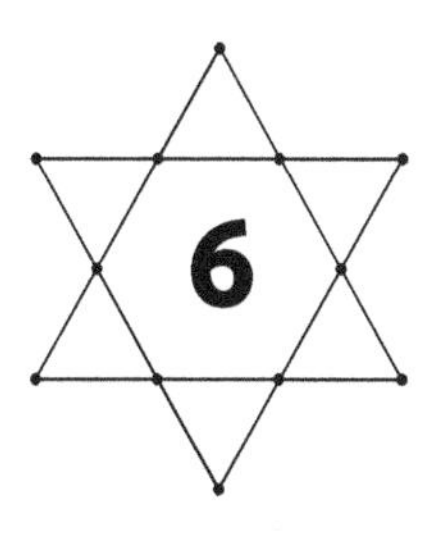

Culture and Stories in the Jewish Mind

As we sat in the circle each student stated their name, where they came from, and whom they represented. The group was anything but diverse on the surface, as everyone in the room was a Jewish college student, most often from a middle-to-upper-class family, with enough financial support to come to a conference on youth programming. However, upon discussion about the issues and philosophy behind the various movements in attendance, the apparent similarities began to fade to the background. Representatives from every denomination of Judaism had various things to say about how they approach God and how they feel the concept can or cannot be taught. Others were offended to even mention God at all, and still others were offended that one would even consider God to be just a concept, as it was clear to them God was intimately involved in the world and all their lives. I recognized that the major points of difference stemmed from significant theological differences between the parties—the different stories they understood about God and about Judaism.

—David Halpern

Overview of Cultural Judaism

A fully informed Jewish neurotheology must take into account that various Jews may not have a unified neural approach and may tap into various mental processes—cognitive, emotional, and experiential—as part of their religious and cultural beliefs and myths. For many Jews, the cultural connection is as important, if not more important, than the religious one. And neurotheology would have to pay particular attention to the relationship between the cultural elements of Judaism and the religious and spiritual ones as they relate to the brain. Jews frequently identify with specific ethnic or cultural constructs such as suffering, guilt, study, history, and the effects of the Diaspora on those constructs. Importantly, culture has an influence on the functions of the brain, and there is a reciprocal influence of the functions of the brain on the individual members of the community that subsequently lead to a sense of culture. In this chapter, we will consider the reciprocal relationship between Jewish culture and the Jewish brain.

Until now we have been discussing the various positions in Judaism about many specific theological beliefs and common practices. However, culture also strongly influences how a person thinks and behaves, and Jews are a people with significant and yet quite disparate cultural heritages. Depending on the culture one lives in and is exposed to, as well as the culture one has come from in the past, one may interpret events and occurrences quite differently, both religiously and theologically. When people are confronted with a personal spiritual experience, their culture may have specific ways of interpreting that experience. Much of our own experiences and discussion in this book itself may be due to the fact that we have studied these subjects in America, a decidedly Western and modernized country, with specific ethical, cultural, and philosophical beliefs. These beliefs may subtly alter the way we interpret and analyze Jewish texts and neurological studies, and particularly their relationship to each other. This would most certainly have been the case for the Jews throughout history and may even help explain many cultural differences between groups of Jews as well.

Initially in Jewish history, with the suppression of the Bar Kochba revolt under the Roman Empire and final quenching of the Second Temple period of Jewish attempt at self-rule, the Jews began to shift from a people who identified more as a nation with a land to more of a people in constant exile. In and of itself, it might be interesting to consider how the Diaspora process would affect a person's brain. Our brain is typically designed to try to establish a stable environment. When confronted with an unstable environment, the brain works toward establishing some type of stability, whether through social interactions or development of a consistent ideology. Exactly where this might occur in the brain is unknown. Instability in the environment would typically be regarded as a danger to personal survival and elicit a fear response in the limbic system. Frontal lobe functions, along with rational processes in the cortex, would be invoked to create ideas and behaviors designed to reduce the stress and anxiety associated with an unstable or dangerous world. Most likely, any opportunity to establish a stable community would be an important process of the brain in dealing with something such as the Diaspora.

There were already considerable Diaspora communities at this time in Babylonia. Most likely, during the Roman period a large group of Jews emigrated to regions of present-day Germany and established communities there as well. These two large areas, Babylon and Germany, become the starting points for two very diverse groups of Jews, largely classified as Sephardim (lit. "Spaniards"), who lived in the North African and Middle East areas, and Ashkenazim, who settled in much of Germany, France, and over time Eastern Europe. These two groups also lived during the medieval time of the wars between Christian Europe and Muslim Africa/Middle East. Much of the tradition and culture is often seen as borrowed in part from the areas and peoples that they were surrounded by. However, the various differing relationships between the Jews and the ruling people at that time complicated this exchange and eventually led to very different outlooks and understandings within subgroups of these two larger Jewish groups.

Culturally, the Middle East and North Africa had a technological and philosophical advantage over Christian Europe, being far more advanced in scientific study and philosophy. As a result, much of the initial work in Jewish philosophy came from Babylon, Egypt, and eventually Spain. Thus, the rationalist Greek tradition was passed on through the Ottoman Empire to the Jews themselves and led to great works like Rav Saadiah Gaon's *Book of Beliefs and Opinions* and later Maimonides's *Guide for the Perplexed*. What then occurred was the result of cross-cultural exchanges between the two main Jewish groups. Jewish scholars in both groups began to interact and absorb ideas from the other's philosophy and theology. One main difference, however, was that the Jews of Ashkenaz at that time were still met with significant discrimination and were forced to live separately from the rest of the community in what would come to be called "Jewish ghettos." These areas made it possible for the Jews to maintain a unique culture and continue to develop independently of the general culture, but still in tandem and interchanging ideas with it.

However, with the rise of the Enlightenment in Christian Europe, the fall of the greater Ottoman Empire, and the rise of more fundamentalist regimes in the Muslim-controlled areas of North Africa and the Middle East, the Jews of Europe began to entertain the possibility of different styles of education. The breakdown of the ghetto walls at that time caused a significant shift in what had been the status quo for some time for many of the Jews. This resultant exposure to general society after years of community unity was met with very different responses from different leaders. Some leaders felt that the community had built itself on ideas and principles that would be best maintained by keeping the general knowledge and interaction with the world to a minimum. The Chatam Sofer (Rabbi Moses Schreiber) best articulated this response in his line "*Chadash asur min haTorah*" (taken from the ruling regarding new grain being forbidden until the Omer offering time; see *Responsa Chatam Sofer*, pt. 1, *Orach Chaim* 28), meaning that new innovations and changes to the tradition are prohibited according to the Torah. This position, while articulated in a strong fashion, was not unique to Rabbi Schreiber, as many Jews felt their existence was threatened not physically, but now spiritually by the new threat of full assimilation. From a brain perspective, such concerns are justifiable, since the brain had found its stability through the close-knit Jewish community and the ideological stance differentiating Jews from other individuals. The notion that these stable forces could be jeopardized by integration into a larger society would theoretically elicit the same fear reaction in the brain as that experienced initially through the Diaspora.

This experience led some Jews to maintain their insular communities with strict rules about outsiders and outside influences, which creates an incredibly distinct and unique cultural perspective about the world, God's relationship to the rest of the world and the Jews, and humankind's place in that world. These communities often took a stricter interpretation of Jewish law as well, and many continued in the spiritual traditions of the Hasidic movement that had begun prior to the Enlightenment period and did not seek to engage in any of the new philosophy or learning from the secular world. A strong spiritualistic interpretation of Jewish philosophy ran through this segment of Jewish life, with rabbinic scholars studying Kabbalah in secret, as well as serious in-depth Talmudic study and expanding response on Jewish law.

In contrast to this approach, other rabbinic leaders believed that there was something to be gained from engaging and becoming a role model for the world as Jews. This movement was initially called *Torah im derech eretz*, or "Torah study with the ways of the world," modeled by Rabbi Samson Raphael Hirsch. It meant that there was a way to engage with the general world in which the Torah and Jewish life could enrich and be enriched through general science and philosophy. However, the traditions and beliefs of traditional Judaism were still upheld to be true and now could be openly kept and discussed in this new world of openness and acceptance. This model allowed for a reopening of the intellectual perspectives within Judaism that had formerly been maintained by those Sephardic scholars generations prior. As above, this approach would likely be embraced by those individuals with a brain that found greater comfort being integrated in a more stable manner into the rest of society. Of course, it was also essential to maintain their traditions and Jewish beliefs, law, and worship of God while still integrating to some extent with the general society. The beliefs of the general society then became much more culturally ingrained in this group than in others before them, and therefore one of the main themes in the discourse of Jews from this segment was, aside from the intellectual encounter with God, wrestling with exactly how the Jewish tradition should appropriate or distance itself from the general society regarding all things intellectual, aesthetic, and otherwise.

Such was not the concern for many Jews after the ghetto walls were removed. Basing themselves off of the works and leadership of Abraham Geiger, many began to institute reforms in the temples, adding traditional choirs and organs like those of the churches, and arguing against much of the traditional beliefs of the past. As the movement consolidated and gained momentum, the Reform movement was born, taking off first in Germany, but subsequently in America more than any other country. The movement officially maintained a belief in God but rejected the historical accuracy of the Bible or binding nature of specific commandments.

However, in the lands of the Middle East and North Africa, the Sephardic Jews did not remain dormant. Instead they formed their own cultures and groups depending on location and were often very distinct from each other and significantly from other Jews who might come from Europe. The Sephardim began to engage more strongly in the Kabbalah and utilized the communal leader, known as the *chacham*, "wise man," to understand these mystical rules and practices. Perhaps through this type of spiritual and mystical approach, the brain of these individuals found greater comfort. Much of the surrounding environments influenced the Jewish culture in the Middle East, perhaps even more so than in Europe. Specific cultural taboos and superstitions arose, as well as a reliance on strong family ties and a mistrust of those from outside the community. These same responses could also be seen in the Jewish ghetto communities in Eastern Europe as well as Russia and may have also contributed to the widespread popularity of the Hasidic movement as well as the ability for the Jewish community to remain relatively stable during this time period despite anti-Semitic attacks on various communities over history.

However, all of these responses were then faced with the rise of Nazi Germany, World War II, and the Holocaust, where the entire Jewish existence in Europe was decimated. Jews once again became victims and suffered significant trauma at the hands of a supposedly enlightened people. The effects of this experience cannot be overstated on the mind-set and culture of people who came from that time period. All of the previously mentioned distinctions continued to exist but were now built on a background of severe destruction and desire to rebuild or forget what was lost. In addition, during this time period the State of Israel was created, and for the first time in two thousand years the Jews were considered a nation again in addition to a religion/culture. Practically, this caused the removal of almost all of the Sephardim from the neighboring Arab countries into Israel and created a unique culture in Israel, with the mixing of Ashkenazic and Sephardic Jewish rituals and customs. Many times these cultures clashed and were recreated, especially considering that the initial Jews who settled and pushed for the State of Israel were almost entirely secular Zionists, with their own set of values and practices far different from many other Jews who were fleeing Europe.

Obviously, this is a seriously truncated and less than complete examination of the Jewish cultural development leading up to the present, and much is missing in the above discussion. However, what is clear is that significant differences between groups can continue to have effects on the development of what might be considered "the Jewish brain" and additionally what might be the different patterns different cultural elements might take.

From the Sephardic tradition, cut off from the previous intellectual empires, the new cultural elements of the "Eastern" perspective of the family unit began to have supreme importance. Spiritual impressions and prayers also became much more significant with the reversal of the focus from intellectual endeavors to more kabbalistic and mystical experiences. Often, the *chacham*, or wise mystic, was sought out by leaders and community members for advice and to help guide and lead the community through the *chacham*'s knowledge of the esoteric. This is all the while that the communities still revered and were quite loyal to the intellectual Sephardic authors of the medieval period such as Maimonides. However, viewing such leaders through a different cultural perspective can lead to very different religious experiences. Prayers became much more of a communal event than an individual intellectual or even emotional endeavor. Much of the prayers were said out loud by the entire congregation, and communal events were a time laced with much familial and spiritual meaning. Neurologically, feelings associated with the family are related to the limbic system, as well as the social areas in the parietal lobe. Those individuals with intense family relationships are more likely to have significant neuronal activity in these regions, which might be different from the abstract reasoning areas of the brain. And as we will explore in chapter 7, spiritual practices and experiences are associated with a variety of neurological changes.

For the Ashkenazim, we break down the cultural implications based on what we have discussed above. Those who view themselves as no longer bound by tradition or its beliefs will have a very different experience mentally, both educationally and in their day-to day-actions. If one believes in God but does not accept other tenets of Judaism, the unique perspectives based on prayer and specific practices may be lacking. However, there is still a significant effect of emotional influences on how people orient and make decisions in their lives. For example, post-Holocaust Jewry struggled with much of the guilt and hardship of remaking life after such loss, and much of this guilt has been passed on to children of survivors and the community as well. But guilt also has a neurological correlate. Brain imaging studies have suggested that guilt is associated with negative emotions associated with moral transgressions or breaking of cultural norms. It is an important self-referential process that appears to utilize the prefrontal regions that are involved with perspective taking.[1] The prefrontal cortex also is associated with initiating and modifying behaviors, and thus guilt changes the ways in which we behave. In Judaism, this has led to a variety of cultural behaviors with the goal of alleviating the guilt and remembering the events that led to such emotions.

Living in a post-Holocaust world can also invigorate communities to "never forget" by creating memorials and specific activities of remembrance. Such concepts heavily affect the amygdala, which identifies important things, especially negative ones, and also the hippocampus and memory centers of the brain to ensure that there are strong and persistent memories. There is an urge to preserve some form of "Jewish life," which may also relate back to the emotions associated with responsibility and guilt. In any situation then, one might create in one's mind a "Jewish" response by simply imagining what one's own prewar parents or grandparents would have responded in such a situation. Another point to ponder is whether this perspective may be a generational one, and with the passage of time, the further generations get from the initial cultural separation and with continued cultural blending, the effects may diminish. If the creation of the constructs and belief system was due to the unique neurological makeup of "the Jewish brain," then they may remain and uniquely influence someone even when the person believes and practices nothing related to Judaism whatsoever. This is also true for many Conservative Jews as well. Conservative Judaism having been formed as a break from Reform for violating too many traditions, the Conservative Jew faces the same dilemmas as the Reform Jew but at a slower pace, as well as the dilemmas of the Orthodox, depending on the person's level of education and religious commitment.

All portions of the Jewish population have seen significant cross-cultural exchanges both with the general secular world and between the different groups of Jewish thought. Particularly since the strains of spirituality and intellectualism were not confined to any specific groups in the past, with the continued development of communities in Israel and the Diaspora, this cultural shift is clear. Many Hasidic Jews, who would have formerly been thought to have strong spiritualistic understandings and leanings, may actually subscribe to a much more formulaic model of tradition. They continue in their community to do what was always done, and the emphasis on individualistic spiritual experience may be lost in some way. Other Jews have a slightly more moderate stance and are not as isolated from the world but still maintain a separate community, believing that secular knowledge is mostly important only for earning a living. This group is able to have both members who are mainly focused on textual study and others who also seek mystical/spiritual experiences. The continuation of the more moderate Orthodox group, now often called "Modern Orthodox" Jews, has also undergone a transformation. Students now study much more of the spiritualistic and kabbalistic teachings than they did in the past, and there has been a push for more stringent legal observance while trying to balance modern cultural issues regarding the place of women, as well as the treatment of gay and lesbian members, in the Orthodox community. Whatever subsect or group one would subscribe to, if one continues to practice and pray with the above-mentioned beliefs, one is able to access many different perspectives from what were originally separate camps. This means that the perspectives on the commandments and prayer may no longer line up with the appropriately designated group and can even be distinct within individuals.

In any event, the cultural elements of separateness, guilt, and study all play into just how "Jewish" one may culturally feel. However, recent history has added another element that also has fundamentally changed the Jewish cultural experience.

The creation of the State of Israel has brought up a plethora of theological and neurotheological issues among the Jews as well. For those who have no official ties to Israel more than any other country, as they have previously revoked the need for Israel to exist, the country serves as an interesting unique Jewish cultural heritage repository. By traveling there and seeing one's history, the non-identified individual may become fascinated with Jewish history and want to learn more about the Jewish culture, ultimately leading to more self-identification as a Jew.

For those who view the State of Israel as religiously significant, this can have serious cultural implications for how to respond to it as a Diaspora Jew. Are Jews to be loyal to their religion first over country? And does that religious loyalty extend to a national existence of the Jewish people? Dual loyalties are complex issues that Jews never before had to entertain in this specific way. For all types of Jews with complex backgrounds, Israel and one's relationship to it and the Diaspora will certainly be something that continues to shape Jewish culture for the foreseeable future.

How a given individual's brain manages these potentially concordant or discordant issues likely involves both emotions and experiences along with abstract processes that set up various oppositional concepts. The brain is then taxed with how to resolve any potential paradoxes. Neurotheology will need to be able to evaluate the various cultural influences in Judaism. Various cultural traits such as guilt, separateness, or study could be early targets of study, since they more specifically relate to different brain processes. Whether such brain processes are more or less prominent in Jews in general or within various groups or individuals is an important area of future neurotheological research.

The Stories and Myths of Judaism

Myths are stories that explain essential aspects of a given tradition such as Judaism as it relates to humanity and ultimately reality.

Myths frequently help to explain these fundamental aspects of reality in terms of efficient causality (creation or foundational myths), final causality (salvation or apocalyptic myths), or both. Although myths are regarded as crucial ways of understanding how reality works, it is never clear if they have any bearing on the truth of the ideas contained in the myths. This does not mean to imply that myths are false, only that it is very difficult at times to ascertain the validity of any story that human beings use to understand the world around us. This includes science itself, which also provides a story about the world based on certain *a priori* principles.

The methods of science are clearly different from those of religion, but from an epistemological perspective, as we shall see in future chapters, neither necessarily has a lock on the true meaning of reality. The basis for this problem is that all human beings are using their brain to interpret the various stimuli and inputs from the external world to produce a workable model of how the world actually works. Since we can never get outside of our brain, we have no way of knowing whether the stories that our brain helps to provide us are accurate with respect to the true reality. Regardless, we need these myths and stories, whether they are based on religion or science, to understand how we as human beings interact with reality. Therefore, to understand the neuropsychological basis of myth formation is to begin to understand the basis of religion and ultimately of theology.

The general structure of myth appears to be based on a presentation of opposites. Oppositional concepts are probably related to neuronal activity at the junction of the temporal and parietal lobe, which is an area particularly involved with language and abstract thought processes. These binary ideas include things such as (1) inside-outside, (2) above-below, (3) left-right, (4) in front-behind, (5) all-nothing, (6) before-after, and (7) simultaneous-sequential.[2] Such basic spatiotemporal relationships are then combined with either a positive or a negative emotional value arising from the limbic system. Thus, "within" is usually identified with good and "without" with bad, "above" with good and "below" with bad, "right" with good and "left" with bad, and so on. These emotional responses are not absolute and may vary depending on the culture and society.

However, some have a more universal connotation. For example, "above" is usually safer than "below" because one can look out for predators more easily when one is situated high up. The result is that "above" is considered good while "below," which may be more dangerous, is considered bad. The emotions that are connected with these concepts are essential for understanding our world and how to interact with it. We might even argue that these basic concepts and their associated emotions arise in the brain as the result of evolutionary forces that lead to enhanced survival. We would prefer to be up high and with other people than down low and alone. This may also explain why spiritually minded individuals have a tendency to describe their experiences as "ascending" and why heaven and positive religious experiences often take place in an abode "above" humanity in some way, be it physical or metaphysical in nature. This is geographically why Jerusalem became such an important city. The elevation of the Old City is approximately 2,500 feet, and the city is generally surrounded by valleys. This location made Jerusalem a "good" place to live. But for Jews, Jerusalem of course is far more than this, since it is even considered to reside in heaven itself, as described in the book of Ezekiel (chaps. 40–48), as well as the general tradition of there being a Jerusalem of this world and a Jerusalem of the next: "The Jerusalem of this world is not like the Jerusalem of the world to come" (Talmud, *Bava Batra* 75b). This notion of being "up in heaven" is a powerful idea and emotionally positive. And of course, "down low" is usually associated with hell (though Judaism does not appear to subscribe to the same hell that is usually thought of as part of Christian thought).

There are other opposites that grow from these initial spatial relationships that are expressed in myths.[3] The moral oppositional question is about the difference between right and wrong.

What is right behavior, and what is wrong behavior? And certainly when it comes to the Jewish people, the perpetual question has been "Why do bad things happen to good people?" This question was addressed directly in the book *When Bad Things Happen to Good People*, by Rabbi Harold Kushner.[4] This question, for many Jews, lies at the heart of trying to understand the Holocaust, leading to works such as those by Victor Frankl striving to find meaning in such a world.[5] Another fundamental question addresses the opposites of life and death. Why do we die? And what are we supposed to do while we are living?

Perhaps the most important opposite in religion is the difference between human beings and God. How are human beings, who are finite, limited, and faulty, able to have any relationship with a being who is omniscient, eternal, and perfect? To address such questions, the myth often describes an individual (e.g., Abraham or Moses) or group of individuals (e.g., the Jews) who have come to some greater realization or revelation about the divine nature of the world. From there, various practices, beliefs, and rituals arise as part of the myth. These practices and beliefs in turn support the myth in a cycle of mutual interaction. Examples of these kinds of myths abound in every tradition, and thus almost any religious text can be considered as presenting some type of problem that must be resolved.

The second aspect of myth is to unify these opposites to resolve the mythic problem. The ability to constitute a mythic problem and its resolution involves a number of brain structures and functions. Areas of the brain that support abstract thoughts, causal reasoning, binary opposition, and ultimately holistic union all play a key role in the formation and resolution of myths. To begin with, myths are created using named categories of objects, such as good and evil or human being and God. Brain scan research has shown that there may be individual groups of neurons that support specific concepts, particularly in parts of the temporal and parietal lobes that house memory and abstract thought areas. Myths, like all other rational thoughts, also involve causal sequences of events, such as the need to build an ark because of the flood because of God punishing the evil in humanity.

The resolution of the myth involves finding a holistic union of the dyad presented in the myth as the original problem. This resolution, therefore, appears to employ the brain's "holistic" abilities to unify the opposing sides of the myth. This is the critical function of myths in general—they provide a way to help us understand the opposites. Through myth we can understand how God can allow bad things to happen to good people. And in the context of the primary religious myth, we can understand how to interact and connect with God. The resolution of the opposites most commonly makes use of the holistic processes of the brain, thus uniting the opposites. As mentioned previously, the holistic processes likely involve the parietal lobes, which help us connect to other people and feel at one with the universe. By engaging the holistic processes, we come to understand how good and evil are intimately interrelated. Or perhaps we understand that by doing certain actions, we can join with God. This idea is most clearly paralleled in kabbalistic theology by the *Zohar*'s understanding that ultimately the darkness and *sitra achra*/other side, which originally came from the Godhead, will return to the side of light. Only then will everything and everyone be unified in serving God, and God's oneness will be complete.

Other examples of mythic solutions to the God-human antinomy are a solar hero, a Christ figure, Mother Earth, or divine kingship. Similarly, good and evil might be unified into a single entity such as Absolute Good that encompasses both good and evil. From the neuropsychological perspective, these resolutions may be caused by a shift in cognitive dominance from a more binary or reductionist way of thinking to a more holistic way of thinking. This might be argued to represent a shift in brain activity from the dominant (left) hemisphere to the nondominant (right) hemisphere of the brain, but this is still speculation at this time. As mentioned, the cognitive functions of conceptualization, abstraction, causal thinking, and antonymous thought typically arise more from left hemispheric functions, particularly near the junction of the temporal and parietal lobes. The right hemisphere generally is associated with more holistic ideas and perceptions. The cognitive unification of logically irreconcilable opposites presented in the myth structure certainly represents a shift from one mode of thought to another. This is not merely a cognitive shift, but involves a transformational experience and understanding of the myth or the world. The cognitive resolution also evokes a very powerful emotional feeling. The emotion actually provides a sense of support for the overall mythic resolution, since it usually feels extremely positive when one resolves irreconcilable opposites.

In this particular aspect Judaism remains unique in the fact that its unifying elements that sought to create a complete resolution via "absolute unity" were never the mainstream of Jewish thought. The tension seen between the finite human and the infinite God was interpreted to be the goal in many strands of Jewish thought, and the ultimate resolution seen only after one had passed from the finite world and could understand God's perspective in the metaphysical realm. The intellectual branches of Jewish thought do not even focus on this end piece at all and see no reason to live in a peaceful holistic way, with everything making sense. In fact, the dialectical tension of living a finite life for an infinite creator, a life of spirituality and a life of strict observance, a life of simplicity and a life of aesthetic pleasure, may be exactly what Judaism demands of its adherents, a lack of peace of mind and a constant striving toward creating a somewhat antinomian truth that may never be fully realized. This is often understood as the message from the book of Job, where Job's life is destroyed without reason and he ultimately questions God. God in response does not give him an answer that makes sense to human logic. Instead, God demands Job admit that his knowledge of the entire world pales in comparison to God and that he therefore cannot in good conscience assume to understand God's plan for the world

(See Job 38). But it is in this realization that there is a resolution to the mythic problem. The union is not within the human realm per se, but the unifying answer is set in the realm of divine reality.

In contrast, much of Hasidic and kabbalistic thought is geared toward understanding that the ultimate, as well as current, goal is to unify humanity's finite world with the infinite and bring everything into one larger holistic perspective. If one takes these two diametrically opposed perspectives and places them in the larger camps discussing myths and rituals as we have discussed, then there appears to be a strong split in Judaism between those who subscribe to the traditional myth-resolution model of most world religions and those who see humanity's struggle as something to be recognized and respected, and never fully finished as a defining feature of the physical or human world. Unity is only for those who have already left this world, to experience the oneness that is God.

Considering various brain functions and their possible correlation with unique religious experiences may also be helpful in understanding many of the various interpretations and interesting stories found in the Bible and throughout rabbinic literature. According to many of the commentators throughout history, the biblical stories are meant to emphasize specific spiritual messages in addition to the explicit legal rulings found in the Torah. If the Torah was simply a law book, then the entire first book of Genesis would be largely irrelevant except as background. Why all of the stories? It appears that according to many of the commentators and most certainly the rabbis of the Talmud, the stories were meant to be messages for how all of us can act currently. Many debate the actual historicity of some of the stories. For example, Maimonides is famously quoted as stating that anything that has God actually doing something physically is to be taken allegorically (since God operates on a different level from the physical universe) and that certain highly spiritual encounters were in fact visions that the founding fathers of Judaism had. This assertion is hotly contested by many within the Orthodox camp, as the Talmud (*Shabbat* 63a; see *Yevamot* 24a) and commentator Rashi (also known as Shlomo Yitzchaki) in Genesis and Exodus state, *Ein hamikra yotzei midei p'shuto*, "A verse is never taken out of its literal context." This would seem to imply that the verses themselves may have additional allegorical messages but that the historicity of the events should not be doubted either. This is a rather complex topic, since the traditional understanding also recognizes that not every verse is meant to be entirely literal. The point here is that, either way, the stories themselves should be used as moral lessons for one who is studying them.

Where things get interesting from a neurotheological perspective is how the different groups and denominations of Judaism might interpret these stories and what neurological mechanisms might be involved. Is the story simply about personal individual self-improvement? Are their moral lessons for whole societies and cultures? Are these messages meant to be taken in a personalized, spiritual sense, or can they be simply about externally doing the right thing? A thorough analysis of how some of these stories are interpreted would be very telling in this situation. Let us explore a few stories in more detail to see how we might apply neurotheology.

The creation story itself is a very telling example. The basic opposite of the creation myth is existence versus nonexistence. As human beings, we have a fundamental problem to resolve, namely why there is something rather than nothing. It is through the resolution of this opposite that the creation myths help us to understand the true nature of reality, as well as our place and God's place within that reality. When we look at the first verses in Genesis, it appears that the world is created twice (further engaging the binary processes of the brain)—first in Genesis 1:1–2:3 and then again in Genesis 2:7–25. Various traditional commentators have sought to explain this interesting phenomenon. Rabbi Menachem Mendel Schneerson, the late Lubavitcher Rebbe, quotes the argument in the Talmud (*Chagigah* 12a) between the schools of Shammai and Hillel about whether the heavens or earth was created first, and he shows that they quote verses from the two accounts to back each of their perspectives.

Instead of leaving it as a basic argument in textual analysis, the Lubavitcher Rebbe states that these are two very different perspectives on life's purpose in general. The school of Shammai stated that heaven was created first, with the ultimate goal to be the physical reality that we live in. Using the kabbalistic concepts discussed in the last chapter of *Tzimtzum* and the slow evolution of humans in the universe, the Lubavitcher Rebbe states that this position suggests that God made a spiritual world, with the ultimate goal for it to descend into a physical state. From this perspective, humanity's ultimate goal is to reunite with the higher spiritual purpose and to use this world and God's commands to ascend and transcend it. However, the school of Hillel believes that the physical world was made first. Their approach is that God never needed the physical world, yet God made it for us to implement the will of God stated in the rest of the Torah. The purpose of creation in their eyes is ultimately unknowable, but at the same time God wants us to live in this world and bring the truth out here, through the study of the Torah.

What is fascinating about this understanding of the dichotomy of the creation story is that it cannot be true. How can two differing opinions about a piece be true at the same time? The Lubavitcher Rebbe then states, however, that in this case, since it is a "spiritual" argument and not a legal/halachic one, both perspectives can be simultaneously true. That is how he explains the double story of Genesis, as a way to have both perspectives illuminated.[6]

This perspective has the elements of dichotomous living imbued in it, as we touched upon in our discussion of the binary function of the human brain. However, it also has elements of the general spiritual camp's perspective, where one's purpose is in both situations to reach some sort of spiritual sensation of having God's presence in the world and for one to "ascend" in some way. That sort of emphasis on spiritual and out-of-body experience can only be associated with the general spiritual camp and kabbalistic teachings. These potentially "deafferentation" events may be similar to the spiritual experiences we discussed previously. However, as we have also discussed, the camps in Judaism are not so clearly divided. For every spiritual push from the parietal centers in the brain supporting a mystical connection with God, there is also a pushback from the practical this-world experience. Even the spiritualist, if he or she wishes to remain in the Orthodox camp, must accept the position of doing actions in this world. The dichotomy remains, but it is synthesized. This also demonstrates the power of myth to resolve binary opposites via the holistic processes of the brain. Exactly how this happens is dependent on a given individual's brain and the approach one takes toward resolving the basic mythic problem. However, what is important is that the resolution is achieved in some way that seems satisfactory to the individual and provides an ideology for deepening our understanding of the world.

This, then, is the major difference between the approach of the Hasidic and kabbalistic camps and that of the more rationalistic models. The rationalist looks only for the *peshat* (basic understanding or literal meaning) in the *pesukim* (verses). The rationalist will seek only to translate those verses rather than try to synthesize and resolve any seeming contradictions.

What is even more interesting is that ultimately those in the rationalist camp utilize contradictions in the verses to explain the events, but from a fundamentally different perspective. Rabbi Mordechai Breuer explains that the two stories are depicting God's relationship to humanity and the world. The first story, which has the creation take place from the outside in, as it were, is where God is seen as the one who created nature. God is removed in some way from humanity and an external source of power to be feared. However, the second story is stated separately to emphasize the second element of any religious or spiritual person: the revelatory component. This dichotomy allows for the passages not to be contradictory, but to speak to two distinct elements in humanity's relationship to God.[7] Rabbi Joseph B. Soloveitchik used the two stories to emphasize the two components of humanity's existence itself, that of the absolute conqueror and analyzer of worlds, the "majestic man," and that of the humble lonely man who searches for a relationship, the "covenantal man."[8] These two Adams are seen as the components of the human personality and are meant to be blended into one human being whose purpose is to serve the will of the creator.

This story is a highlight for exactly how the different approaches view humanity's purpose and the human brain. Is it the spiritual pursuit of achieving some form of union with God, or is it the simple service of God and forming a relationship with God simply by doing God's commands? These two approaches, as we have shown, significantly overlap in much of their understanding both textually and theologically. This would make sense, as they all try to remain within the framework of an Orthodox theology and existence. The question then becomes: which processes are these stories emphasizing? Spiritual disassociation is present in both, but in one it is the be-all and end-all of existence, and in the other it is an important component of a relationship with God, but not the final step.

Another telling story is the interpretation of the creation of the founder of monotheism, Abraham. Genesis 12:1 states that God told Abraham, "*Lech lecha*," which can be translated as "Go for you." This is the first we hear of Abraham after a brief note from the previous section at the end of the Torah portion *Noach*, where we are told that Abraham went with his father and brother on their way to the land of Canaan and then stopped on the way. God's words appear to Abraham out of the blue, and we are told nothing about his past from the verses. The Lubavitcher Rebbe states that these words are meant to mean that Abraham was told not only to leave physically, but to go on an internal spiritual journey we all must take. He was told to leave his family, his home, and his mental place of comfort for a new start. However, the Lubavitcher Rebbe goes further, stating that the symbols of "land," "birthplace," and "father's house" represent more than just specific instances of comfort, but are symbols of natural desires—home and society, and even intellect. Thus, the call is for Abraham to forge a new identity by relinquishing everything in his existing identity.[9] The emphasis on the spiritual longing and mode of immersion in a new experience is clear and very telling, as it underlies the goal of spiritual and religious experiences of that group and of the kabbalistic system in general.

However, this is not how others have understood the story. Rabbi Soloveitchik explains that Abraham was told to leave his family after years of self-discovery and work. Abraham is depicted in this setting as persistent in his search of reality and not something otherworldly at all, but practical and relevant to him and to others. He is seen as the seeker of truth but unwilling to leave the world behind, even if it would shun him.[10] These characteristics are much more consistent with what one might view as the ideal political leader, not necessarily a spiritual one. However, this perspective of Abraham as being told to go in order to rebuild himself as an iconoclast who has tolerance only for truth is one that is appealing to the rationalist in search of religious meaning. Whichever perspective on Abraham one takes, it is telling that the founder of monotheism himself is depicted with differing emphasis on components of his character, depending on which religious mode one uses in approaching the text.

From a brain perspective, this mythic story presents the fundamental dichotomy of being a person with God or without God. Abraham has to contend with the two parts of himself as he ultimately becomes a person with God, incorporating everything that he had been into a new way of being. From the perspective of those evaluating the story, our brain presents the opposites so that we can understand the challenges involved not only with Abraham, but with ourselves. Through the process of the mythic story, we can come to understand how to bring the disparate aspects of our own self into a more holistic, spiritual person. We might even ponder the challenges affecting Abraham's brain and whether he had to contend with the complex rational and emotional processes associated with becoming a man of God. The implication from the Bible is that it truly changed every aspect of himself, including his name from Abram to Abraham.

The myths above represent a small sample of the stories of the Jewish tradition but demonstrate the power and structure of myth. By presenting a fundamental question or problem as a binary opposite, the myths can help bring together a holistic solution that provides knowledge and understanding about the world, about how to believe, and about how to be human. Once such an understanding arises, not only is there a cognitive resolution, but an emotional one as well. The resolution of myth appears to elicit powerful positive emotional responses that can include feelings of love, of compassion, or simply of peace with this fundamental understanding of the world. However, a cognitive and emotional understanding, while valuable, does not necessarily get us all the way to a true spiritual understanding. For a deep, visceral understanding of the meaning of myth, it is most effective to be able to introduce a behavior or practice that is attached to that myth. The behaviors and practices associated with myth are primarily in the form of rituals. These rituals form the basis for how people interact with each other and also how they engage with their religious and spiritual traditions. The next chapter takes a more detailed neurotheological look at the rituals and practices of Judaism.

The Neurophysiology of Jewish Spiritual Practices

The crowd rushed in, and I was pushed up hard against a large black velvet coat. The man in front of me was enormous, at least a foot taller than me. He wore a large felt hat and had on high black socks and shoes, dressed in full Hasidic garb. Pressed against my back was another man, pushing to get in to see Rabbi Shimon bar Yochai's grave. I was pushed tight enough to pick my feet off the floor and continue moving. The experience was extremely frightening and invigorating at the same time. Everywhere people were dancing. In one section the men lined up in eight rows on benches and rhythmically moved back and forth to the music, singing in motion. Thousands of Hasidim and others flock to Meron, Israel, the reported burial place of Rabbi Shimon bar Yochai, who kabbalistic tradition names as the author of the *Zohar*. It was the night of Lag BaOmer, the thirty-third day of the Omer count from Passover to Shavuot. It was on this day that he is reported to have revealed many of the kabbalistic secrets before he died, and the day is commemorated with huge memorial bonfires and singing and dancing late into the night as the mourning period of *Sefirat HaOmer* [Counting of the Omer] comes to a close. The experience left me wondering, just what was it about that night with all of those people together that made it so memorable? What was even religious or spiritual about it for all of those people to feel such a strong need to return each year to that small mountaintop town?

—David Halpern

Introduction to Myths and Rituals

As we develop a neurotheological approach to Jewish thought and experience, specific practices in Judaism have great potential to be mapped onto their neuropsychological correlates. To begin our analysis of Jewish practices, it is important to expand upon two fundamental elements of these practices—myths and rituals. Of course, myths and rituals are a part of every religious and spiritual tradition throughout the world, and Judaism is no exception. And as with each tradition, there are unique qualities that differentiate Judaism both in terms of its essential elements as well as the physiological processes that support them. To begin, we can consider the basics of how myths and rituals affect us and then turn our attention to Jewish myths and rituals.

Myths and rituals have been an essential part of religious and spiritual traditions throughout history. As we described in the previous chapter, it is also important to note that the use of the term "myth" is not meant to be pejorative. We are not using myths to denote something that is false, but rather a fundamental story that is critical for helping human beings understand their world and how to interact with it. Neuroscientifically speaking, making myths may be seen as a behavior arising from the integration of the function of various parts of the brain as we try to understand the world around us. Rituals are a more physical expression of myths, and they help create powerful experiences and feelings for participants as well as help resolve the mythic problem. It is through myth and ritual that all religions, including Judaism, elaborate their beliefs and foundational doctrines.

The Ritual Solution of Myth

It is one thing to resolve a myth cognitively and emotionally, but there is a third element that can help to forever engrain the mythic solution in the mind of the individual—and that is to experience it throughout the body. The best way to feel the myth throughout the body is through the enactment of the myth through ritual. Properly performed ritual, in and of itself, produces a powerful unitary experience and, when coupled with a myth, helps to resolve the myth on both a cognitive and intensely experiential level. The Jewish people have realized the importance of ritual for thousands of years including one of the oldest perpetually held rituals in the world, the Passover seder.

From a Jewish perspective, many myths that make up the story of the Jewish people are enacted in rituals: the creation story, the dialogues and stories of the forefathers, the Exodus, and the ultimate promises of redemption and a messianic era, to name a few significant ones. It should also be noted that many of the Jewish rituals, or mitzvot, act as an explicit "remembrance." By eating the matzah, one reexperiences the Exodus, and by performing the seder, one remembers what it means to be slaves and how God led the Jews out of Egypt.

Other rituals have the goal of forming a deep relationship with God. Reconciling the dyad of physical and spiritual existence is a key component of many kabbalistic interpretations of Jewish rituals. This reconciliation is utilized even more often as a basic teaching model for understanding the reasons for mitzvot. The idea of *teshuvah*, Jewish repentance, can be understood as a self-reconciliation with one's ideal self-confronting one's actual or physical self and forming a new person who knows who he or she is and what he or she can now do. However, one way of understanding *teshuvah* actually has the individual transforming with intense self-reflection into a "new" person who can thus not be judged for the previous person's actions. The ritual is believed to spiritually transform the person and form a new mythic concept of the nature of the human being as related to God. On a more "everyday" level, when Jews pray and recite the *Shema* prayer, they are asking for God to be both relatable and all-powerful and to either grant their request or form some sort of relationship with them.

Rituals are the experienced aspect of myth. But rituals have an extensive history that goes far beyond human religious rituals. Rituals appear to have originated within the animal kingdom primarily for mating purposes. The work of Eugene d'Aquili and his colleagues in the 1970s identified a number of elements of rituals. This work was elaborated further in books such as *The*

Mystical Mind and *Why God Won't Go Away*.[1] The primary purpose of mating rituals was to bring two animals together for mating. It is a unifying behavior that also has underlying neurophysiological correlates. Particularly, the autonomic nervous system that regulates both our calming and arousal functions of the body becomes rhythmically stimulated by a given ritual. Rituals that have many frenzied movements result in activation of the arousal system of the body, while rituals that are slow result in activation of the calming or quiescence system.[2] But these changes also affect higher parts of the brain and, in humans, appear to induce powerful feelings of connectedness or oneness that often have religious and spiritual overtones.

Many scholars have struggled with the best way to define rituals and their specific components. From a neuropsychological perspective, rituals appear to have several common elements:[3]

1. Rituals are structured or patterned.
2. Rituals are rhythmic and repetitive (to some degree at least), that is, they tend to recur in the same or nearly the same form with some regularity.
3. Rituals act to synchronize emotional, perceptual-cognitive, and motor processes within the central nervous system of individual participants.
4. Rituals synchronize these processes among the various individual participants.

The last component necessarily refers only to rituals performed in groups and not to individual rituals such as individual prayer or meditation practices.

Human rituals work on both an individual and group level, helping people feel a greater sense of connectedness to the group, tribe, society, the universe, or God. Importantly, human ceremonial ritual should probably be considered a "morally neutral technology," which, depending on the myth in which it is imbedded, can either promote or minimize particular aspects of a society and promote or minimize overall aggressive behavior. Thus, if a myth incarnated in a ritual defines the associated unitary experience as applying only to the tribe, then one ends up with only the unification of the tribe. It is certainly true that aggression within the tribe has been minimized or eliminated by the unifying experience generated by the ritual. However, this may only serve to emphasize the special cohesiveness of the tribe vis-à-vis other tribes. The result may be an increase in overall aggression (specifically inter-tribal rather than intra-tribal).

The myth and its associated ritual may, of course, apply to all members of a religion, a nation-state, an ideology, all of humanity, or all of reality. As one increases the scope of what is included in the unitary experience, the amount of overall aggressive behavior decreases. This exact description of feeling responsible for one's self and concentrically expanding circles of belonging fits perfectly with Rabbi Abraham Isaac HaCohen Kook's description of how one's level of religious consciousness expands with personal development (See Arpilei Tohar pg. 108, Orot HaKodesh 3:101). At first people are concerned only with themselves, then their neighbors and fellow Jews, then all of humankind, and finally the entire animal kingdom in ever-increasing unitary experiences.

In Judaism in particular, the ceremonial ritual is far more significant than the individual meditative or prayer experience. While the individual experience may be personally enlightening, the ceremonial ritual is prescribed and dictated from God, the rabbinic authorities, or communal custom to be something that everyone must participate in to be considered a part of that Jewish community. The ritual of communal prayer is a great example of something that was formatted to induce cohesiveness and unity in the wake of the national tragedy of the destruction of the Temple. For the Jews to live on in solidarity in the Diaspora, where there was no longer a unifying Temple service, they needed to have similar prayers and practices that would be reproducible by Jews wherever they went. Thus, when one goes to a new Jewish community and walks into a synagogue, he or she will be able to pray with the group, participate in their rituals, and keep the Sabbath with them even if their practices differ slightly based on locale and heritage.

Perhaps the most well-known Jewish ritual is the Passover seder, which embodies the myth of the Exodus out of Egypt, the end of Jewish slavery, and the special covenant between God and the Jews. It is an extremely important ritual on many levels and certainly bears consideration in the context of the underlying neurobiology. Part of the power of the Passover seder is that it has rituals layered upon rituals. The seder itself is a grand ritual in that it is highly structured. In fact, the word *seder* itself means "order" or "arrangement." And it describes patterned behavior from the first prayer, to the four glasses of wine, to the four questions, to the conclusion with the afikomen. It is also repetitive and rhythmic, containing songs such as *Dayenu* or the singing of the four questions. There also is repetition in specific elements, such as drinking a glass of wine four times throughout the meal. These repetitive physical behaviors signify to the amygdala that something important is happening. And the repetition primes the brain for each subsequent appearance of the idea or concept. These synchronize the brain of everyone present to experience the importance of the Jews being led by God out of Egypt. This is especially true as the four cups are linked to four specific verses describing God taking the Jewish people out of Egypt: "I shall bring you out . . . I shall deliver you . . . I shall redeem you . . . I shall take you." (Exodus 6:6–7)

As we mentioned, the seder also has another layer, which has to do with its rhythm and repetition over the course of a person's life. Passover is a holiday that occurs at the same time of year every year in a person's life. This annual repetition, especially as it coincides with spring, is another powerful reinforcement of the ritual and its meaning. It is embedded into the memory at a person's earliest stage of development in infancy and childhood and occurs until death, continually reinforcing the story. And there is yet another neurobiological layer, which is the connection that occurs across generations, since this ritual has been occurring for thousands of years. This connects the story to the group of people—the Jews—throughout time. The memory, emotional, and social areas of the brain all become part of this process.

Neuropsychological Models of Ritual

How does all of this ritualizing happen from the neuropsychological perspective? We might consider both the "top-down" and "bottom-up" processes involved. By top-down, we are referring to practices that start with an activation of the brain's cortex such as the frontal and parietal lobes that eventually lead to activation in the limbic system and autonomic nervous system connecting with the body. Practices such as meditation and prayer are excellent targets to study in this regard, since they can incorporate mythic concepts in a powerful ritual experience.

During certain spiritual practices, the frontal lobes are initially activated as the person brings attention to a particular belief, word, or prayer. Simply by repeating this concept, the limbic system begins to pick up that this belief is very important. The person begins to feel powerful emotions such as joy, awe, or love. As the process continues, the parietal lobes are deafferented, and the person experiences a loss of the sense of self and a sense of connection with the story or with God. This also results in a powerful autonomic activation that might combine either a strong arousal response accompanied with awe or joy and/or a strong quiescent response accompanied by a sense of bliss or peace. Logical paradoxes or the awareness of polar opposites may appear simultaneously, both as antinomies and as unified wholes. During intense meditative experiences, the experience of the union of opposites includes the total union of self and other. Similarly, in the *unio mystica* of the Christian tradition, the individual experiences of the union of opposites, or *conjunctio oppositorum*, or in the kabbalistic ultimate experience of *achdut* or complete *yichudim/devekut* also includes the experience of the union of the self with God is described.

On the other hand, many rituals are more typically a bottom-up phenomenon in which the rhythms and movements of the body evoke changes in the autonomic nervous system and subsequently the limbic system and eventually the higher parts of the brain. Neuroscientists such as Antonio Damasio, who wrote *The Feeling of What Happens*,[4] have suggested that many thoughts and feelings are interpretations of bodily and physiological processes in more of a bottom-up phenomenon rather than top-down. In rituals, this may be manifested by the presentation of polar opposites by the analytic or left hemisphere (i.e., the presentation of a problem to be solved in terms of the myth structure) and the simultaneous experience of their union via the activation of the holistic functions derived from the parietal lobe in the right hemisphere. This could explain the often reported experience of the resolution of unexplainable paradoxes by individuals during certain meditation states on the one hand or during states induced by ritual behavior on the other. In fact, there may be significant similarities from a neuropsychological perspective between meditation and group ritual in terms of the resolution of opposites, since both types of practices essentially meet in the middle. The neuropsychological similarity rests in the final activation of the holistic process in the brain, enabling a unitary experience that reconciles opposites. It should be obvious that the initial neuropsychological mechanisms underlying meditation on the one hand and ritual on the other are actually quite different, but perhaps they become more closely aligned in the latter stages.

These two perspectives have also been the focus of some debate among the Jewish leadership over time. As was mentioned previously, in the past some Hasidim would jump with joy and ecstasy in prayer, and intense movements were seen as part of the process of spiritual enlightenment. However, there was significant pushback from the larger Jewish community against such practices, as moderation in general was considered the norm and refinement in prayer and in person became the ideal for most rabbinic leadership.[5] This led to a sublimation in many Jewish circles of the original practice to a more simplified version where singing and dancing, while not ecstatic, was still performed. Still others opposed the practice and desired the more traditional prayer service without said song and dance. Most recently, the controversial and polarizing figure of Rabbi Shlomo Carlebach is a good indicator of this trend. He drew many individuals interested in that sort of song-and-dance spiritual connection to God. However, he drew sharp criticism from many rabbis in the Orthodox community for breaking from the traditional services and methods of service to God.

In terms of the more specific neurophysiological changes associated with various rituals, they begin with repetitive motor activity or visual, auditory, or other sensory stimulation. This rhythmic stimulation strongly drives the autonomic nervous system. With prayers and chanting, we might see an increase in sympathetic or arousal activity if they are particularly vigorous and active. Alternatively, if the prayers and ritual involve very slow rhythms, then the parasympathetic or quiescent system can turn on. The myth's meaning may be presented within the ritual prayer, thereby activating the cognitive functions of the left or analytic hemisphere, while the ritual rhythms might activate the holistic processes of the right hemisphere. Even more so, it seems that rhythmic or repetitive behavior synchronizes the limbic system activity and concomitant emotional states in a group of participants. It can generate a level of arousal or bliss that is both pleasurable and reasonably uniform across the individuals so that necessary group action is facilitated. Rhythmic activity likely causes these effects, in part, via its ability to function as a form of communication.

If the ritual is effective and persistent, the one side of the autonomic nervous system becomes so highly activated that this may result in a subsequent activation of the other system, referred to as spillover or breakthrough.[6] This unusual physiological state is associated with other aesthetic-cognitive effects besides a sense of the union of opposites. Many religious traditions indicate that such states yield a feeling not only of union with a greater force or power, but also an intense awareness that death is not to be feared and a sense of harmony of the individual with the universe. This sense of harmony with the universe may be the human cognitive extrapolation from the more primitive sense of union with other participants that ritual behavior generates in prehuman animals. This can help build the case that ritual evolved from the behaviors of primitive animals to the most complex human religious rituals (at least from an evolutionary perspective).

It is interesting to note that some Jewish rituals associated with prayer do not always lead to a specific rhythmic synchronization. Sephardic prayers have the group reading everything aloud together, and Hasidic groups often sing and repeat prayers together. However, in many congregations, the prayer leader simply sets the speed of the prayers with the congregation, and most of the prayer is in silent individual meditation and recital of prescribed words. The concept discussed earlier of "*davening b'avodah*" also appears to be a strongly individualized service, where people focus on their own inner mental state before and during prayer. This may also be why when the Talmud speaks about the formalization of the prayer service after the Temple's destruction, it states that the main reason was that individuals could no longer find the words to appropriately pray on their own. With prayer maintained as an individual experience, the group component of prayer becomes a unifying communal component rather than an additive piece to the generation of holistic feelings. However, that does not diminish the fact that there are many Jewish groups that use song and group dancing and singing to help increase cohesiveness and to create a "more meaningful" prayer experience for the attendees.

Ecstatic versus Blissful Experiences

Let us explore this basic neuropsychological model of ritual behaviors in more detail and consider not only a broader array of rituals, but especially the more powerful, ultimate spiritual experiences. For example, ritual can now be considered in terms of being rapid or slow and with different phenomenological characteristics as well as different associated neuropsychological processes. There is likely an initial parasympathetic drive associated with "slow" rhythmic rituals, like the slow *niggun* sung by many Jewish groups (particularly the Hasidim). This can be contrasted with the mechanism associated with "rapid" rituals, such as Sufi dancing or voodoo frenzy.

As any ritual continues, one arm of the autonomic nervous system becomes more and more active, leading to spillover and breakthrough of the other arm of the autonomic nervous system. It would seem unlikely for human ceremonial ritual to go beyond this point. However, various rituals and more likely spiritual practices such as intense prayer or meditation might yield several fascinating neuropsychological states and associated experiences.[7]

1. **The hyper-parasympathetic state:** *When parasympathetic activity is exceptionally high, the result would be an extraordinary state of relaxation and peace.* It may occur during spiritual practices such as chanting or prayer or during "slow" rituals such as the slow immersion in the mikvah. (For women the ritual of mikvah is done after waiting for the requisite "clean" days after cessation of menses, while for men it is only voluntarily.) In extreme form, the hyper-quiescent state may be experienced as a sense of oceanic tranquility and bliss in which no thoughts or feelings intrude upon consciousness and no bodily sensations are felt. In Buddhist psychology, this state is called access concentration, or *upacara samadhi*. This may possibly mimic what is described by some texts as the method many of the prophets would undergo to prepare themselves for receiving prophecy. In order for a prophet to feel relaxed enough to be receptive to the Divine Presence, he often would require music to help him prepare (see 2 Kings 3:15). It is possible that this is similar to a slow parasympathetic relax-

ation state necessary in order to perceive the divine flow, with one's mind empty except for thoughts of God.

2. The hyper-sympathetic state: *When sympathetic activity is exceptionally high, the result would be a powerful state of intense arousal and energy.* This state may occur under various circumstances in which motor activity is extremely "rapid," such as frenzied dancing or drumming. While Judaism seems to have nothing exactly parallel to this particular approach, the closest might be the continued rhythmically increasing dancing of the Hasidim of old. At times they would reach a state of spiritual ecstasy in their trances and higher spiritual states of prayer.

3. The hyper-parasympathetic state with breakthrough of the sympathetic system: *When parasympathetic activity is so extreme that "spillover" occurs and the sympathetic nervous system becomes mutually activated, the person would not only experience great bliss, but a sudden experience of a tremendous release of energy.* Thus, someone in deep meditation or prayer may experience an "active" bliss or energy rush. Occasionally, this may occur in people during "slow" ritual behavior, which can result in a brief eruption of the sympathetic system with consequent brief states of altered consciousness. This may mimic what many of the early medieval Jewish philosophers explained as the path to prophecy. One first meditates significantly, and only after much work and spiritual endeavor, one merits the gift of prophecy. However, all agree that it is not something that is naturally given, but something that God can freely give or withhold even if one has reached the requisite state of spiritual preparedness. Such a concept is typical of other enlightenment states in which the individual feels as if the experience is being given to him or her rather than any purposeful attempt at making it happen.

4. The hyper-sympathetic state with breakthrough of the parasympathetic system: *When the sympathetic activity is so extreme*

that the parasympathetic system becomes mutually activated,a person would experience an overwhelming sense of peace and bliss at the height of an intensely energetic feeling or practice. This state may occur during frenzied dancing, intense exercise, or perhaps during sexual climax. Such a state might be similar to the way that the holiday of Purim is explained by some of the commentators, who maintain the position stated in the Talmud that on Purim people must become inebriated at the festive meal so that they cannot tell the difference between "Blessed is Mordechai" and "Cursed is Haman," the Purim story's hero and villain respectively (Talmud *Megillah* 7b, See also *Shulchan Aruch Orech Chaim* 695:2). This state of absolute drunkenness, so antithetical to the moderation that is ordinarily required of the religious individual, is explained to be due to the unique nature of the Purim holiday. On this day, God turns the world on its head, as it were, and allows the unification of the bad and the good within human understanding to all be under God's will. The great unification exists even after intoxication, which may be similar to the parasympathetic spillover experience after sympathetic stimulation. However, the opposite argument could also be made that overstimulation of the parasympathetic centers via alcohol is what creates the stimulation to bring a sympathetic spillover and leave one in a spiritually enlightened state. Which pathway occurs in actuality is something that theoretically could be studied scientifically by observing markers of autonomic activity during these practices. Evidence of hyperstimulation of either the sympathetic or parasympathetic sides could help identify the basis of such experiences.

A fifth and final state would be one in which there is mutual activation of both the sympathetic and parasympathetic systems. In such a state, the person experiences maximum arousal and maximum peace at the same time. It is likely that this represents a powerful altered sense of consciousness in which the experience takes over the cognitive and emotional processes. This state is perhaps the ultimate mystical experience, in which the entire brain is experiencing something enormously intense in terms of the emotions, beliefs, and sense of unity—often described as an enlightenment experience. Mystical kabbalistic experiences are also of this type, in which the person becomes fully unified with the light of God.

The relationship of these five sympathetic/parasympathetic states to a whole assortment of altered phases of consciousness and mystical states is an important element to a neurotheological perspective of ritual. Such states may be related to various levels of Jewish rituals and especially kabbalistic experiences. Of course, there is more to ritual than autonomic nervous system processes. The cognitive and emotional processes associated with myth and ritual formation are also crucial. But the autonomic nervous system activity is an important mediator of such experiences and also aids in the transmission of the brain's activity to the rest of the body. The ability of the autonomic nervous system to connect the brain and body provides the means by which ritual can result in very visceral feelings.

In addition to the direct effect of rhythmicity, there are other neuropsychological components of ceremonial ritual that might augment the effect of rhythmicity and help cause changes in the autonomic nervous system and other brain areas during rituals. First of all, rituals frequently incorporate "marked" actions such as a prostration, a bow, or specific movements of the arms and hands. These marked actions draw attention as being different from ordinary baseline actions and would likely produce an orienting response in the amygdala. The amygdala acts to perform environmental surveillance and can direct attention toward something of motivational importance in the environment. Activity in the amygdala might result in a mild fear response, which in a religious or spiritual context might be called "awe," or *yirah* in Judaism. This feeling of awe can be ascribed to a particular story, belief, or God.

In addition to the amygdala's response to ritually marked actions, it should be noted that the sense of smell or sound can function as a driver of the sympathetic nervous system. The middle part of the amygdala receives fibers from the olfactory tract that are the neurons for the sense of smell. If we experience a strong smell, there is concomitant activation of the amygdala. It would make sense that the use of incense or other fragrances, such as the *besamim* spices smelled at the end of the Sabbath, might cause direct stimulation of the amygdala. Other Jewish smells such as of the Passover meal or the burning of candles similarly can cause a strong reaction in the brain.

In Judaism, the sounding of the shofar might be the greatest amygdala stimulator of all. The blast of the shofar likely activates the "acoustic startle reflex" that signifies something of importance to pay attention to by activating the amygdala and getting the sympathetic nervous system fired up to respond.[8] It gets our heart pounding and our blood pressure up as we respond with a sense of fear and awe. It is no wonder that the shofar is set aside for the most important holidays. During Rosh Hashanah the shofar is blown repeatedly, both as a call to repentance and as a symbol of crowning God as king over the world. Both of these symbols would necessitate feelings of awe and fear in their ideal states. Additionally, the final shofar blow on the holiest day of Yom Kippur, at the end of the day when the Jews achieve atonement, is a final call of joy and awe mixed together. Particularly when one's body has been weakened and made more susceptible to the stimuli by spending the entire day in fasting and prayer, the final blast can be quite powerful and awe-inspiring, giving meaning and significance for someone to make plans for true change.

Evaluating Specific Jewish Myths and Rituals

Each Jewish mythic story and ritual has its own specific elements and way of presenting opposites and then resolving them. The specific elements of the myths and rituals are what ultimately define a given tradition like Judaism. Let us look with more detail at a few such myths and rituals in Judaism from a neurotheological perspective.

The practices such as the four species—the mezuzah, matzah, prayer, and charity—as well as the Sabbath and kosher laws, can all be considered from the perspective of different neurological processes as described above. For the religious individual, the added perspective of neuroscience probably means very little, since it does not change the inherent religious basis. However, we hope that this is the beginning of a Jewish neurotheology that provides an added perspective that might actually contribute to our understanding of the many different Jewish beliefs, laws, and practices. Perhaps we will understand their meaning in deeper ways by knowing how they affect the brain and body. Perhaps we will find a deeper relationship between the brain and our religious selves. We might even find more effective ways of being Jewish.

The waving of the four species is a physical action that follows a specific set of movements. Such movements are typically initiated by the frontal lobe function for willful behavior and particularly the initiation of motor activity. The orientation of the body and a sense of spatial orientation are typically attributed to parietal lobe functions. Additionally, the shaking is often done by groups of people, which can cause what are called "mirror" neurons to fire. Mirror neurons essentially reflect or mirror what we see when another person does something. Thus, if you observe a person raising her hand, your brain's mirror neurons will fire in response and send a signal to raise your hand. You can stop this process consciously, but there is some amount of synchronization with another person that occurs using these neurons. Thus, during the group component of waving the four species, the mirror neurons help synchronize the shaking in the brain of all the participants. If the congregants believe that while shaking the four species they are also requesting rain from God (See *Mishnah Rosh HaShana* 1:2 & Talmud Taanit 2b, *Sukkah* 37b), the group action becomes one of a larger communal prayer, and members might feel that they are participating in something larger than themselves.

As we have described before, deactivation of the parietal lobe during such rituals has been shown to be associated with a sense of connectedness with others involved in the ritual or with God or the universe. However, if the individual only performs the ritual without substantial interest in the larger experience, then perhaps just the frontal lobe would be activated, and it might be no different than what is activated during regular motor activity.

The difference between individuals performing the four species ritual is most clearly expressed in the mystical approach that allows shaking to be done individually in the sukkah before prayers. When an individual shakes the four species, one is engaging in a personal synchronization of one's movement and thought centers. By thinking about the union with God and with the person's entire body (spine, lips, eyes, and heart, which correspond to the four species, according to the kabbalistic literature), one might experience something similar to a trance-like state with increased frontal lobe activity and decreased parietal lobe activity correlating with how much the individual felt a part of the larger community. Such an analysis supports the potential importance of this ritual as one that not only has theological importance, but literally utilizes the body to engage Judaism to its fullest extent.

The act of hanging the mezuzah on the door, while important as a ritual, is in fact secondary to having the mezuzah itself on the doorpost. When one passes by the mezuzah, it can be a reminder that God created the world (See Nachmanides commentary on Exodus 13:16), or it can be seen as additionally protecting the house from dangerous events (See *Zohar Parshat V'etchanan* 264a). The reminder one notices when passing by a mezuzah would trigger emotions and a memory of the *Shema* prayer recited every day, possibly through the limbic system. The general emotion would be one of safety and positive feelings associated with protection, potentially mediated by the limbic system's positive emotions along with an increase in serotonin that stabilizes negative emotions. The more kabbalistic understanding of protection from evil would be a stronger association in this case up to the point where one felt that the mezuzah itself gave off some sort of power. This sense of agency empowers the mezuzah and further connects us to God. We have previously suggested that certain "marked" movements or objects stimulate the amygdala, thereby notifying us of their importance.

To further support this point, we recently performed an fMRI study evaluating various visual symbols that had religious and emotional significance.[9] The symbols were taken from world traditions and were initially evaluated for their religious and emotional impact. We identified the five symbols that most clearly were categorized as either religious/positive, religious/negative, nonreligious/positive, and nonreligious/negative. Examples of these include, respectively, a dove, the devil, a smiley face, and a gun. The brain scan results showed that religious symbols had an impact on the primary visual areas of the brain, even before our consciousness became aware of the symbols. And our religious beliefs actually affected the way our brain perceived these visual stimuli. So, if our study of religious symbols means anything, perhaps the symbol of the mezuzah stimulates basic brain functions associated with our perception of reality.

The eating of matzah is also an important ritual, which we briefly discussed above. If we are reliving the former Exodus, we must activate our emotional and memory centers in the limbic system, especially the hippocampus. However, if we would then move to a higher level of feeling, like we were excising the bad "leaven" from our body in the act of eating (See Sfat Emet Book of Exodus: Parshat Bo: 20), then a more complex set of correlates would be expected. The physical act of eating is also an intentional one, potentially mediated through the frontal lobes as well as coordinated motor messages. This orientation of memory and action might actually involve the dopaminergic system of the basal ganglia, mediating the glutamate neurotransmitters to allow for an increased amount of communication between the different parts of the brain. If that was the case, then the ancient experiences that are read about would now be "experienced" by the individual seeking to participate in the reliving of the Exodus myth. Additionally, when trying to excise the "leaven," one could also potentially be entering a state of self-dissociation, perhaps mediated by a decrease in parietal lobe activity. Matzah, as well as other foods that carry particular meaning, also stimulate the taste centers of the brain. The taste brings up memories in the hippocampus, which reinforces the meaning of holidays or specific religious concepts such as the freeing of the Jewish people from bondage. This may all be another reason why there is a specific command for everyone on Passover to consider as if they themselves have gone out of Egypt, as well as it being the only Jewish holiday to have a specific food that Jews are commanded to eat (the matzah).

Charity likely requires a complex set of interacting neurological functions. Giving charity requires action, specific action that uses one's own money or time for another. Thus, there is bound to be some frontal lobe involvement in planning the act. However, there appears to be an important emotional element to charity, particularly as it relates to the understanding of another's emotional feelings. Several research studies have demonstrated that the insula, which helps us to interpret feelings, plays a prominent role in prosocial behaviors such as giving charity.[10] Maimonides (*Mishneh Torah*: Laws of Gifts to the Poor 10:7) states that helping another person become self-sustaining is the highest form of charity, and that requires a lot of planning and preparation, as well as emotional investment. However, the dissociative feeling of spiritual healing and forgiveness only occurs when one gives the charity as an act for *teshuvah* (repentance). In such a case, one may again have the feeling of communicating while doing a simple action. In this situation, the ritual act of giving charity elicits the emotional limbic system as well as the positive feelings of protection and security offered from the memory of the previous mystical experiences. One final point about charity stems from an interesting study of children who were asked to donate money to a charity.[11] For the most part, factors such as genetics, socioeconomic status, feelings of attachment, and other prior environmental factors did not appear to affect whether they would donate—in fact, most did not. However, if they were prompted to donate, they were much more likely to do so. This supports the importance of religious concepts such as *teshuvah* as a way of encouraging charity. Such responses also implicate the importance of the emotional and social areas of the brain in relation to works of charity.

The laws of *kashrut* combine myth and ritual and also derive from both intellectual and spiritual reasons. *Kashrut* is actually a collection of various dietary laws, many of which have their basis in rabbinic literature, but no reason is explicitly given in the biblical texts. The main reasoning for those who do not seek a higher spiritual sensation with their observance of *kashrut* is that God gave these laws and their reason may not be known, but they will assuredly increase fealty and obedience to God, as well as distinguish those who choose to keep them. Another option is that the laws are geared toward engendering compassion. As Rabbi Haim Halevy Donim, the former Sephardic chief rabbi of Israel indicated, kosher laws make one's table like the altar in the Temple and creates an atmosphere in which the food is sanctified to God as one eats.[12] From this perspective, every bite one eats becomes a spiritual ritual. But this ritual can go even further because it represents earlier rituals that one has not actually participated in, such as offering sacrifices in the Temple, even though it has been destroyed for over two millennia. The neural systems involved in maintaining the myth and ritual of kosher laws likely would be similar to those discussed above in relation to eating matzah, with frontal lobe involvement and dopaminergic system activation. Additionally, the *Zohar* and other kabbalistic works see nonkosher food as having some sort of spiritual taint, so that when one partakes of nonkosher food, the person's spirit is actually affected.[13] This concept of a physical act having spiritual consequences would be an intense integration of the frontal lobe's actions and the emotional response of the limbic system, as well as the spiritual dissociative state of decreased parietal lobe activity.

Shabbat is another uniquely Jewish experience that has multiple mythic and ritual facets. The Bible states that the purpose of Shabbat is to remember that God created the world and took the Jews out of Egypt (See Exodus 20:8–11 and Deuteronomy 5:14–15). On the Sabbath, "work" is forbidden. Tradition defines work as whatever the Jews did to maintain and move the Tabernacle during the time in the desert on the way to the Holy Land from Egypt. This work is categorized into forty-nine *malachot*, or "works," that are prohibited biblically on the Sabbath. Traditionally, the way the *melachot* are explained is that their work was defined as creative work, and by taking a break from creativity, one can come to realize that God is the one with true creative power (See Rabbi Samson Raphael Hirsch's commentary on Exodus 35:1–3). Additionally, by recognizing that God is the one who took the Jews out of Egypt, one can reorganize one's priorities to be in line with spiritual beliefs. Thus, if God is the one who took the Jews out of Egypt and therefore has earned the Jews' loyalty, one may recognize the need on the Sabbath to think about one's own spiritual and physical commitments to God in order to be "a better Jew." This creative break could potentially correlate with decreased frontal lobe activity, as well as a naturally increased "pre-meditative" state that would allow for easier self-reflection, perhaps even similar to what happens with some mindfulness techniques. Through decreased brain activity and forced relaxation from all other creative processes, the mind is allowed to relax and form novel connections and thoughts. In the end, Shabbat may ultimately yield greater creativity by helping the brain "take a break." In fact, research suggests that creativity most likely occurs during states of daydreaming and mind wandering, something that can occur when one is not focusing on basic daily activities, but rather resting.[14]

Theologically for Jews, Shabbat is even more than just resting the mind. The Talmud (*Taanit* 27b) states that one is given an extra soul on Shabbat and that this soul increases one's level of spiritual sensitivity and understanding. This position can be interpreted in light of what was just stated, namely that Shabbat is a time for introspection, when additional cognitive abilities are unlocked. An interesting neurotheological perspective might explore whether unique areas of the brain are activated or deactivated during Shabbat. And what would such findings mean in the context of the soul or even a second soul?

Hasidic writings view this Talmudic concept of a [15] second soul as indicating that one has additional abilities on Shabbat that will aid in the meditative and introspective process. If this is the case, then the changes in neural activity on Shabbat might be assumed not to derive from simply a lack of overstimulation, but a definitive change in neural connections associated with the assumption that the day will aid in spiritual endeavors. When one anticipates something, the amygdala, along with the frontal lobe, can activate in unique ways to push someone in that direction, thus contributing to the potential increased ability for spiritual meditation on Shabbat. Another possibility is that changes in certain neurotransmitters such as dopamine or serotonin occur during this relaxation that predispose the individual to intense spiritual experiences. Some data from our own research has suggested that spiritual retreats alter dopamine and serotonin levels in such a manner that might predispose individuals to intense spiritual experiences.[16]

Jewish Prayer and Introspection

Prayer is the most general, but also can be the most complex of all the rituals. This is particularly the case since prayer comes in many forms, some heavily prescribed and some very spontaneous. In addition, prayer can be done individually, in a small group, or with an entire congregation or community. In order to understand how prayer works, one must understand the basic background of the origin story (i.e., the "myth") of the Jewish prayer experience. The Talmud (*Berachot* 26b) states there are two background stories to the origins of the current prayer system of three times a day. One, which appears to be the earlier chronologically, is linked to the *Avot*, the forefathers. Each one of the three (Abraham, Isaac, and Jacob) instituted the prayers (*Shacharit*, *Minchah*, and *Maariv*, the morning, afternoon, and evening services), respectively. Each of these was seen as an individual outpouring of emotion that the forefathers felt during their respective lives and tests they underwent. However, that is not the only opinion. The second position is that the daily prayers correspond to the three daily regular sacrifices in the Temple. The *korban tamid* refers to the "sacrifice" brought each morning and afternoon to the Temple. Additionally, there was an overnight burning of the afternoon *tamid* as well. These three offerings that were brought on the altar came to an abrupt end with the destruction of the Temple. As such, the rabbinic assembly felt that something was needed in place of the offerings and instituted the three regular prayers in place of these sacrifices. Thus, prayer became a proscribed, specific custom done with regularity and little variation in practice. These two varying interpretations of the origin of Jewish daily prayer speak to the two different perspectives on prayer being either a highly ritualized group event or a highly personalized individual spiritual experience.

There is also the issue of whether the daily prayers are said aloud in a group, as many Sephardic Jews do, or if there is an oscillation between saying prayers as a group and saying them as an individual, as is the norm for most Ashkenazic Jews. If one would say all the prayers together out loud, then the mirror neurons and frontal lobe system for action, as well as the temporal lobes for language, would likely be activated. Prayer often involves some movement and bowing, and this movement would be initiated via the frontal lobes and amygdala for the orienting response. The combination of private and communal prayer among Ashkenazic Jews might allow for individuals to oscillate between the feeling of being lost in the group and that of being alone in their thoughts in front of God. This could result in a type of potential meditative mindfulness combined with a unity experience both with the group and with God.

The option to say the prayers in any language one understands means that there can always be activation of the temporal lobes for language understanding. However, if the prayer is recited in Hebrew, even if the person does not understand Hebrew, Jewish law states that he or she can still recite it along with the congregation (See *Shulchan Aruch* Laws of Prayer 101:4). Such a mimicking of what is said without understanding could be a very interesting type of prayer to evaluate, since it would seem to activate only the mirror neurons related to the phrases said and the actions performed. It would be fascinating indeed to determine how such illiterate prayer compares with prayer that one understands fully.

Other types of prayer could be conversational, in which one views prayer as a means of asking God for things as if one were simply speaking to another person. One study of conversational prayer in a Christian context showed that the same language and social areas of the brain were activated as those involved with speaking to another person.[17] Thus, conversational prayer appears to be a way in which a person can interact with God on a very personal level, but not one aimed to induce intense spiritual or mystical states.

Prayers can even be used before praying and other religious activities. Before every religious act, many Hasidim and kabbalists recite a prayer called *Leshem Yichud*, which states that the purpose of the act they are about to do will be to unify God's name, thereby having them recall the larger theological purpose behind the commandments in general. This framework is created by the general understanding that the kabbalistic system has about the nature of the world and humanity's place in it. Many kabbalists meditate on the names of God before prayer and fulfilling commandments, since this will unify God's light through the *sefirot*, the spheres of existence. This idea became more popular in medieval times through Abraham Abulafia, a prominent kabbalist, and was turned into a system of *partzufim* by Rabbi Isaac Luria, a prominent Sefat kabbalist (See Pri Eitz Chaim Gate of Teffilin Chapter 16). The *partzufim* are arrangements of the spheres into separate worlds that lead into each other and create the metaphysical flow that ultimately creates the world. Luria believed that by meditating on the flow before prayer, one can enhance one's own prayers and achieve divine inspiration.

This process may have interesting neurological correlates. As one meditates or prays, there is an increased focus on a specific object such as a word or name. As the process goes on, the focus on the name leads to a reduction in sensory input into the parietal lobe. And as the parietal lobe begins to shut down, the individual will lose his or her sense of self and feel a deep sense of oneness with the object. Applied to the notion of the *Leshem Yichud*, focusing and repeating the name of God might allow the person to feel deeply connected to God and perceive God's oneness.

The emphasis on God's names in prayer continued until the time period of the Hasidim, when the Baal Shem Tov, the potential founder of the Hasidic movement, spread the idea that joy, dancing, and singing are the true elements that create emotional closeness to God (See Likkutei Yekarim 1b & 2b). This evolution allowed for the spiritualistic elements within Judaism to become more accessible to the masses and reduced the amount of time needed to be spent on intense meditation in order to achieve the *devekut* state. At the same time, the same Baal Shem Tov emphasized a concept of *tfilah* that requires significant *avodah* "work" where one meditated significantly on what they were going to say during prayer as well as how to conceive of the Godhead prior to engaging in prayer. The combination of intense happiness coupled with the universalization of internal preparation allows for a much broader and more emotional prayer experience. Among the Hasidim of Breslov, Rabbi Nachman explained (See Likutei MahaRan—Collections of Rabbi Nachman #54 Miketz) that when one has achieved total *devekut*, then one will realize that God is the essence of reality as we know it. This total absorption in the

Godhead is indicative of a strongly spiritualistic and meditative orientation to religion and was totally at odds with many other Jewish thinkers at the time, who sought for a more "this-world" practical approach to the religious teachings. Nevertheless, the mainstream thinkers still utilized much of the Hasidic thought on meditation and introspection.

Rabbi Israel Salanter, a nineteenth-century rabbi, founded the Mussar movement, which focused on introspection and self-examination, among the non-Hasidim. He taught a method of self-evaluation similar to what Rabbi Elya Lopian (one of the leaders of the Mussar movement) *defined as mussar,* or "making the heart feel what the mind understands." In this method, you simply sit and watch how your mind wanders, thus allowing you to view yourself in a "beyond-self" state. This idea is similar to the notion of "mindfulness" in which you pay attention to your thoughts and feelings without judgment. You simply rest within the present moment. But the *mussar* approach also may be a variant of "passive meditation" currently used by many to clear the mind of thoughts. The notion of being absorbed into the Godhead could have similar neurological correlates as with other mystical experiences in which the individual becomes one with the universe or ultimate reality. As we have described, current data suggests a decreased activity in the parietal lobe associated with the loss of the sense of self and the sense of absorption into the object of the meditation—in Hasidic or kabbalistic thought, this would be God. A study that could explore whether such neurological changes occur during these highly spiritual Jewish practices might provide unusual insights into the power of these experiences.

No matter how one looks at prayer and its varieties, there are still two frameworks of thought—the intellectual and the spiritual—that play out differently in the neurological and theological realms of Judaism. The interaction between these two frameworks means that the cognitive and neurological processes that are engaged in the religious actions of the Jew will always be complex and individually differentiated. We might expect certain parts of the brain such as left temporal and parietal lobes that enable abstract, rational thought processes to become more active during rationalistic approaches to Jewish practices. And we might expect the superior parietal lobe and emotional centers of the brain to interact more during spiritualistic approaches.

Every individual may at times feel more or less spiritual and may also engage intellectually in different religious events. Which perspective dominates for a particular individual depends on his or her personal background, education, and perspective on the religious acts being performed, as well as the assumptions and expectations about what the act will do. As such, there are those who have few mystical experiences and feel no real need to pray in a meditative state, but who regularly attend prayers and fulfill all of the required Jewish rituals. These people do not feel lacking in some way or feel that their religion has left them without meaning, but in fact find much meaning in their encounters and feel that they are doing God's will. In comparison, there are those who go through the motions but feel nothing at all while doing the rituals. These individuals do not have true "religion" from a theological or neurological perspective and in fact appear to be using the religious movement for other reasons, be they personal, communal, or emotional. Lastly, the people who yearn for some sort of higher spiritual connection may end up getting more heavily involved in kabbalistic and Hasidic practices as a way to fully engage their spiritualistic side.

Permanence and Impermanence

An important characteristic of rituals, as well as religions in general, has to do with the balance between permanence and impermanence. It is the rhythm between permanence and impermanence that appears to define that which is adaptive both within a religion as well as for a religion with respect to the outside world. In other words, for a religion to persist, it must maintain certain specific myths and rituals that form the foundation of its belief system. These permanent myths and rituals help produce the memory and the cognitive and emotional attachment of the adherents to that religion.

In the example of the Passover seder above, the same story is told year after year. There is a permanence to the content of that story. Each year, the same four questions are asked. However, there is also impermanence, since the melody may change to make it more modern sounding. But the actual four questions remain the same so that participants can understand the meaning behind the holiday. But what if the answer to those questions would change? What if Moses failed to lead the Jews out of Egypt? Not only would the ritual change, but the entire religion of Judaism would change. Over time, various myths and rituals of a given religion may be modified or updated as new scholars apply different perspectives to their meaning. As humanity progresses, various ideas and technologies may affect the way we think about specific religious beliefs. Take for example the creation of the universe. As science has enabled us to understand the cosmological workings of the big bang, it has forced a rethinking of the creation story in Genesis. Thus, religious concepts need to continually adapt themselves to subsequent generations in order for them to remain viable. But if the specific beliefs of a religion change to such a great extent that it no longer is recognizable, then the religion itself has changed. The balance between permanence and impermanence arises as part of the rituals of a given religion and help determine the specific elements as well as the existence of the religious belief system itself.

Judaism itself has adapted and morphed to new periods of history. However, there is significant debate among the various sects of Judaism as to how this adaptation should proceed. The creation of the Reform movement itself was in response to a growing desire from many Jews in the newly emancipated Germany to blend and be more similar to the larger European culture. When the Jews had been relegated to the ghettos and forced to maintain their separateness, being unique and maintaining their traditions were far easier, as there was nowhere for someone from the community to go if they would even want to leave. However, with the Enlightenment in Europe and the option to leave the traditional religion behind, many Jews began to change their practices. The Conservative movement came about as a kind of counter-force in response to feelings by many of the Reform-type leadership at the time that the traditions important to maintain Judaism were being ignored to the great detriment of the continuity of the religion. The story is told of a banquet honoring the first graduating class of Hebrew Union College where nonkosher seafood was served, angering some guests. However, the Reform leadership of the time dismissed it as "kitchen Judaism," which led to greater rifts within the larger Jewish community. Issues and differences such as whether to include kosher laws led to the creation of the Conservative movement within Judaism. Many Reform rabbis argue that Judaism should continue in the spirit of the law. This means that the purpose of the rituals and specific laws of the Bible and Talmud should be understood and then utilized in a way that fits into the modern era, with more modern ethics and sensibilities.

A Conservative rabbi would agree in principle that the spirit of the law is incredibly significant but would also believe in there being more continuity of the law, stemming from the belief that the biblical laws are in fact divine but subject to the evolution of time and could be changed depending on circumstance. This has led to much theological debate within the Conservative movement about what exactly still must be kept in Judaism and what is no longer needed.

Among the Orthodox, there is widespread agreement as to the divine origin of the Torah and the divine authority of the rabbis of the Talmud and later codifiers in determining Jewish law. This means that there is far less change in the practice of Judaism between generations and that the legal system, once espousing an opinion that is accepted on a certain matter, cannot simply change. Orthodox theology does believe in a meta-halacha that applies for all time. The great debate among Orthodox scholars today centers around whether the halacha, the Jewish legal system, can adapt to new situations that have not been encountered before.[18] Many consider the system to have been set up and finalized already and that until there is a new *Sanhedrin*, a full legal body sitting in Jerusalem with worldwide Jewish recognition of their authority, no rabbinic legal enactments of the past that have been accepted can be changed. New situations need to simply be viewed from the lens of the tradition, and appropriate responses to these new situations can be formulated. Others believe that that the legal system has certain rules and applications, and these should be thoroughly analyzed to determine how to best apply them to new technology and situations that never existed before, as well as to older *halachot*, Jewish laws that are inconsistent with the meta-halacha of the community. If there is an ethic independent of the Jewish legal system, and this ethic contradicts the halacha, then this camp believes that they should do what they can to help the system adapt to the ethical realities. In practice, the situation is far more complicated, and these two camps are often blurred, and individual situations often lead to different resolutions.

Despite this debate, all would agree that the basic Jewish laws given by the Torah and kept for generations would be unchangeable and immutable against the forces of time and history. A good example of this application would be the issue of the application of the international dateline to the Jewish legal system. When one goes to Hawaii, the international dateline puts them in a certain place. However, Jewish law may also place the island on a different location timewise, leading to a situation where the Sabbath does not fall on Saturday. There are differing positions on the matter, and no definitive answer has been reached. However, there are some Orthodox Jews who will do no work for two days while in Hawaii to ensure no Sabbath laws are broken. Others, because of their halachic position, believe Saturday and the Sabbath are the same day and will keep only the usual Sabbath from sundown Friday to sundown Saturday.

Rituals in Judaism in general also speak to different people in different ways. When someone lights Shabbat candles on Friday night using the candlesticks their grandparents used, the emphasis may be on the need to bring the Sabbath in with lighting the candles, but the intergenerational continuity is clear. One can have a spiritual one-off experience while lighting candles one Friday night and praying, but if one does it consistently, the ritual becomes ingrained in the neuronal connections in the brain. If it is seen as a chore that must be completed, the ritual can lose its meaning. However, consistency also allows for one to maintain a ritual despite non-spiritual moments and events in life. When the opportunity again presents itself, the deeper meaning behind the ritual can be awakened. The ritual can also take on meanings for identity and defiance. Many of the Jews in Eastern Europe, all of Europe during the Holocaust, and in Communist Russia were significantly persecuted for even trying to carry out their rituals. However, this made them feel like the rituals were an even more necessary part of their identity. One might not even be thinking of any higher spiritual goal when carving Hanukkah candles out of potatoes in order to fulfill the mitzvah of lighting them, but that act elicits strong feelings of belonging and a desire to link back to the intergenerational connection one has to the *masora* (generally translated as tradition) of the Jewish people.

Thus, there is an ongoing struggle between permanence and impermanence that is part of every religion and part of every myth and ritual in that religion. From the perspective of the brain, we see a similar interaction. An eighty-five-year-old woman is still the same person she was when she was five years old, twenty-five, and fifty, but she has also adapted and changed. The brain through its neuroplasticity can grow new connections in order to take on new ideas, experiences, and beliefs. Of course, if the person becomes too different, she would lose her identity.

Hence, our brain is always balancing permanence and impermanence, and this is subsequently reflected in our religions.

Improving Ritual and Liturgy

Given the ability to begin evaluating myth and ritual from a neurotheological perspective, it may be possible to consider the general applicability of this approach to Jewish rituals. A more specific question to be addressed is, can neurotheology have anything to say about current Jewish liturgical practice? Liturgy refers to the particular arrangement of religious rituals. There are many familiar liturgies in Judaism, such as the Passover seder or the holidays of Rosh Hashanah or Yom Kippur. Each holiday has a certain liturgy that consists of specific prayers, songs, and stories that are put together in such a way as to provide a consistent, spiritual, and religious event. What is particularly important is that each holiday may have its own set of prayers, songs, and stories that create a ritual specific for that holiday.

The different aspects of the liturgies give each holiday, and hence each ritual, its own meaning. However, certain prayers, songs, and stories may be performed at every ceremony, since they refer to some primary tenet of the religion. These might consist of prayers or songs that praise God or refer to God's infinite oneness. For Judaism, the usual daily prayers in fact make up most of the ritual prayers on any given day, the only exceptions being the High Holy Days of Rosh Hashanah and Yom Kippur. All the other Jewish holidays do have additions and changes to the prayers/*davening*/*tefilah*, but the basic structure remains the same throughout. The universal elements of a given religion's liturgy help make all of the rituals part of the religion in general. Thus, Hanukkah and Sukkot do not exist on their own, but they exist within the context of Judaism and have their own specific practices and prayer rituals.

On a fundamental level, liturgy must strike a rhythm between the theology or mythic elements of the religion in general and the significance of an individual story. Judaism provides an outstanding example of this in which each holiday has specific stories and themes in order to express a particular component of Judaism. However, the prayer called the *Shema* is said at every ceremony in the Jewish religion. This is because the *Shema* represents the primary tenet of all of Judaism:

שמע ישראל, ה' אלוהינו, ה' אחד

בָּרוּךְ ,שֵׁם כְּבוֹד מַלְכוּתוֹ ,לְעוֹלָם וָעֶד

Hear, O Israel, Adonai is our God, Adonai is One.
Blessed is the Name of the Honor of God's Kingdom forever and ever.

This prayer expresses what distinguished Judaism from all other religions in the ancient world, an insistent and uncompromising monotheism. This prayer has been preserved and repeated for thousands of years. Similarly, the personal *Shemoneh Esrei*, or eighteen blessings (though now there are nineteen), recited every day by Orthodox Jews, has remained an integral component of the Jewish prayer service across denominations. No matter what the holiday, the opening and ending of the *Shemoneh Esrei* remain the same: opening by praising God and concluding by thanking God for everything one has and requesting for ultimate peace. These themes are not specific to the themes brought out in any given holiday or event, but are overriding principles that Jews seek to include in the prayers each day. In fact, it was this prayer that we specifically decided to explore using brain imaging to assess what exactly might be going on in a rabbi's brain.

Imaging the Rabbi's Brain

As we got into the midst of writing this book, and after all of our discussions about various aspects related to the rabbi's brain, we realized that one very important thing was missing—an actual scan of a rabbi's brain. So we decided as a small experiment to have both of us go into the scanner to check out our brain while performing one of the most fundamental prayers in Judaism—the *Shema*. As with all of our studies, we had to determine a reasonable comparison state that matches everything about the prayer stay with the exception of doing the prayer itself. The *Shema* is actually sung so we decided to make the comparison state singing a simple song of *Row, Row, Row Your Boat*. We figured that both states involved singing a song that we knew well, only one has deep spiritual and religious meaning, while the other one is simply a song. In order to avoid movement of the head, we decided that we would sing only in our mind rather than out loud.

Our overall hypothesis was that, similar to other types of prayers, doing the *Shema* would involve activating the frontal lobe attention and language areas. Given the emotional importance of the *Shema*, we also figured that there would be activation in the limbic system. And perhaps if we were lucky enough, even though we were doing the prayer for a short period of time, we might see a decrease of activity in the parietal lobe if either of us felt an intimate connection with God during the prayer.

The results of the scans were quite interesting to us for several reasons (see Figures). Perhaps most importantly, when we compared the scans between us, we were quite surprised to find a pretty dramatic difference. Only, we were not really surprised at what the difference was. David, the Orthodox rabbi, had all kinds of changes going on in his brain while singing the *Shema*. Andy, who is a Kohen but raised as a reform Jew in a family that was not particularly religious, had virtually nothing going on in his brain while singing the *Shema*. In Andy's brain, singing the *Shema* was not much different than singing, *Row, Row, Row Your Boat*. To him, this was somewhat surprising because he was trying to do the prayer has fervently as possible. But obviously, the brain scan revealed otherwise.

So what did we actually see in the rabbi's brain? Starting with the frontal lobes, we saw a very unique pattern. A number of areas in the front lobe, particularly on the right, were substantially activated. This is consistent with what we have observed in many other practices that also requires focus, attention, and language. That David was focusing intently on doing the prayer is revealed by the widespread increases of activity in his frontal lobe. What was particularly fascinating is that in his left frontal lobe there were a number of areas that revealed a decrease. We have observed decreased frontal lobe activity when people are engaged in spiritual practices that lead to a sense of surrender in which people feel as if it is happening to them rather than them making it happen.

What was unique about David's brain is that one side of the frontal lobe went up while the other side went down. This has not typically been observed in the other spiritual practices that we have studied, and we have looked at a wide variety of practices including Christian and Islamic prayer, Buddhist meditation, mediumistic trance states, mindfulness meditation, Sikh meditation, and speaking in tongues. Thus, at least with this rabbi's brain, we find an interesting and unique pattern of activity in the frontal lobes. Our interpretation is that the process of doing the *Shema* involves a combination of intense focus and surrender at the same time. Of course, it is also possible that we captured David's brain in a dynamic process between the increased and decreased frontal lobe activity. In other words, perhaps he was transitioning from high frontal lobe to low frontal lobe activity. We have previously proposed that this is an essential part of the process of intense spiritual experiences including mystical or enlightenment experiences.

Not surprisingly, David showed increased activity in the limbic system, particularly the amygdala. These areas are well known to be associated with intense emotional responses and since this is the most important prayer in Judaism, we would expect him to have such a response. Finally, David had a mild decrease of activity in his parietal lobe, the part of the brain that helps us to feel connected, in his case connected to God. So for our rabbi, the brain actively engages many different parts associated with many different aspects of the Jewish prayer experience. And it appears that there are similarities and differences between the *Shema* and practices from other traditions. And finally, the results indicate something we have argued for some time, that the more you are engaged and believe in the practice you are doing, the more you activate your brain.

While certain elements of Jewish prayer are religiously obligated, much of the "form" of the prayer, i.e. who leads, what is

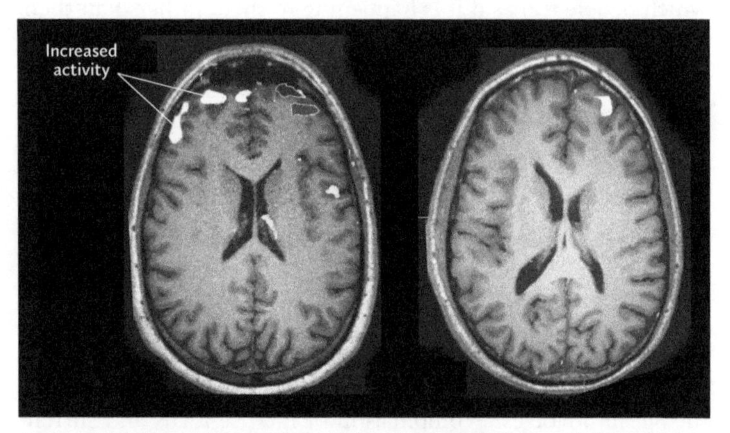

Increased right frontal lobe activity (white) and decreased left frontal lobe activity (gray) in David's brain (left image) during the Shema. Andy's brain is on the right showing minimal increased activity.

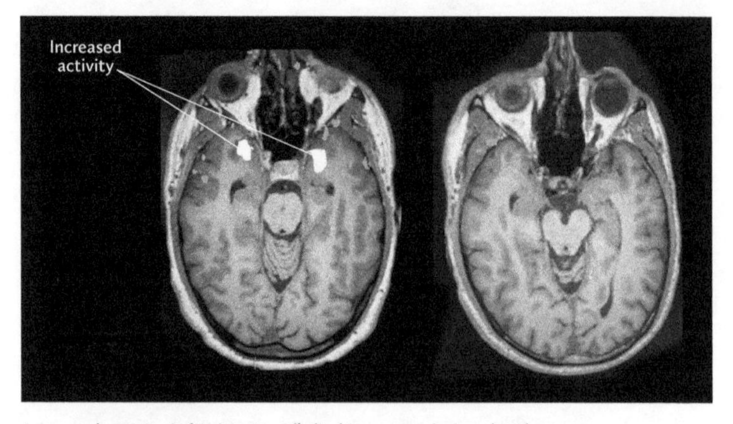

Increased activity (white) in David's limbic system during the Shema.

said outloud and what is said quietly, what tunes people use and whether there is dancing or singing, or quiet contemplation, is often up to the past rabbinic leadership and their perspectives on what is appropriate/the best way to behave during prayers. These leadersduring the prayer

From a neurotheological perspective, we might consider how important it is for a given individual's brain to enable him or her to follow rituals to a greater or lesser extent and whether this information can be used to develop future liturgy. It might also be possible to more actively engage some of the modern perspectives on how one is to engage God through various acts or beliefs. For example, Martin Buber, a 20th century German philosopher who moved to Israel after Hitler came to power in Germany, famously described the need to not define God, for such is an "I-it" relationship, and instead to simply exist with God in an "I-thou" relationship.[19]

For Buber, it was essential to develop the I and thou relationship with God by letting God come to you. Theoretically, there may be neurological correlates that could focus on how the brain processes the I and thou relationship with other people or with God. Remember the study mentioned above that conversational prayer with God activates similar brain areas as those involved in having a conversation with another person. Perhaps such processes might help develop more effective rituals and liturgies to optimize this I and thou relationship.

Or perhaps we might find neurobiological patterns associated with powerful emotional experiences such as that associated with the experience of "radical amazement" discussed in the writings of Rabbi Abraham Heschel.[20] Heschel was a 20th century Rabbi raised and studied under Orthodox institutions in Germany, but became a leader of the Conservative movement and wrote extensively on the need for moral correctness and the fact that while the law is important, it must not become solely focused on legalistics and lose its inner meaning. This desire to maintain some halachik observance but not accept the gamut of rabbinic restrictions is one of the foundational distinctions between Conservative Judaism and Orthodoxy (as well as Reform Judaism in the other direction). And it was through enacting the right behaviors that Heschel felt a person could still have the experience of radical amazement. We might wonder whether it would it be possible to characterize the neurobiological basis of such an experience or even find the most effective ways of helping people to experience it? Could neurotheology help to demonstrate the types of religious experiences attained by different approaches?

One interesting argument regarding the overall evolutionary basis of religion might bear directly on this issue. Several scholars, such as Scott Atran and Pascal Boyer, have argued that religions function as a "signaling mechanism" by which individuals indicate to each other who is a true member of the group., The rituals require a certain degree of personal (or financial) investment that signals a true commitment to the group. Within Judaism, there is much debate surrounding the different denominations and their differing ritual observances. Some Orthodox groups will not recognize members of the Reform community as "observant," because they do not adhere to the same rituals and beliefs. However, Judaism also has the distinction of being a nationality, and its members are therefore always considered Jews by default. Nevertheless, the status of those converted under the auspices of a Reform or Conservative rabbi can be called into question by those who are Orthodox due to a lack of similarity on ritual observance and what this indicates theologically. This further raises the issue of permanence and impermanence. How much of the liturgy and rituals does one have to do in order to be considered part of the tradition?

Another example might be whether a story should be told in the original liturgical language or in the secular language. If that story is brought to another culture, then it would make sense to change the language so that the new audience can understand the story within the rhythmic flow of the liturgy. Insistence that the story be told in the original language would necessitate various awkward strategies for translation, all of which would interrupt, in one way or another, the rhythmicity of the ritual, inevitably resulting in a diminished effect. An interesting example of this in Judaism is the law surrounding prayer: though prayer is traditionally said in Hebrew, the Jewish law/halacha states that people may pray in any language that they understand or in Hebrew even if they do not understand the words specifically. The uniqueness of this approach is that it allows for the need for the individual worshiper to maintain the connection to God in personal prayer, but also maintains the fealty and continuance of the religion to the traditional flow of the liturgy. For even though one can translate prayers and say them in any language, the original beauty and meaning behind the prayers are diminished somewhat, and a little is "lost in translation."

The rhythmic aspect of liturgy involves the liturgy itself. As described earlier, various songs, hymns, chants, or prayers can all drive the sympathetic or parasympathetic systems and concomitant brain structures. Slow, rhythmic hymns may result in a sense of peace and tranquility. If certain stories or ideas are narrated or sung during this time, then they will be associated with powerful emotions of peace and tranquility. For example, during a slow rhythmic chant that would activate the parasympathetic nervous system, a narration about the love that God feels for human beings might leave the individual participants with the sense of overwhelming peace and an all-encompassing, powerful love.

In the end, a liturgical sequence that employs both aspects of sympathetic and parasympathetic activity—some fast songs, some slow hymns; some words of love, some words of fear; stories of glory, stories with morals; prayers exalting God and prayers asking for help—will allow for the participants to experience their religion in the most powerful and integrated way. They will experience the profound peace of the love of God, and fear and awe of the power of God, and a strong sense of what is right and moral. And if a ritual has the proper rhythms, then the participants may briefly experience something a step further. If both the sympathetic and parasympathetic systems are activated during ritual, then eventually they may experience a breakdown of the self-other dichotomy and a sense of being more intensely united to God or to whatever the religious object of prayer or sacrifice may be. In Judaism, the only time when this experience would be found regularly might be in the High Holy Day prayer service, which is mixed with strong slow prayers filled with ideas of repentance and pouring one's soul out to God, as well as exciting communal prayers based around reimagining the high priest's service on the holiest days. The combination of intense meditative prayers with rousing singing and excitement would parallel these ideas of religious ritual sympathetic and parasympathetic responses. The timing additionally makes sense because the High Holy Days are the time traditionally set aside when Jews will attempt to create lasting religious change in their personal observance and lives, a change that can only be brought about after significant introspection and perhaps religious experiences.

Given these aspects of liturgy, and the opportunity to evaluate them from a more scientifically oriented perspective, we might be able to get beyond what may have been a more "trial and error" approach to developing liturgy. Most likely, the impact of various parts of liturgy was determined by interactions between those creating the liturgy (i.e., priests or rabbis) and the participants. These interactions may have been formal, in terms of planned discussions, or informal and arising simply from the reaction of participants and congregants. Studies into the practice of ritual by liturgists also helped them determine what were the important stories, songs, and so on. Then, based on these reflections and studies, new liturgical sequences were developed and tried. There seems to have been a gradual selection process occurring that helped human ritual progress to its current state. As more effective (more adaptive) parts of ritual were discovered, they were included in the ritual, and less effective parts were excluded. However, with regard to permanence and impermanence, the religion had to have its basic tenets preserved in its ritual so that the religion could maintain its identity and vitality.

The *Anshei Knesset HaGedolah*, the "Men of the Great Assembly" after the destruction of the Second Temple, recognized that the individualized prayers would no longer be enough to allow everyone to feel their connection to God, as the Temple they used to travel to at least three times a year had vanished. Instead, the Men of the Great Assembly made a bold move to transition the people from a national center to a communal one based around service to God, wherever they went. The Talmud (*Berachot* 8a) speaks of God's presence itself going into exile with the Jews, and after the destruction, it was only found "in the four cubits of the halacha/Jewish law." The Jewish system of study and communal prayer was crystalized in set texts and rituals that beforehand had been more fluid. Much spontaneity was lost, but the existence of a Masoretic community that linked itself consistently and constantly to the Temple service and the dream of a return to Israel with the *Beit HaMikdash* was a must given the historico-religious circumstances. Thus, Judaism maintained its spontaneity within a certain framework to allow its basic tenets and constituency to survive.

In the future, it might be argued that there could be a more "scientific" approach to developing ritual and liturgy. The early liturgists had no background in research or neurophysiology to be aware of why the rituals they were creating had the emotional impact that they did. However, the early developers of these liturgies eventually got a sense of the rhythm needed for their rituals. Now that neurotheology can provide a theoretical framework for how the rhythm works, as well as encouraging more research-based methods, perhaps future rituals could be developed or current rituals could be tailored to specifically affect either the parasympathetic or sympathetic systems, depending on the goals of the ritual. If the purpose is to try to give the participants a sense of awe, then a liturgy needs to activate the sympathetic system with music and words. By directing the excitation of the sympathetic system toward God, a sense of the awesomeness of God might be elicited. Conversely, if a myth is attempting to describe the immense love that God has for everyone and of God's infinite goodness, then a ritual that directs the parasympathetic system and its associated emotional response toward God may help to give people this sense. Further, if the purpose is to try to achieve a sense of unity with God or the universe, then a liturgy would need to be constructed that incorporates both sympathetic and parasympathetic components fairly intensely. It would also be essential to "test" new liturgies, as well as specific ways of implementing older liturgies, to determine if the desired effects are indeed happening. Surveys of congregations and rabbis can help determine whether a given liturgy succeeded in eliciting certain emotions or beliefs. And these liturgies could be changed if they fail to do so. We might even consider neurophysiological studies to try to observe how specific rituals or prayers affect the brain and body.

From the perspective of Jewish prayer, the idea that the prayer services can change significantly is one that is hotly debated among the different sects of Judaism. While Conservative and Reform Judaism do believe that some of the prayer services can be edited and changed, the Orthodox service maintains that the prayers instituted by the Men of the Great Assembly are the authoritative prayers of Judaism in general. One aspect of what makes Orthodox Judaism unique among other religions is that it has a set determined number of practices that must be followed by the observers of the religion, and the prayers are no exception to this rule. However, there are additional elements within the prayers that allow one to explore additional experiences in prayer beyond what the formal words state. Prayer is not simply repeating the words as a group in a rhythmic way. In fact, the rote recital of the words without meaning or understanding has been constantly criticized by rabbinic scholars throughout the ages. This notion is supported by some of our recent research that shows that the brain is not significantly activated when one merely says the prayers without engaging in their meaning. When a person is fully engaged and believing in the prayers, the brain becomes much more active, and the areas of the brain involved in our sense of self and our emotions become substantially altered. This mirrors the concept of "*davening b'avodah*" discussed earlier. Without focus and contemplation prior to and during *davening*, a significant amount of the neural network that would be activated in prayer would remain dormant and underutilized.

The future of liturgy, especially as developed utilizing a neurotheological approach, may find ways of more effectively linking the cognitive, emotional, and experiential aspects of myth and ritual. Although higher cognition may have evolved as a very practical, adaptive, problem-solving process, it carried with it—indeed it requires—the formation of myths that present problems for which the ancient rhythmic motor behaviors help generate solutions. In other words, when ritual works (and it by no means works all the time), it powerfully relieves our existential anxiety. Furthermore, when ritual is most powerful, it relieves us of the fear of death and places us in a sense of harmony with the universe or with God. Ritual allows individuals to become incorporated into myth and, conversely, allows for the very incarnation of myth. One can understand why so powerful a behavior has persisted throughout the ages and is likely to persist for some time to come.[21]

The Body's Response to Spiritual Experiences

Our physical needs are great, we need a healthy body. We dealt much with "soulfulness"; we forgot the holiness of the body. We neglected the living/healthy, strong and physical (needs); we forgot that we have holy flesh; no less (valuable) than our holy spirit. We abandoned the life of action, and the clarity of our senses and the connection of our concrete physical existence, because of a decrease in our fear, because of a lack of faith in the holiness of the land. "Faith—this is the Order of *Zeraim* [Seeds]—for one believes in the life of the (physical) world and plants." All of our *teshuvah* [repentance] will succeed by our hands only if it will be—with all the wonder of its spirituality—also a physical return, which produces healthy blood, healthy flesh, mighty solid bodies, a hot spirit radiating over powerful muscles, and the power of the holy flesh will enlighten the soul that has been weakened, as a remembrance of the revival of the physical "dead."

—Rabbi Abraham Isaac HaCohen Kook
in *Orot, Orot ha-Tehiyah* (Lights of Renascence), chap. 33

The Importance of Religion and Spirituality in Health

The Torah indicates that good physical health will follow those individuals who are favored by God. The Bible actually goes on to try to prove this by describing one of the first examples of a controlled medical trial. It turns out that even the authors of the Bible understood the need for a good research study design. In the first chapter of the book of Daniel, we read about something that looks a lot like a controlled nutritional study. Daniel is talking to an official in Nebuchadnezzar's court. The official is worried about Daniel's diet and Daniel says:

> "Please test your servants for ten days, and let us be given some vegetables to eat and water to drink. Then let our appearance be observed in your presence and the appearance of the youths who are eating the king's choice food; and deal with your servants according to what you see." So he listened to them in this matter and tested them for ten days. At the end of ten days their appearance seemed better and they were fatter than all the youths who had been eating the king's choice food. (Daniel 1:12–15)

The point of this parable was to show that those people who are more religious tend to do better in terms of their overall health and well-being, regardless of what they actually eat. It is interesting that even in the Bible, it was realized that proof of that concept required a comparison group, a nonreligious group. From the religious perspective, such a health benefit would be considered a natural consequence of being more closely connected to God.

When it comes to the physiological effects of religious and spiritual practices on the body, there is a growing amount of information. Although there are few studies evaluating specifically Jewish practices and experiences, we can explore the implications for existing research in terms of the more prominent elements of Judaism. Clinical data on illness and mortality also shed light on the relationship between a person's overall sense of religiousness or spirituality and physical and psychological well-being. This is actually an important area of study for neurotheology because it becomes very helpful to understand how religious and spiritual practices can affect the human body, and subsequently human health.

A number of studies have demonstrated that those individuals who are more religious tend to have better overall physical health. This appears to be the case for the general population as well as for people who have specific disorders. For example, various systematic reviews and meta-analyses demonstrate that religious involvement is associated with decreased morbidity and mortality[1] and that high levels of religious involvement can result in living up to seven years longer.[2] A study from Israel demonstrated this effect by exploring mortality rates in different types of kibbutzim.[3] Over sixteen years, from 1970 through 1985, mortality rates were evaluated in eleven religious and eleven matched secular kibbutzim in Israel. The study team found 268 deaths occurred among 3,900 men and women thirty-five years of age and older. The results revealed a significantly lower mortality rate in religious kibbutzim compared to the secular kibbutzim. Importantly, the finding was observed in both sexes, across all ages, and consistent throughout the sixteen-year period of the study. The findings also held across different causes of death, including heart disease, cancer, and other causes. Even the usual female mortality advantage over men disappeared, with the secular women living about as long as the religious men.

Some results have suggested that mortality and morbidity may vary by religion but also seems to be related to how that religion is viewed and integrated into the current society. The more accepting a society is of a given religion, the more benefit those religious individuals may obtain.[4] For instance, a study of contemplative monks in the Netherlands showed that compared with the general population, mortality varied with time during the 1900s depending on whether the monks were viewed favorably by the society. Greater morbidity and mortality have been reported among Irish Catholics in Britain, which may be a reflection of their disadvantaged socioeconomic status in that country.[5] Such an issue may be particularly relevant with regard to Jewish people given the many times throughout history that they were persecuted. During such times, Jewish mortality was likely worse than when Jews were able to fit into a specific society.

However, it also appears that in the past, some of the religious rituals themselves may have offered physical protection from some diseases. The accusation that Jews somehow were causing the plague was in part based on the realization that in some towns, the Jews living in their isolated areas were less affected. This has been theorized to be due to the increased hygiene of the Jews resulting from their ritual requirements of hand washing and immersion in ritual baths more regularly than what was previously standard in the medieval period.

Religiosity also appears to protect people from various illnesses. In the United States in Maryland County, an early study of 91,000 people found that those who regularly attended church had a lower prevalence of cirrhosis, emphysema, suicide, and death from ischemic heart disease.[6] Several studies have implied that religious participation and higher religiosity may have a beneficial effect on blood pressure.[7] Studies have indicated that people who are more religious may have better outcomes after major illnesses or medical procedures. In a study of 232 patients following heart surgery, lack of participation in social or community groups and an absence of strength and comfort from religion were consistent predictors of mortality.[8] Another study of heart surgery patients revealed that stronger religious belief was associated with shorter hospital stays and fewer complications. And in a small study of 30 elderly women after hip repair, religious belief was associated with lower levels of depressive symptoms and better ambulation status.[9] A study of African American women with breast cancer showed that those patients who did not belong to a religion tended to have poorer survival outcomes.[10]

Not all studies have found this positive relationship between religiousness and health,[11] so much more data is needed to understand how religious and spiritual beliefs interact with health.[12] However, there is a substantial amount of data suggesting an overall health benefit. The benefit may extend to both physical and psychological health. We will focus here on the physical health aspects and consider psychological health in the next chapter.

Indirect Mechanisms

When it comes to understanding the physical health benefits that appear to be tied to being a religious or spiritual person, we can explore a number of possible mechanisms. Understanding the mechanisms is also part of the overall field of neurotheology. Neurotheology recognizes the intimate link between the brain and body. Thus, if we are to understand how religious phenomena are associated with the brain, evaluating the effects on the body provides information regarding this overall relationship.

The reason behind the physiological changes and health benefits associated with being religious can typically be divided into what we would call "direct" and "indirect" mechanisms. The indirect mechanisms refer to various practices and guidelines offered by any given religion that might have an effect on health. For example, in Judaism, the special dietary laws of kosher provide an overall health benefit for those following it. For example, avoiding shellfish protects people from acquiring various illnesses from such food. Preparing food in a careful manner also protects people from food poisoning. Washing one's hands prior to having a full meal has the added advantage of preventing any particulate matter or bacteria that had been on one's hands from their work to enter the gastrointestinal system. Inspecting fruits and vegetables carefully to ensure they do not have any bugs on them also may decrease the incidence of disease transfer.

While health benefits may derive from following kosher laws, the laws were developed for non-health reasons. For something to be "kosher" it must simply adhere to the various dietary rules and regulations codified in Jewish law. As such, there can be very unhealthy food that is kosher as well. You may not be able to get a "kosher cheeseburger," but a kosher burger or hot dog with fries and an extra-large soda are all completely kosher even if they are not great for your physical health. Ultimately, the tradition does not rely on the reason that keeping kosher is better for one's physical health to explain why one should do so. The dietary laws fall under different categories and have multiple explanations for why each specific ruling is stated. The famous restriction not to eat milk and meat together is derived from the three-time repeated verse of "Do not cook a goat in its mother's milk." (Exodus 23:19, Exodus 34:26 and Deuteronomy 14:21). The tradition explains that need to repeat this verse to prohibit all milk and meat for consumption, cooking, and benefit (Talmud Hulin 113b, 115b).

Maimonides claims the reason for the literal verse may have been due to an ancient idolatrous practice of cooking a baby goat in its mother's milk (see *Guide for the Perplexed* 3:48). This explanation is one of many from the early commentators' attempts to explain the kosher laws as a form of separation from idolatry. Rabbi Samson Raphael Hirsch explains kosher laws as a way to increase one's recognition of passivity (in contrast to bullishness and aggressiveness) and to learn to have a more refined moral character by only consuming those things that are more passive.[13] This explanation relies on symbolism and may not mean that one is actually internalizing the "nature" of what is consumed. However, some kabbalistic perspectives actually take literally the maxim "you are what you eat" on a metaphysical level, and explain that the nonkosher food is spiritually damaging to the soul and must be avoided for one to achieve appropriate spiritual connection to God (see Isaac ben Moses Arama's commentary on the kosher verses in his work *Akeidat Yitzchak*).

It is worth noting that other traditions also have specific dietary guidelines that can be beneficial for overall health. For example, Hindus are typically vegetarians, since they believe all life is precious. Irrespective of this metaphysical or theological perspective, there is a growing amount of evidence to support eating a plant-based diet as a way of engendering good health.[14] Avoiding red meats and dairy products helps to eliminate saturated fats from the diet that can lead to heart disease or stroke. This was similarly found in a study out of Jerusalem in which secular subjects consumed more total fat, more saturated fatty acids, and less carbohydrates than religious subjects, most likely related to a reduced intake of dairy products among religious individuals.[15] Thus, a healthful, spiritual diet can be an indirect way that religion contributes to overall health.

Other health-related guidelines promoted by religions, including Judaism, espouse avoiding high-risk behaviors, promiscuity, and overindulgence in bad habits. Most doctors would encourage patients to follow the same guidelines. Being promiscuous, especially in times before antibiotics, could lead to a variety of serious illnesses, including syphilis or gonorrhea, that could cut somebody's life short. And in today's world, global epidemics such as HIV and other sexually transmitted diseases result from sexual activity and affect more people in promiscuous societies. A study of over three thousand adolescent girls found that religious beliefs were associated with having fewer sexual partners.[16] Limiting sexual activity and striving for a monogamous relationship are more likely to reduce health problems and mortality associated with such diseases. In fact, marriage, especially happy marriage, carries significant health-related improvements.[17]

Physical activity, which has a number of important health benefits, is an essential part of many spiritual pursuits. Maimonides provides a long discussion in the *Mishneh Torah* (Laws of Temperments 4:2) about the importance of exercise and healthy, measured eating. For example, Maimonides states that a person "should engage one's body and exert oneself in a sweat-producing task each morning." This makes sense for Maimonides, who followed the Platonic principle of the golden mean in all things. The physical body was no exception and required a strict regimen and balance to maintain its homeostasis. Judaism in general has a positive outlook on exercise and maintenance of the body, when not taken to the extreme. Genesis 2:15 states that God placed Adam in the Garden of Eden *le'avda u'leshomra*, "to work it and to guard it." Humanity has always been tasked with being involved in the physical world and its unique God-given beauty. Maimonides was correct on a physiological level, since exercise does a lot of good things for the body.[18] It stimulates the autonomic nervous system and increases heart rate variability, which makes the body better able to handle stressors. Exercise also improves the immune system's functions. Most importantly, aerobic exercise improves the function of the heart and blood vessels, reducing blood pressure and lowering weight. Overall, these effects reduce the risk of heart disease, stroke, and atherosclerosis.

But exercise might even help improve prayer. By improving the function of the body and reducing stress, exercise may enable better mental focus on prayers.[19] Furthermore, since we have already considered the importance of the autonomic nervous system in meditation and prayer practices, exercise may prime this system to allow for more intense experiences. Many traditions, including Judaism, have realized the potential value of physical movement in relation to spiritual practices. In Judaism, *davening* can be a kind of exercise when done with great zeal. The *Sefer Chasidim* (57; *Book of the Pious*, written by Judah ben Samuel of Regensburg, founder of the Jewish mystical movement of Germany in the medieval period) states, "A person needs to shake his entire body during *tefilah* [prayer] since the verse says, *Kol atzmotai tomarnah HaShem mi chamocha* [Psalm 35:10: "All my bones shall say, 'God, who is like You?'"]." The Rivash (Isaac ben Sheshet, Spanish medieval scholar and rabbi), quoted by *Mekor Chesed* (by Reuvein Margolies, Israeli scholar and author from the Ukraine) commenting on the *Sefer Chasidim*, states, "When a man is drowning in a river and making many movements to remove himself from the water, surely those that are watching him will not laugh at him and his movements; so too, when one prays and makes many movements, one should not laugh at him." This clearly indicates that there was a tradition of significant movement for some supplicants during prayer. The *Kitzur Shulchan Aruch* (12:11) states that in general it is praiseworthy to run to do mitzvot, and as such one should run to the synagogue. However, it also states that when one does reach the synagogue, one should stop running and enter with humility and seriousness, as one enters a king's chambers.

Many Jews are also aware of the daily morning prayer that focuses on health and posture: "Blessed are You, Adonai our God, Sovereign of the universe, who straightens the bent." Is this just a metaphor, or would participation in exercise that straightens our bodies so they are not hunched, stooped, bent, or subject to skeletal pain not help people be true to the profound words of the prayer?

Other traditions such as Yoga or Tai Chi have more definitively combined movement with spiritual practice. Yoga, not what most people do in the United States, but actual spiritual Yoga, is designed to bring the body into the best state possible for meditation in order to induce deep spiritual experiences. Similarly, Tai Chi is designed to help bring the person into harmony with the universe through movement and meditation.

Another important indirect mechanism by which religion can help with a person's health has to do with providing a strong social network. Being part of a synagogue or a religious community provides a tremendous amount of social support for people dealing with various issues in life, especially health issues. Rabbis frequently visit patients in the hospital and help patients feel they are not alone, which makes it easier for them to cope with their particular problems. Theoretically, it makes sense that a strong social support network would be beneficial for any person. There is also a great deal of data supporting this contention. Patients with strong social support networks cope better with illness, are more compliant with treatment, and typically have much better outcomes. For example, several studies of breast cancer patients found that those with strong social support networks and those who are married have significantly better overall survival rates compared to those without good social support or who are unmarried.[20] Marriage itself, and its stability, may be associated with higher religiousness. In a study of religious and secular kibbutzim in Israel, the divorce rate was eleven times higher in secular compared to religious kibbutzim.[21] Religious communities, particularly those that are more Orthodox, often place strong social pressure to preserve marriages and reduce divorce.

Bad behaviors such as drugs and alcohol are also typically frowned upon by religious traditions. For example, compared to the general population, Mormons and Seventh-day Adventists have lower mortality rates from cancers that have been linked to reduced intake of tobacco and alcohol.[22] Living "the good life" and avoiding such high-risk behaviors would no doubt help to enhance a person's health and well-being. And while it is helpful for a doctor to tell a patient to avoid such behaviors, the added impact of a religious tradition or even God espousing the same ideas makes it more likely that a person will follow these guidelines. Religion has been shown to affect whether a person begins using drugs and alcohol, how significant the use becomes in a person's life, and how easy it is for a person to quit and recover.[23]

Judaism has a complex relationship with alcohol. On the one hand, alcohol is used regularly for religious ceremonies, such as the *Kiddush* every Sabbath and holiday meal, and it has traditionally been associated with the holiday of Purim, on which the Talmud states one has an obligation to get drunk (*Megillah* 7b). However, the way this law is codified is indicative of the general attitude toward alcohol in Judaism in general. Alcohol is viewed as a tool, which can be utilized to great spiritual effect or be misused. The *Shulchan Aruch* (by Rabbi Joseph Caro) cites the law from the Talmud, and the *Rama* (Rabbi Moshe Isserles), the great halachic codifier of Ashkenaz (European Jewry in the medieval period), states that on Purim one must drink only a little more than usual and then sleep (*Orach Chaim* 695:2). Additionally, the *Chayei Adam* (155:30) states, like the earlier commentator on the portion of the Talmud, the Meiri, that one who gets drunk and is unable to perform mitzvot undermines the entire purpose of the rabbinic command, which was to enhance service to God. Thus, alcohol is consumed from a young age, but ideally its consumption is regulated to the point where one is used to drinking in moderation and is both halachically and socially restricted from binge and alcoholic-like behaviors. However, this does not mean that someone who has a drinking problem or someone predisposed to an addiction would not be more likely to succumb due to the repeated use of alcohol in religious ceremonies. For these individuals, however, Jewish law is clear: they are not to partake of alcohol, and if they do so, it is to their detriment and against the religious practice. Alcoholism and other addictive behaviors are entirely contradictory to a religious mind-set, which views one as dedicating one's physical endeavors toward God.

Another indirect mechanism may have to do with providing people a sense of meaning and purpose in life. When people feel that they have direction and control over their life, they are much more likely to be successful. They typically have less anxiety and less depression, and they will enact behaviors that are beneficial to their overall well-being.[24] If you feel that God is behind you, whether in everyday life or dealing with a particular illness, you are going to do the things you need to do in order to survive as best as possible. But the larger issue of reducing stress and anxiety may arise out of religion in the context of "ontological anxiety." Since religions provide a sense of meaning and purpose, as well as a sense of understanding of the world, the great ontological anxiety that humanity faces tends to be reduced. And whenever we have less stress and anxiety, our body tends to work more effectively.

There are likely other indirect mechanisms by which Judaism, and other religious belief systems, can be a benefit to a person's overall physical and mental health. These can include approaches toward rearing children, forming communities, approaches to living, and even potential genetic and environmental components that have yet to be fully explored.

Direct Mechanisms

The direct effects of religion are related more to specific practices such as meditation or prayer that directly have an impact on the brain and body. We have already considered some of these effects but will focus here more on the bodily effects. As it turns out, meditation and prayer practices appear to have a number of effects on the body, based in part on the autonomic nervous system, but also on the body's stress system and immune system.

We know that prayer and meditation have a significant impact on the brain and autonomic nervous system. We have already considered how these practices can either increase or decrease activity in the sympathetic (arousal) or parasympathetic (quiescent) systems. The most common impact of prayer and meditation is a reduction of the sympathetic nervous system. Such an effect reduces the stress-related changes in the body, thereby decreasing heart rate and blood pressure, along with a number of other changes.[25] One can only imagine the effect of persistently lowering blood pressure and heart rate on an individual's overall health and well-being. High blood pressure is one of the biggest problems in the health-care field and is a risk factor for many diseases, such as heart disease and stroke. A practice such as prayer, especially in a synagogue with other people creating strong social support, should be very effective at reducing blood pressure and heart rate.

In addition, a reduction in stress is associated with lower levels of cortisol, the main stress hormone produced by the adrenal glands. We need cortisol to live and respond to everyday stressors, but too much cortisol can actually damage neurons and suppress the immune system. In fact, people with autoimmune diseases take cortisol or a cortisol derivative to suppress their immune system and lessen the disease symptoms. A number of research studies have shown that the stress-reducing properties of prayer and meditation practices lower cortisol levels in the body.[26] Lowering cortisol levels would likely have the effect of improving brain function and also improving immune system function.

Studies have shown that meditation and prayer practices boost immune function. For example, one study found that people who meditate have a better response to the flu vaccine in terms of turning the immune system on to fight the flu virus more effectively.[27] There is also a growing body of evidence to suggest that cancer forms when the immune system is not functioning well. A long-term practice of prayer or meditation might enable the immune system to work as effectively as possible to keep cancer at bay.

Of course, there are certainly plenty of religious people who die of heart disease and cancer, and there are plenty of atheists who live to one hundred years old. Prayer and meditation by themselves are unlikely to be able to maintain a person's health indefinitely without other medical interventions. However, in terms of maintaining an overall healthy lifestyle and keeping the body functioning as best as possible, religious and spiritual practices can often be beneficial.

On the other hand, we must be careful not to presume that a religious practice like prayer should be considered a medical intervention. No doctor, in fact no person, should advise another person to pray more *in order* to be healthy. Prayer should be done because it has deep personal and spiritual meaning for the individual. If it happens to have additional health benefits, that is all the better. But prayer is a religious practice and not something that can be done only for the purpose of improving one's health. Telling an atheist to pray because data supports its health benefits would not make any sense. However, a religious person might be encouraged to engage his or her faith—including individual as well as group practices—as part of an overall healthful lifestyle.

In a similar manner, we must always be careful not to accuse a person who develops a disease such as cancer of not being religious enough. There is no scientific evidence that not being religious or not praying is a cause of health problems. In fact, studies have shown that those people who feel that God is punishing them with a disease because they were not spiritual enough earlier in their life typically have a much poorer outcome. After all, why would you follow your doctor's advice if God is just going to punish you anyway? This would seem a bad path to go down for both a healthcare provider and a religious individual. Judaism's approach to suffering in general is not to focus on the specific reasons why something is occurring to someone on a theological level, but rather to approach the issue practically: is there something I can do to improve my situation physically and/or spiritually from where I am currently? Is there something that I can learn from this situation in order to grow as a person? Those are the questions that Judaism feels are worth asking, from both a personal religious and theological perspective. Questions laced with emotion about why specific circumstances occur to someone are very difficult to answer satisfactorily and are best left to the realm of the unknowable. Rabbi Yannai says, "It is not in our hands—not the rest of the wicked nor the suffering of the righteous" (*Ethics of the Fathers* 4:15).

The overall point is that when religion has a positive influence on an individual such as providing support, coping opportunities, and meaning in life, religion can also have a health benefit. If religion is viewed negatively or the person views God as a punishing God, then religion can be detrimental to a person's health.

A study of older individuals suggests that people who felt that they were struggling with God also had poor overall outcomes.[28] When people become old and disabled, sometimes they cannot engage religious practices or participate in a religious community the way that they would like to. This can cause great angst for an individual, leading to higher stress and poor physiological function. A good example of this in Judaism is the loss of ability to learn Jewish text as people get older. This may be due to loss of vision, hearing, or sight or due to a stroke or neurodegenerative disease like Alzheimer's. Whatever the cause, these people are aware that what they used to find religiously fulfilling and meaningful is now being cut out from under them as a result of illness and frailty. For this reason, pastoral care in the health-care setting can be extremely beneficial to at least enable people to optimize their religious and spiritual beliefs as they confront a given health problem.

Is Religiousness Inherently Healthy?

The final direct effect of religion may be simply the act of being religious. Samir Becic notes that "spiritual awareness is one of the key components of a healthy lifestyle that impacts the whole body and rejuvenates the spirit. Physical fitness and healthy nutrition allows that spirit to flourish to new dimensions and many people experience a closer relationship to God."[29] The problem with determining the effect of being religious on health is that there are so many other variables that excluding everything else is often difficult. We would have to consider designing a study on mortality in such a way that we factor out every possible variable—age, gender, socioeconomic status, disease status, psychological status, and so on. We would even have to factor out genetic effects and environmental influences. If we can factor every variable into our analysis and still find that religious people live longer on average than nonreligious people, we might actually be able to identify some inherent quality of religiousness that confers a health benefit to people.

It certainly is not unreasonable to think that a person might derive some health benefit from being deeply religious, in and of itself. Of course, some might argue that there is still an ancillary effect, such as being more optimistic. Data show that those people with the highest levels of dispositional optimism have the lowest rates of heart disease and death.[30] But there still could be something inherent about being religious or spiritual that provides a direct benefit to the person. Perhaps having a healthy soul is sufficient for keeping a healthy body. As Maimonides said, "Since keeping a body healthy and whole is from the ways of God, for it is impossible for one to understand or know knowledge from the knowledge of the Creator and be ill" (Maimonides, *Mishneh Torah*, Laws of Temperaments 4:1).

Additionally, when one is deeply religious, the mind-set one has in how they interact with the world fundamentally changes. Self-esteem has been significantly correlated with mental health, and mental health and wellness clearly correlate and have a reciprocal relationship with the rest of one's physical health. The brain controls our physiological function as well as governs our mental world, and the two are inextricably linked. When people have a religious outlook, their own mental health can clearly benefit. If someone truly believes the sayings of the mussar/Jewish ethical leaders of Navahrudak (pronounced Novardik) that this world is not something to hold in high esteem and that God is the only thing a person should be worried about impressing, then the attitude one has when faced with life's struggles fundamentally changes. Rabbi Nachman of Breslov suffered from severe depression his entire life. However, it is clear that his strong faith and religious convictions were an incredible source of strength for him and allowed him to continue despite sometimes crippling symptoms.[31]

Another important point to remember is the stress that being religious can place on an individual. Specific societal expectations and demands beyond what the basic tenets of the religion require can often place practitioners in stressful situations. The stereotypical anxiety-ridden Jewish practitioner who puts bleach on the window shades before Passover and is constantly worried could have a diagnosis of generalized anxiety disorder, which is not something that should be ignored. There are clearly strong cultural and sociological pressures that have pushed modern Jews to perhaps have an unhealthy burden of mental illness, most particularly anxiety and depressive symptoms, in their attempts to follow their religion. But since science has no way at the moment of accurately measuring and delineating the religiousness of an individual, relative to other factors, future neurotheological scholarship will need to carefully evaluate the impact of inherent religiousness on a person's life.

One final point about whether religion is inherently healthy or not might be found in a unique set of studies designed to explore whether specific religious holidays are associated with lower mortality. Many people have observed that a person might survive long enough to make an important event and then die shortly thereafter. A great-grandparent might survive to a child's bar mitzvah and then die a few weeks later. And some have noted that people seem to survive to important holidays such as Yom Kippur.

But what does the data show? It seems that there may actually be a relationship between important religious events and people not dying.[32] An early study of death rates in the 1900s in New York and Budapest, which both had high Jewish populations, found a decline in mortality in September during the years when Yom Kippur fell between September 28 and October 3.[33] A study of Passover in California from 1966 to 1984 found a dip in the death rate of men the week before the holiday,[34] although an interesting opposite pattern has been found in Jewish women, with an uptick in deaths before important holidays and a decrease right after. Such a finding may be related to the different roles that men and women play in these holidays that end up providing protection for men and a lack of protection for women. The most recent study showed that for Jewish men and women in Israel from 1983 to 1992, there was a clear and significant drop in the number of deaths around the Sabbath (Saturday), but no consistent drop around other holy days.[35] This pattern was found to be stronger for men than for women.

These studies are interesting but also have a number of methodological challenges with regard to how mortality was measured and who was actually being studied. However, the larger point is that such studies raise interesting questions. Even if death takes a holiday, is it an indirect or direct mechanism at work? Is there something about the brain that can somehow help the body keep it together and functioning when it needs to? Or is there something more supernatural at work? Either way, neurotheology might bring a unique perspective to how our brain, body, and soul come together for the purposes of human health and well-being.

The Role of Religion in Health Care

Recently, physicians and others in the health-care industry have recognized the importance of patients' religious beliefs in the context of their health care. There is an overall realization that people are not just biological animals, but have a psychological, social, and spiritual side that each can play an important role in a person's health. Ignoring that part of the person may actually be damaging both physically and mentally for a patient. For this reason, many health groups are emphasizing the importance of religious and spiritual beliefs for patients. For example, the *Diagnostic and Statistical Manual of Mental Disorders* recognizes religion and spirituality as relevant sources of either emotional distress or support.[36] Also, the guidelines of the Joint Commission on Accreditation of Healthcare Organizations (JCAHO) require hospitals to meet the spiritual needs of patients.[37] The medical literature reflects this trend, with an almost exponentially increasing number of studies on religion and spirituality in health.[38] There are specific areas such as the treatment of addictions (e.g., drug and alcohol abuse) in which religious concepts can be very useful. For example, 12-step programs invoke the notion of a higher power as part of the recovery process.

Some proponents of including religion and spirituality in health care recommend that physicians and other providers routinely take religious and spiritual histories of their patients.[39] A religious history not only helps the doctor understand the patient better, but helps clarify if religion is an important source of coping for the patient, or may actually guide medical decision-making on topics such as abortion, organ transplantation, or end-of-life issues. A religious history may also provide information about whether a given patient feels that his or her religious beliefs are supportive or deleterious. Dealing with a major health problem can move some people closer to God and religion and others further away. There also has been greater emphasis in integrating various religious resources and professionals into patient care, especially when the patient is near the end of life.

In addition to doctors recognizing the importance of religion, many patients are interested in integrating religion into their health care. According to surveys, approximately 75 percent of patients want physicians to include spiritual issues in their medical care, and almost 50 percent would like physicians to pray with them.[40] In fact, almost every physician has been asked at least once to pray with a patient or patient's family. However, there is often a gap between what patients want and what physicians are prepared for. Many physicians have not been adequately trained to address religious issues, and limited time is cited as a crucial factor. There are other issues as well, since discussing religious beliefs can be uncomfortable for physicians, especially when the physician and patient have different beliefs.[41] Future studies will be needed to help better determine what role, if any, religion should play in formal medical care.

On the other hand, religious communities could be more widely utilized to promote good health and even provide access to health care. For example, religious communities like the synagogue could hold health fairs, in which members can be screened for diseases such as high blood pressure or diabetes. Simply making sure that synagogue events utilize healthful foods and beverages could be an important way to make people healthier. Instead of having a bake sale, the synagogue could have an organic vegetable sale. As Dr. Daniel Amen put it after seeing the donuts, cakes, and ice cream at one of his church's functions, "Not only does the church prepare you for heaven, it sends you there more quickly." Thus, religious groups and communities can utilize their substantial resources to positively influence people in ways that many secular organizations cannot. If the rabbi gets up and speaks from the pulpit about significant physical or mental health challenges in the community, he or she may have opened the door for those who are struggling to come out and speak about their own struggles. And if rabbis can visibly show the path toward leading a healthy life, both physically and spiritually, they can be major contributors in their congregant's lives towards better health practices.

Psychology, Neurology, and Religion from a Jewish Perspective

I was standing in the patients room one Sunday morning. "Excuse me doc, but are you a Pastor?" She asked. "Well not really" I responded, "I am a Rabbi but I don't work with the pastoral care services here, I am just your doctor today." "That's good enough for me!" the elderly woman responded. She asked if I would care to pray with her, and I of course said I would be happy to. "It's nice to know there are people here who care about God" she said. "There certainly are," I said, "and God cares about you and wants you to stay healthy as well." We proceeded to talk for an hour about her strength and how she drew on her lifetime of prayer and faith to get her through hard times (an abusive husband, losing her home, and continuous conflict with her children, to name a few), and how now, as her health problems became more complex, her faith was even more important to her. Tears streamed down her cheeks at the end of our conversation. "I am so used to going to Church every Sunday, I thought I might die today from how upset I was that I could not go, but you saved me." At times, being a physician means only caring for specific medical needs of patients, but at other times, and I would argue most times, we should not forget that being a physician means being "one who cares" and that unique ability to physically and mentally ease someone's suffering is a common goal and burden/gift for the Pastor (Rabbi in this case) and the Physician. No matter how it is done, the healing is often profound and impacts people deeply. Perhaps with these connections highlighted, we can help both physicians and spiritual leaders guide those under their care to live more healthy, full, and fulfilling lives.

—David Halpern

The history of Judaism and psychology is extensive and complex. As Woody Allen once quipped, he was able to take a tax deduction for his psychoanalysis by declaring it a religious contribution! But religion and psychology, especially in the context of Judaism, can lead to a number of fascinating issues. Neurotheology encourages us to take a long look at the relationship between mental illness and religious and spiritual phenomena. Some have even argued that everyone who has had an intense or unusual spiritual experience has had some kind of delusion or, worse, a psychosis. On the other hand, some studies have suggested that religious people have the lowest risk of psychological problems such as anxiety or depression. Still others have tried to integrate traditional therapeutic interventions with religious or spiritual concepts, yielding generally successful results. And finally, we might consider religion in the overall context of the psychological development of the person. This spiritual, or faith, development can help us understand not only how human beings move from infancy through old age, but how religion can become a part of our psyche and our brain processes.

Some work has been done on describing the psychological experiences a person can have with others and the relationship they have with the "other" that they know as God. Some have noted that the relationship with God can in some cases be parallel to other sorts of attachment relationships.[1] Many religious perspectives also overlap with various psychology theories as well. One who thinks of God on a regular basis and integrates religion into one's state of mind and everyday actions can have a perspective on life that is filled with schemas of religious content.[2] Additionally, Judaism has for a long time emphasized that within a person there are multiple desires, known as *yetzer*s, which influence the person to do both good and bad actions. This perspective is one that is in line with psychological understandings of a person as a complex organism with many different competing desires and value systems that must be integrated to create the sense of self as we know it. How that self develops and how various psychological and neurological factors might influence these *yetzer*s are important to consider from the neurotheological perspective.

Psychological Disorders and Religion

It may be helpful to briefly review several common psychological disorders, especially those that have a specific relationship to religious and spiritual phenomena. From a neurotheological perspective, it is further useful to consider how the various brain structures and functions may work abnormally in these disorders.

Schizophrenia is perhaps the most bizarre of all psychological disorders and the most well known with respect to unusual religious or spiritual beliefs. In general, schizophrenia is characterized by positive and negative symptoms.[3] Positive symptoms refer to hallucinations, delusions, or agitation. Negative symptoms include a flatness of emotions, social withdrawal, apathy, and decreased thought content. It has become clear that the problem lies in the neurophysiology of the person's brain, and therefore treatment has consisted mostly of antipsychotic medications. Because schizophrenia is caused by abnormal brain function, it would seem likely that there might be a genetic predisposition to acquiring schizophrenia. This has been shown to be the case in twin studies. If one twin has schizophrenia, there is a much greater chance of an identical twin having schizophrenia than a fraternal twin. Studies of the disorder in families have not been as clear. This suggests that the neurophysiological cause of schizophrenia is probably related to more than one gene but may also have other environmental factors.

People with schizophrenia can experience something called looseness of association. This refers to their difficulty in maintaining attention and jumping from one idea to another without any coherent transition. Often this occurs because they are unable to exclude irrelevant material or competing ideas from their primary thoughts. They may also think with a bizarre or "magical" form of logic in which causal connections are assumed to exist when no connections are perceived by others, as in the example given at the beginning of this chapter. They may not be able to think abstractly or to understand metaphors. Thus, they may take sayings such as "people who live in glass houses shouldn't throw stones" in a literal way, without deriving the deeper meaning. In addition to these symptoms, schizophrenics may also have language difficulties in that they may make up new words, repeat the same phrases over and over, repeat every word someone else says, or be completely unable to speak. Another common aspect of schizophrenia is the perceptual changes, which can be either hallucinations or illusions. Hallucinations are sensory experiences that do not have an apparent counterpart in the external world as determined by others. These hallucinations can be experienced in any of the senses. Thus, they might hear or see things that are not really there. Illusions are misperceptions of actual objects or events in the external world. They may experience distortions in the flow of time, movement, or perspective.

The brain dysfunction underlying schizophrenia is important to evaluate from a neurotheological perspective because this provides potential information regarding the brain mechanisms associated with religious experiences. If hearing the voice of God is similar to having an auditory hallucination, then we might expect the auditory areas and the receptive language areas of the brain to become abnormally activated. If on the other hand, brain imaging studies could show a distinction, then perhaps there is a true difference between schizophrenia and religious experiences that have similar elements. Schizophrenia is also associated with abnormal dopamine function such that the treatment typically includes medications that affect the dopamine and related neurotransmitter systems. Thus, we can similarly explore how the dopamine system changes in people with intense spiritual experiences or differs between religious and nonreligious individuals.

With regard to schizophrenia, the most well-known aspect related to religion is the people who claim to be important religious people, such as the Messiah or Jesus. This delusion can overtake such people's entire life, as they act and believe they are truly the Messiah. Because of my (AN) background in neurotheology, I have had a number of "Messiahs" come to me requesting a brain scan as a way of proving that they are different from the rest of humanity. Their hope is that the brain scan will look so distinct from any other person's brain that the only reasonable explanation is that they are the Messiah. I have even been visited by one person claiming to be the reincarnated King David. While this was more original than claiming to be the Messiah, it still is a similar type of phenomenon. A rare presentation is the "Jerusalem syndrome," as some have come to call it, in which people come to the city of Jerusalem and begin to develop thoughts that they are the Messiah or King David or another biblically oriented savior. There is some debate whether this happens only once people come to Jerusalem or whether the city just attracts people with religiously oriented psychosis. However, one thing is clear: religion can be a focal point for psychosis to manifest.

Psychosis itself has a complex place in the halachic literature of Judaism. The Talmud (*Chagigah* 3b–4a) clearly states that there are cases of people who are labeled a *shoteh*, one who has "lost their mind" in the colloquial sense. These individuals are described as those who demonstrate bizarre behaviors. They might go out alone at night, sleep in the cemetery, tear their clothing, and lose/destroy everything that is given to them. David experienced a patient having a strong negative psychotic reaction one time when she screamed at him as he walked into the room, "Keep him away from me!" She was restrained to the bed and had a wild look in her eyes. As she stared at David, she immediately looked at the *kippah* on his head. "He's with the Mossad; they sent him to kill me!" she cried. David was asked to leave the room and was only able to return to see this patient the next day when her psychosis was more subdued with appropriate medication. She was a young Iranian immigrant who had been pulled over for reckless driving, which had clearly been due to a psychotic episode. What was going on in this woman's head? She recognized the *kippah* as a symbol of Judaism and linked that to Israel and, thus, the Mossad. However, many cognitive steps were missing in between because of the significant psychological disorder. Such individuals are exempt from Jewish laws and requirements. However, when it comes to Jewish legal capacity to make decisions and execute documents, there is some debate about how exactly the mind of a person works and how a psychotic individual should be viewed. Some authorities quote the examples cited in the Talmud to mean that a person must have one of those above-mentioned defining problems to be considered mentally incompetent (*Mordechai, Gittin* 421). Others (Maimonides, *Mishneh Torah*, Laws of Testimony 9:9–10) believe that these examples only reveal the larger issue: how do we assess the mental competency of the individual and whether he or she is deemed sane or insane, and what happens when the status of mental impairment changes? For example, if someone with psychoses is on medication and can manage life well, he or she may be deemed halachically sane and obligated in all of God's commands. However, should one experience a relapse and no longer be able to manage his or her life and behaviors properly, then Jewish law does not require him or her to continue to fulfill any commands while in such a state. It would appear that the law recognizes that the human mind, and brain, is something that is in constant flux, and judgments made at one moment cannot be assumed at another point in the future without reevaluation.

Judaism also makes clear distinctions between what it views as "pathological psychosis" versus true prophecy and visions from God. Prophecy in Judaism is not inherently assumed to be psychosis and an untrue experience. However, the concern for psychosis always lingers and is particularly concerning in the modern era, with Judaism's theological principle that prophecy ended with the destruction of the Temple (See Talmud *Yoma* 9b). Surely there are many who believe that it will return in the messianic era, but the individual who begins to receive that prophecy would have to have been chosen very specifically to carry out that message.

Therefore, the sages instituted that any prophets themselves will be tested by the high rabbinic court of the time that they appear in. With the destruction of the Temple and the move from a centralized religion to one that spanned the world, Jewish sages recognized the need for the law to maintain the Jews and keep them connected. The legal aspects of Jewish law prevent even a prophet from abrogating that law, one of the major points of contention between Judaism and Christianity. As such, the messianic individual who claims to have received prophecy changing the legal system will immediately be rejected by traditional Jews, since the tradition itself acts as a buffer for such attempts, aided by the previous failed attempts in Jewish history (See Maimonides *Letter to Yemen* and *Mishneh Torah*: Law of Kings 11:4 Venice Edition 1574).

Of course, one always wonders whether any of these self-purported Messiahs might actually be the true Messiah. How would we ever make that type of clear determination? Could we use science? Maybe a brain scan? And do we even need to? Would a Messiah be so obvious that science would not be required to make that determination? The problem with these questions is that they start from a particular bias—that anyone claiming to be the Messiah is by definition crazy until proved otherwise. And there are scholars such as Richard Dawkins, who has labeled everyone who believes in God as being delusional in his book *The God Delusion*.[4] Neurotheology would remind us to be careful about such a quick conclusion. After all, how do we assess what is normal? Many people whose spouse has died report seeing or hearing the spouse. This bereavement hallucination has been referred to as a "normal" hallucination, since it happens so often and apparently leads to better coping with the death. And if people who claim to be the Messiah are crazy because they are so intense and devoted to their belief, what are we to make of nuns, monks, and rabbis who might also follow a particular belief to the exclusion of a "normal" life (e.g., not getting married or spending all day in meditation or prayer). We might also ponder whether a person with known schizophrenia is capable of having a normal religious experience and whether an otherwise normal person is capable of having an abnormal religious experience. Again, we would have to determine what we mean by normal and abnormal with respect to both the person and the experience.

In Judaism there have been a few historic cases of individuals who claimed to be the Messiah and received large followings. The two most notable cases are the ancient Roman rebellion led by Bar Kochba, which was quashed by the Romans and led to increased destruction and persecution of the Jews. The next more recent phenomenon was that of Shabbetai Tzvi in the seventeenth century. Riding around the Middle Eastern Ottoman Empire, he began to create a fervor and messianic revivalism across much of Jewry at the time. He was brought before the sultan and given the choice of conversion or death, and he chose the former. His conversion ended much of the messianic fervor and left the Jewish communities rocked with the realization that he had been a false Messiah. Some continued to believe in Shabbetai Tzvi and created a religious theology that required the "raising of the sparks" previously mentioned in the Lurianic kabbalistic system to refer to spiritual sparks hidden in the world revived not only by the keeping of Jewish law and spreading God's name, but by actively doing things contrary to the law to bring "sparks from the darkness." This perspective, while espoused by a minority of the Jewish community, was enough to create strong condemnations of the group and helped cement a wariness of anyone claiming to be the Messiah.

Current Orthodox tradition regarding the Messiah runs the gamut between viewing the Messiah as a physical person who will come and produce clear miracles that will prove beyond a doubt that person is the Messiah, to there being a specific person but that much of the change will take place in a "messianic era"[5] in which the kingdom of God will be restored to Israel and everyone will live in peace and prosperity, working to solve the world's problems.[6] These two extreme perspectives often fall out theologically along the lines discussed previously, with the emphasis on clear miracles and a specific individual (or two, with the first being "Messiah son of Joseph" to wage war and create the era for the second, being "Messiah son of David") often taken up as a certitude by those of a more traditional theology as well as those with a more kabbalistic/Hasidic tradition. The other perspective is often ascribed to Maimonides due to the way he explains the messianic era in his *Mishneh Torah,* Laws of Kings. This perspective has also become popular among the Religious Zionist movement under the writings of Rabbi Abraham Isaac HaCohen Kook, who described the creation of the Jewish state as one of the first steps of the messianic process.[7] From this perspective, the Messiah may be a man who comes with miracles, but rather, "the messianic era" is a concept which exists beyond the Messiah's lifetime, in which the ultimate goal of God for humanity is achieved in this world: for us to live keeping God's commands and working toward perfecting the world. As a neurotheological aside, we might also ponder what brain functions might lead a person to believe in one messianic approach or the other. Are there cognitive, emotional, or experiential components that support one belief over the other?

Another significant mental disorder is depression, which is believed to affect as much as 10 percent of the world's population. By depression, we are referring not to the day-to-day depression that all of us might feel if we have a negative experience, but to a depression that comes to represent for that person a total approach to life. In other words, people who are depressed view all, or virtually all, aspects of life in a negative manner. Reality to them appears in every way to be a negative experience. They may feel that they just do not care about anything or that nothing makes a difference to them. They lose interest in their usual hobbies and activities and generally feel lethargic and disinterested. In the end, they may have thoughts of death or even suicide.

Religious and spiritual attitudes can be modified by depression, as people will sometimes forgo participation in religious activities when they feel deep depression. Although depression is focused on various life aspects, sometimes depression can be directed toward God either as a kind of existential crisis or as a move away from God. This has prompted some therapists to develop psychotherapy approaches that incorporate religious and spiritual concepts. The goal is to help a person heal his or her relationship with God or find a way to incorporate religion back into the person's life as an added method for coping and recovering. These combined approaches have been very effective, particularly when used with religious individuals.[8]

In a similar manner, religion is sometimes an integral part of treatment for substance abuse with drugs or alcohol. The most well-known is Alcoholics Anonymous, in which a higher power plays a prominent role. The individual needs to begin early on to accept that healing comes from a power greater than his or her self. This reliance on a higher power is believed to provide another important source of spiritual support that the person is lacking.[9] Some have even suggested that the person transfers the addiction from drugs or alcohol to religion or spirituality. However, such an actual psychological transference has not been firmly established through research.

Jewish individuals might be able to rely on the complementary approach incorporating the concept of *teshuvah*, or repentance. Most people in psychological distress realize or come to realize that they must enter into a process that requires some time to resolve long-standing issues and work through them in a constructive manner. Of course, for those people with major depression, there is a physiological process that can overwhelm a person in spite of taking various approaches to alleviating it. *Teshuvah* can be understood as the process through which "one's emotional pain is linked with the requirement and inevitability of change. In *teshuvah*, the change takes the form of eradicating delinquent behavior according to the precepts and ideals of Jewish law."[10] Using neurotheology to incorporate brain processes along with psychological and spiritual measures might yield the most comprehensive evaluation regarding whether a religious, psychological, or combined approach might be the most effective.

Neurological Disorders in Religion

When it comes to neurological disorders, several have been considered prime suspects in relation to intense religious or spiritual experiences. Patients with seizures in the temporal lobes have occasionally been found to harbor extreme religious beliefs or have distinct religious and spiritual experiences. But other disorders such as stroke, tumors, or Alzheimer's disease may all have an impact on religious experiences and beliefs. The actual evidence for the relationship between temporal lobe seizures and religiosity is somewhat controversial. On one hand, there are certainly anecdotal cases of patients with temporal lobe seizures who have unusual experiences. However, as researchers have explored this topic in more detail, it turns out that less than 5 percent of patients with temporal lobe seizures actually have these unusual religious experiences.[11] Further, patients with seizures in other areas of the brain also have expressed unusual feelings of religiosity. Thus, it is not clear whether it might be the seizure itself or the location that might be important.

In support of the relationship between temporal lobe seizures and religiousness, early studies in which surgeons electrically stimulated the temporal lobes produced very intense emotional and sensory experiences for people.[12] These experiences had elements similar to religious and spiritual experiences. Building on this work, Michael Persinger developed what has been referred to as the "God Helmet."[13] This device emits radiofrequencies into the temporal lobes and has been reported to elicit elements of religious experiences such as a "sensed presence." Another group of researchers was unable to obtain similar results, but there has been ongoing controversy whether either group is applying these radiofrequencies appropriately.[14] Either way, it certainly seems that the temporal lobe plays a role in religious experiences, but to what extent is not fully clear.

The potential relationship has led certain scholars to claim that many, if not all, religious experiences are associated with some type of seizure activity in the brain. An article that was published in 1997 in the *Journal of Neuropsychiatry*[15] included a table of the great religious figures of history and their potential diagnosis. It is not surprising that many of the diagnoses were temporal lobe seizures. The overall approach was to look at the various behaviors that were described as being part of these experiences and try to relate them to known disorders. For example, the description of Paul on the road to Damascus suggests that he fell to the ground and heard the voice of God. When he got up, he was unable to see. The act of falling down and having intense auditory experiences can be seen in patients with seizure disorders. In addition, post-seizure blindness can occur and last a few days, just like what happened to Paul. No other Jewish individuals are mentioned in the table, but it is interesting to consider whether any of the other great biblical figures such as Abraham, Isaac, Jacob, Moses, or other prophets may have had any type of neurological condition such as seizures that predisposed them to having intense spiritual experiences. Of course, there are several important problems with this perspective. Most important, patients with seizures, as well as other neurological disorders, typically have repetitive experiences and auras. However, in people with intense religious experiences, their experiences are frequently one-time events or, if they occur multiple times, have different content each time. For example, each time Moses interacted with God, it was a very different experience, from the burning bush to receiving the commandments on Mount Sinai to speaking "face to face." In addition, patients with neurological disorders often have deficits in their cognition and ability to function in society. This does not seem to be the case for the majority of people who have intense religious or mystical experiences. Most of these individuals are able to function at a high level and historically have been regarded as some of the most important human beings in history.

Neurotheology would certainly support a careful analysis of the many individual stories of people who have had intense religious or spiritual experiences and how they may relate to the brain. There are certainly some individuals who do have neurological disorders, and this can be highly informative in understanding how the brain is associated with them. On the other hand, we must be careful not to over-pathologize religious and spiritual behaviors as always being related to a disorder. This was Richard Dawkins's perspective in his book *The God Delusion*.

Other neurological disorders also warrant attention regarding their potential impact on religious beliefs. Specifically, neurodegenerative diseases such as Alzheimer's or Parkinson's disease may have unique effects on a person's religiosity. As a person's brain changes, it could be very important to be able to observe how his or her religiousness changes. In the Torah, we read of many individuals living excessively long lives but maintaining not only their cognitive faculties but their steadfast religious and spiritual beliefs. It would be fascinating indeed to explore whether those individuals with neurodegenerative disorders experience an alteration in their religious beliefs, either in content or strength. There is some evidence for such an effect. In patients with Parkinson's disease, there is apparently a relationship between the loss of dopamine function and certain aspects of religiosity. People with Parkinson's disease appear to be less able to engage religious concepts, particularly if it affects the right side of their brain (symptoms on the left). People with Alzheimer's disease have sometimes been noted to become more creative as the cognitive controls dissipate. The main challenge with studying neurodegenerative diseases is to distinguish whether changes in religiosity are actual or simply the result of an individual's inability to describe what he or she is really feeling. For example, if a woman with Alzheimer's begins to view God in a more simplified manner as basically the being who controls the universe, can we assume that this is what she truly believes, or is it possible that as her language deficits become worse she simply cannot describe what she is thinking in an accurate manner?

One final neurotheological point with regard to neurological conditions is worth considering. There is generally a Western scientific assumption that if a person has a seizure and concomitantly has a religious experience, then the experience was *caused* by the seizure. However, there is an alternative possibility. If we were to assume that God operates on an extremely high level of energy, is it possible that the seizure allows the brain to operate at a similarly high level such that the person can actually interact with God in a manner different from when the person is not having a seizure—that is, in the "normal" state? Of course, this is highly speculative, but it is important to keep in mind that any quick conclusions drawn from some piece of data must be carefully evaluated from all possible scientific and religious sides.

The Rabbi's Neurology

We also need to look at those individuals who hold intense or extreme religious beliefs. We need to understand if the rabbi's brain is truly unique, advanced, or even disordered. What is it that drives people to want to make religion the true focal point of their entire life? Is there something unique about the neurotransmitter set in these individuals? Do they have more dopamine, less serotonin? Are there alterations in their parietal lobe that makes them feel more connected to God? And what about a person's sense of meaning and purpose in life? Many rabbis will tell you that they felt "called" to become a rabbi. How does a sense of calling occur? Is it perceived as literally receiving a message either verbally or via some other form of message? Or is it something that a person comes to understand as part of his or her own cognitive processes? Perhaps becoming a rabbi just makes sense because a person feels deeply connected both to religion and to people and wants to create a life joining those two concepts as actively as possible.

Our Survey of Rabbis found some fascinating results

regarding their beliefs about their own psychology, their desire to become a rabbi, and their perspectives on spirituality and religion. We started by asking questions about their own psyche. We asked questions about how they used emotions and logic and how introverted or extroverted they were. With this information, we can gain some fascinating insights into the brain of a rabbi.

As a whole, 78 percent of the rabbis described themselves as moderately or intensely emotional, with a fairly similar distribution across all denominations and gender. Thus, the emotional centers of the brain appear to play a prominent role in the way rabbis think. This is an interesting response given the emphasis in Judaism on Torah study and rational thinking. In spite of these appeals to the logical parts of the brain, emotions appear to generally rule the day for the rabbis in our survey. In his well-known book *Descartes' Error*, neuroscientist Antonio Damasio argues that emotions are not only essential to human beings, but they are essential to rational thought.[16] Emotions enable us to place a value on various ideas in order to assess how important they are in our ideologies and our lives. It is also interesting to note the extensive interconnections between the emotional centers of the brain and the higher-order rational areas of the brain. A structure called the insula actually sits between the limbic system and the cortex and helps us to bring reason to our emotions and emotions to our reason. Thus, the strong reliance on emotions seems highly appropriate for the rabbi's brain.

Another question we asked was whether rabbis felt they were particularly introverted or extroverted. On one hand, the substantial amount of study that goes into being a rabbi would more likely be associated with someone who is introverted. On the other hand, since rabbis have to speak in front of the congregation and frequently talk with members of the congregation to help with various life issues, one might expect a rabbi to be fairly extroverted. To some extent, the survey results bear this out. There was a fairly even distribution of rabbis who felt more extroverted and more introverted. Only 7 percent of the respondents said that they were very introverted, while 30 percent said that they were somewhat introverted; 18 percent said that they were very extroverted, while 16 percent said that they were somewhat extroverted; and finally, 26 percent said that they were a balance. Rabbis of a specific denomination were not more introverted or extroverted than another and there was a similar distribution in both male and female rabbis..

It appears that each rabbi is able to engage his or her introverted and extroverted sides as part of their profession. It also seems that those rabbis who consider themselves more introverted emphasize their own studies and intellectual development in Jewish learning and legal practice, whereas those who are more extroverted tend to enjoy the time they spend with their congregation, doing outreach and speaking to crowds for inspiration. There is limited data on the areas of the brain involved in introversion and extroversion. Hans Eysenck originally proposed that people who are more introverted have higher levels of baseline arousal in the brain compared to those who are extroverted.[17] Data from a resting fMRI study support this contention, suggesting a relationship between being introverted and having higher resting activity in the brain.[18] A different study found that people who are extroverted have relatively smaller brain areas involved in behavioral inhibition, introspection, and social-emotional processing, which includes the right prefrontal cortex and the right temporoparietal region.[19] Does a rabbi who chooses to spend his time in *kollel* (a place of full-time study for adults whose sole purpose is to increase the number of people learning Torah) or seminary studying and becoming a legal expert have a different brain structure than one who chooses to enter into the world of Jewish outreach, traveling to different communities around the globe to engage Jews in their Judaism? Future studies of rabbis might be interesting to explore how their resting brain level of arousal compares with that of non-rabbis as well as between those who are more extroverted or introverted.

The Negative Psychology of Religion

While most rabbis pursue their religious beliefs toward improving the lives of the people in their community and for a greater Jewish people, religion also can turn negative. Such negative effects of religion occur in Judaism, as well as other religions. Perhaps most relevant to Jews with respect to negative effects of religion has to do with individuals who turn intensely violent as the result of their religion. Many atheists will cite that religion has been one of the leading causes of death throughout history. This is certainly supported by so many religious wars dating back to the beginning of human civilization culminating in religious violence such as the Crusades, the Inquisition, and the Holocaust. It seems that religious beliefs can sometimes promote highly exclusive perspectives on the world that lead to much hatred and anger against those people who are not members of the group. Furthermore, because of the ultimacy of religion, it seems that the emotions become elevated quickly.

One of the important neurotheological questions going forward would relate to how strong negative reactions within religions might be redirected in a more positive or compassionate perspective. While this is no easy task, research might be able to show how such negative emotions and attitudes arise. To begin with, it is important to understand why some people find a hate-filled ideology appealing. Do these people have a history of emotional or physical abuse? Is it possible that their brains are built in such a way that violence is a preferred avenue? There are a number of studies exploring the brains of violent criminals or those with sociopathic psyches. Brain imaging has sometimes shown that these individuals have alterations in their limbic systems or frontal lobes.[20] The result is a person who is either unable to have normal emotional responses or experiences excessive emotional responses, especially negative ones. There may even be certain neurotransmitter effects that might be associated with such negative tendencies.

As we discussed earlier, rituals also likely play an important role in the development of negative emotions. Rituals can produce an intense unitary experience within the participants, but this can cause great animosity toward those not in the ritual, not in that sense of connectedness. In order to shift people into a more positive perspective, more positive-based rituals that focus on love and compassion would be required. The question would be how effective such rituals might be. Again, neurotheological research could help explore which types of rituals might be the most useful for creating a more positive emotional perspective.

There is an entire literature regarding religious extremism and violence, and much has already been learned, but bringing in knowledge about the brain and using scientific methods to address these negative effects of religion might be our best option in helping to find more compassionate and positive approaches within religions and between them.

Neurotheology and Spiritual Development

It is appropriate within the psychology section to consider a developmental spirituality in the context of Judaism. Psychology, in general, has focused a substantial amount of effort on how a person develops from infancy through old age. The focus is typically on the earlier stages in which the growth through childhood and adolescence coalesces into the overall mind-set, beliefs, and behaviors of the individual. This originated with the work of Sigmund Freud, who believed that the early childhood years had a substantial impact on an individual's future psychological structure. Later psychologists such as Lawrence Kohlberg and Jean Piaget explored the development of the mind during childhood through adulthood in great detail.

There has been a relative lack of material on the development of religious and spiritual beliefs throughout the life cycle. One of the most well-known and detailed models of spiritual development comes from the work of James Fowler in his book *Stages of*

Faith, which can provide an excellent framework for a neurotheological analysis of this topic. Specifically, we might wonder how the development of the brain as it grows and changes throughout life might match the development of a person's religious and spiritual beliefs. In fact, it is the ability to change and adapt that gives the brain its power to enable human beings to survive, grow, and learn new things. Enough studies of brain function and structure have been performed to yield an overall model of human brain development from infancy to old age. The brain changes that occur should impact directly upon a person's religious and spiritual beliefs and experiences. Although Fowler's model is more Christian-based, given the close monotheistic relationship with Judaism, it seems reasonable to adapt this model to Jewish religious development. We should stress that this is only a model, and future work focused more specifically on the Jewish mind would be necessary to further advance these concepts.

Spirituality throughout the Life Cycle

Fowler described the stage that precedes the first actual faith stage as "undifferentiated faith," which occurs in infancy. This stage does not contain any specific cognitive content. Even if the person is raised in a highly Orthodox Jewish family, the infant cannot cognitively absorb this information and hold a particular religious perspective. This makes sense from a brain perspective as well, since the higher cognitive centers are not functional until about the age of one year. No higher thought is occurring because the structures and functions that would support that thought are simply not there yet. In spite of the lack of higher cognitive processing, Fowler indicates that this is the stage where the seeds of trust, hope, and love are developed through the influence of the infant's caregivers. Interestingly, animals raised in nurturing environments have substantially more neuronal connections compared to animals raised in restrictive environments. Those raised in nurturing environments have less emotional reactivity and are more likely to explore the world around them, perhaps due to the neuronal complexity and interconnectedness that ultimately lays the foundation for future brain development.[21]

There is another potentially important perspective with regard to the infant brain. It has been remarked by a number of mystical traditions that the ultimate goal of spiritual pursuits is to return to a time in which the mind was at its beginning. From a Jewish perspective, the Bible and rabbinic literature is filled with references to the Jews as God's children. One example is in Isaiah 49:15; speaking about how God will not abandon the Jews, the prophet says, "Can a woman forget her nursing child, that she should have no compassion on the son of her womb? Even these may forget, yet I will not forget you." Additionally, there is much literature specifying that the younger years of the child's life are one of purity and an ideal state; as the Talmud states, "Childhood is a garland of roses," and "The very breath of children is free of sin" (*Shabbat* 152, 119).

Fowler refers to the first stage of faith as the "intuitive-projective" stage, and he describes this to occur between the ages of two and six years of age. A child in this stage begins to develop speech and create meaning. A child is able to sort out and gain some control over the world through the use of language and symbolic representations. Also, this stage is largely characterized by fantasy-filled, imaginative processes that are unconstrained by logical thought processes. It is interesting that during this stage, neurophysiological development is associated with a progressive increase in overall brain metabolism. In fact, the brain has its greatest overall metabolic activity around the age of four years. The initial heightened metabolism is likely associated with the overproduction of neurons and their connections.[22] We would suggest that this may explain the increase in fantasy and imaginative powers of children at this age. Their brain is establishing so many different connections all of the time that there is tremendous expansion and over-connectedness between neurons, with few clearly defined rules and an unusual blending of different experiences and ideas. Children would therefore perceive the world as being composed of many overlapping ideas, experiences, and feelings. They would likely see things in ways that appear to be fantasy-like to older individuals, who have already reduced their neural interconnections and developed more concrete ways of looking at the world with their better defined neural connections. Children in this first stage will likely not see any problem blending ideas about God with very mundane issues. For example, a young child may have no problem stating, "Hashem (God) made the traffic light change green so we can go, Mommy!" or use God to convince their parents they should receive something: "Hashem wants me to have the candy!"

Children during this stage begin to form their first sense of self-awareness, which is most likely associated with a greater maturity of the association areas, particularly the superior parietal region, in which the sense of self, in conjunction with the other association areas, is ultimately formed. However, due to the over-connectedness of sensory neurons with the association areas, the developing self is seen as highly interwoven with the external world, especially as elaborated in fantasies and dream states. On the other hand, with this developing sense of self comes the beginning of experiencing concepts of death, sex, societal taboos, and the ultimate conditions of existence. However, they will not likely be able to make sense of these complex issues in the same way a mature adult would, since they might not be able to clearly distinguish death from life and wrong from right until their higher association areas are able to fully process such ideas.

The Bible does state, as mentioned above, that children are innocent, but it also states that children are *ra minurav*, "bad from their youngest days" (Genesis 8:21). The Talmud (*Sanhedrin* 91) further says that the *yetzer hara*, the urge to follow one's own physical desires, is present from birth (see also *Avot d'Rabbi Natan* 16). However, this is due to the fact that before a child has developed the capacity for abstract thought and true spiritual and moral growth, during the "terrible twos" when everything is "mine," the child can be described from the religious ethos as truly *ra*, or bad. However, while "the child is not a thinker and cannot distinguish between good and evil" (Rabbi Bachya ibn Pakuda, *Duties of the Heart*, pp. 154–55) and imagines everything in a very concrete way initially, he or she has the ability to continue to develop the neurocognitive potential and learn more deeply about Jewish ideas and teachings. As the Midrash comments on the story of Jacob's ladder, when he saw the ladder connecting heaven and earth, it was meant to represent himself, that every human being has a part of them that is basic and earthly and a component of the spiritual that must be developed and "stepped up" to (*Bereishit Rabbah* 68:16, 18).

The educational approach in Judaism recognizes that the second stage described by Fowler, the "mythic-literal" stage, must enable the child to move from a very literal perspective of Jewish teachings and the world itself to a more complex one that develops in adolescence. During the mythic-literal stage, a child begins to internalize stories, beliefs, and observances that symbolize belonging to a community or group, fostering the development of a worldview and ideology. Beliefs are related to literal interpretations of religions or doctrines and are usually composed of moral rules and attitudes. This second stage appears to coincide with a plateau phase in brain metabolism such that the overall activity throughout the brain remains higher than in the adult, but there is no longer an increase in activity.[23] It is believed that during this time, from the age of four to nine, there continues to be a slower overproduction of neuronal connections, and there is a very active cutting back of connections.[24] The removal of inappropriate connections is likely associated with the establishment of specific rules by which neural connections are allowed to continue. If the connection that 1 + 2 = 3 is correct, then the other connections that might lead to 1 + 2 = 2 and 1 + 2 = 4 will be pruned away. In this manner, very specific and literal rules of behavior, language, emotion, and thought are established. While there is still some overproduction of neuronal connections during this stage, the emphasis on the cutting back of these connections may account for the transition from a very imaginative and fantasy-oriented stage to a very literal and rule-based stage.

It is during this stage that Jewish education is traditionally begun, starting with the teaching of the Bible, then moving on to include the Mishnah, and finally the full Talmudic teachings. When students study the biblical text at a young age, they are taught the stories often with the Jewish midrashim and comments by the rabbinic scholars of the time of the Talmud. The rabbinic commentary, while often not the simple reading of the text, are all based on some textual or theological issue that the rabbis picked up on from the text and come to explain a potential religious question one might have. The midrashim are meant to teach very specific religious "truths" and ethical and moral lessons, even though they may not be factually true. Maimonides, in his introduction to the tenth chapter of Tractate *Sanhedrin* of the Babylonian Talmud, says that those who take the Midrash literally are not wise enough to understand what the rabbis really intended. Therefore, one must continue to grow in one's education to understand that the midrashim and stories one learned as a young child, while initially thought to be factually true, are not to be taken at face value. However, when children are in this early phase, they are unable to understand this distinction and initially take much of the traditional teachings at face value.

As spiritual growth occurs, it is important to recognize the impact of the child's environment. If children grow up in a highly rigid environment, it can lead their cognitive and emotional perspectives, as well as faith-based concepts, to become trapped in the "narrative." This may occur because the restricting environment actually leads to a more limited set of neuronal connections. The person is less able to develop new ideas or see things in the abstract. Thus, children whose environment is constantly controlling and judging ultimately will have difficulty formulating their own spiritual concepts and reflecting on the value of those concepts. This can even lead to a distortion of their individual sense of self or their possibly becoming self-destructive if they feel deserving of punishment on the basis of a restricted relationship with the sacred. Thus, people who grew up in a house filled with fear and dread of their parents may superimpose this emotional response on their view of God, giving them a fearful and angry figure in their mind as someone who is coming after them. This type of attitude can lead to hyper-religious observance beyond any normal subgroup's designations or an eventual rejection of religion entirely, labeling it as "bad" in response to the experiences one has had.

The factors initiating transition to the third stage are related to the confrontation with existing beliefs and doctrines. In this regard, the concrete thinking that establishes myths is confronted by new information, exposure to other perspectives, and the development of abstract cognitive processes resulting in a reconsideration of the literal aspect of stage two. The third stage, called the "synthetic-conventional" stage by Fowler, usually begins in adolescence and extends into early adulthood. Neurophysiologically, this corresponds to a time between the ages of twelve and twenty, in which the overall metabolism in the brain begins to decrease at a relatively rapid rate. This decrease is associated with a very active cycle cutting back old connections and growing new connections, with more emphasis on the former. Thus, the person is removing old beliefs from childhood and replacing them with more complex beliefs that will eventually become those that carry forward into adulthood. The newer ideas are still built upon the foundational concepts laid down in childhood. For example, new mathematical concepts are learned, but they are built upon the fundamental laws of quantitation that are already engrained in the person's brain.

From the Jewish perspective, we can clearly see this transition between childhood and adulthood. Given these developmental stages, it is no surprise that the bar/bat mitzvah occurs when a person is thirteen (or twelve for girls, in Orthodox and Conservative communities). This is in between the second and third stages but is a bit more into the third. For this reason, a person can just begin to act like an adult in some ways even though there are still many childhood concepts still in the brain. The bar/bat mitzvah also makes sense at this stage because the person going forward will now begin to develop more complex religious and spiritual beliefs and will need to be a more fully participating member of the community in order to accomplish this. But as we have seen all along, the foundational ideas of Judaism are established early in childhood with the Passover seder and the four questions, the stories of Noah and Abraham, the following of Shabbat or kosher laws, and the celebrating of various holidays. Each of these, and many others, become established within the brain, and hence the mind-set, of the individual. But the simplistic concepts that a child takes from these foundational ideas now grow into more mature and complex ideas.

Furthermore, according to the Talmud, the age of thirteen for boys and twelve for girls is defined as the "coming of age" period, where they become obligated in the commandments and are considered mature enough to handle full responsibility for their actions (*Yoma* 82a; *Niddah* 45b; *Ketubot* 51a). Prior to this, they were considered to only have their own base urges guiding them. However, once they turn twelve and thirteen, children are "considered" to have gained their *yetzer hatov*, "good inclination," which points them toward the morally correct course. As the Midrash states, the *yetzer hara*, or inclination to the bad, is thirteen years older than one's *yetzer hatov* (*Ecclesiastes Rabbah* 69). This is clearly correlated with the adolescent's continued development of a complete outlook at how the world functions as well as builds off of earlier education and spiritual training.

Thus, the third stage is naturally complex due to a variety of factors, including biological ones, such as new hormonal states of puberty, and the growing influence of friends, teachers, and other individuals with whom we come in contact at this age. Sometimes this stage can be characterized as a conformist stage, as we each try to fit in with our peers. But through this process, we each find our own friends with whom we can share ideas. Some of us turn more to science, others to art, and others toward more religious practices. Our individual approach to life and various ideologies are beginning to be solidified.

Likewise, a person's sense of spirituality and religion grows and develops but is still built upon previously established ideas. This is often seen in the phenomenon of the "year in Israel" that many teenagers participate in. Students who travel to Israel after high school or during college may have some background in Jewish studies but are in the midst of forming new identities and beliefs. Upon exposure to a large institution studying Jewish law, philosophy, and the Bible, and coupled with experiences in the Holy Land, many students take on new religious practices and/or identify with their Jewish roots in a more meaningful and stronger way. Independence is then coupled with religious growth, not rejection, and therefore can be an important factor in the spiritual development of the student.

As a person develops his or her overall life's belief system, this can result in defining differences among ideologies and the individuals who adhere to different religious or political approaches. If the social environment is harsh and prone to scapegoating, feelings of anxiety and hatred can be fostered. As negative emotions inundate the developing brain, the neuronal pathways supporting these emotions become stronger. Thus, hatred breeds more hatred from a brain perspective. This is where hatred and intergroup rivalries can arise and where cults and powerful leaders can provide a safe and important context to nurture vulnerabilities relating to the need to conform and feel secure in one's beliefs. In addition, as we have already discussed, rituals can enhance a sense of unity among individuals adhering to the same ideology or myth, whether they support positive or negative emotional vantage points.[25] If a myth and its associated ritual embody a feeling of connectedness only to the group, then aggression against other groups can increase. If, on the other hand, a myth and its associated ritual embody a feeling of connectedness with all of humanity, then aggression against other groups can decrease. These statements are true regardless of the size of the group. A small cult could love or hate everyone else in the world. It just depends on their ideology.

The initiating factor for entering into the fourth stage of faith, referred to by Fowler as the "individuative-reflective," is frequently the growth experience of leaving home and receiving higher levels of education, precipitating a thorough examination of one's own self, values, and beliefs. From a brain perspective, this is the stage in which the higher cognitive and emotional functions become fully established and help in a systematic evaluation of a person's beliefs and doctrines. The overall brain metabolism and its connectivity (new connections versus pruning old connections) are highly stable as well, reflecting this overall mature stage of the human brain. Fowler believes that the emergence of the executive ego occurs in this stage and may be consistent with the fully functioning frontal lobes that are involved in executive processes. The executive ego refers to the internal mental processes that help develop the person's entire belief system and is how the person becomes defined and identified as a fully established individual. There is also the struggle for self-fulfillment as a primary concern versus service to and being for others, a concept well described in Judaism, with an emphasis on performing mitzvot toward others.

This stage may also be when one can be said to have "decided" on certain religious perspectives. Among the responses to the survey of rabbis, it was clear that at some point, the respondents made a decision to join the rabbinate and have a positive impact on others. The average age at which our survey participants stated that they decided to become a rabbi was twenty-six years old. However, Orthodox rabbis apparently made the decision much younger, at about the age of twenty years. Conservative rabbis decided on average around twenty-six years, Reform rabbis around twenty-five years, and Reconstructionist rabbis decided on average about twenty-eight years. Interestingly, rabbis who reported that they were currently unaffiliated made their decision even later, around the age of thirty-four years. Perhaps the more defined the path or the more defined the upbringing, the earlier a person is able to come to the decision to be a rabbi. Alternatively, it is possible that the brain of these individuals matured at different rates or was able to find more definitive paths earlier. After all, some rabbis felt that they were going to be a rabbi almost from birth, while others struggled for a long time before making a decision.

Additionally, when people come to Fowler's individuative-reflective stage, an advantage of Judaism is that they are able to not simply look to different religions and spiritual movements to find what they feel is lacking, but that within the tradition itself there are so many varied understandings and ways to interpret the answers to many of the questions a person may seek. Often this can occur when individuals who were identified as Mitnagdim (a designation given to those Ashkenazic Jews from traditionally Lithuanian areas who were staunchly opposed to the Hasidic movement and its intense emotionality as a severe and unwelcome change to Judaism and their traditions)—highly intellectual and unemotional in their observance of Judaism—begin to feel lost and long for something more meaningful. They then begin to search for specific meaning and ideas from more spiritualistic or kabbalistic thought that they chose to ignore at a younger stage of their religious studies. At the same time, some people who spent most of their time in mystic meditation and studying Hasidic thought may turn to a more analytic approach as they get older. Such developments are common to many members of the Jewish community, from teenage through adult years, who throughout their lives continue to explore the different perspectives within the Jewish tradition regarding the various modes of prayer and theological perspectives on the meaning behind their religious acts. We might reflect on the reciprocal interactions in the brain of the emotional, experiential, and cognitive processes that develop in different ways at different times in each individual.

Likewise, many Jewish theologians found that over the course of their own lifetimes their perspectives and theological attitudes changed, depending on their experiences and life perspectives. A good example is Rabbi Joseph B. Soloveitchik, arguably the progenitor of the Modern Orthodox movement in America, who emphasized intellectual engagement and rigid analysis of the mitzvot/commandments to determine their underlying meaning and significance. At the end of Rav Soloveitchik's life, he was reported to have said that he spent much of his life teaching the intellectual aspect of Torah/Jewish study to his students, and they retained and learned much of that. However, he said, the emotional connection and closeness to God was something that he felt deeply, as anyone who reads his works can tell, but he was unable to convey these feelings to many of his students. Part of the reason may be exactly what was discussed above. If the mind is constantly developing and learning different modes of engagement with religion, myth, and ritual, then the age and physiological development of a student is important in determining that student's abilities to truly synthesize the meaning behind religious concepts. At the same time, teaching those who are at a different intellectual stage of development can force the teacher to reexamine positions they otherwise would have ignored. As the Talmud states, "I have learned much from my teachers, more from my colleagues, and the most from my students" (*Taanit* 7a).

Fowler refers to the fifth stage as "conjunctive faith," which generally occurs around midlife and often reflects a disillusionment with life's compromises and recognition that life is more complex than the ability to use our mental and emotional processes to adequately manage. In addition, the individual gains access to new perspectives as the result of life experiences and encounters with people who hold differing perspectives. The end result is a reclaiming and reworking of one's identity and faith through understanding how it relates to humanity. Neurophysiologically, this stage is associated with a persistent decline in brain metabolism, particularly in some of the higher cognitive centers. This decrease, while frequently unknown to the person, may contribute to the sense of disillusionment and a concern for understanding the true nature of reality. As brain connections are lost, there may be a sense that the answers are slipping away and that it is unlikely that the present path will help the person find those answers. The result may be a concern that the self can no longer face the struggle to know and understand.

In Judaism, this stage may be reflected in the concept of the "dialectical tension" that is often quoted as something one must live with for the rest of one's life. When younger, Jewish students may constantly be searching for truth in the world and religion. However, they recognize that they cannot achieve a true synthesis and harmony between oppositional concepts. God's omnipotence and immediate presence, God's knowledge of the future and yet human beings having free will, are all issues that many in this life stage are unable to effectively "solve." However, the recognition that some issues may not be resolvable is an important one for the religious individual as well. In the whirlwind of Job, God asks him, "Where were you when I laid the foundations of the earth? Tell Me if you understand!" (Job 38:4). This interaction is interpreted to mean that Job was being taught an important lesson by God: We cannot as humans always understand things that appear contradictory to us. There is often a feeling that answers are unattainable, but we should not lose faith. This approach may not necessarily be a bad one from Judaism's perspective and, if one does not have spiritual leanings, may be one's final stage of religious development.

The last stage Fowler describes is called "universalizing faith," in which a person overcomes one or more existential, moral, or religious paradoxes through the apprehension of a more universal perspective. In the stage of universalizing faith, there is a sense of unity between the self and the beliefs and principles of the individual's religious tradition. This may go further and even represent a sense of union with God or ultimate reality. Such a unitary experience might result from various practices such as prayer or meditation or through spiritual experiences that affect the sensory and cognitive inputs into the orientation areas of the brain. As a person ages, there is already decreasing overall neuronal function and interconnectedness, which may contribute to this type of experience. There may also be a realization that various religious traditions share essential characteristics and that all faiths derive from a similar root. This perspective is clearly utilized in many kabbalistic and Hasidic writings as the ultimate goal of finding union between humans and God on a metaphysical level. As Rav Schneur Zalman states in the *Tanya:*

> Now just as in the human soul the principal manifestation of the general vitality is in the brain, while all the organs receive light and strength from the brain that "illuminates" them, so it is by way of parable the revelation of the life force of the world that creates the world, that the creations in the world are clothed and included within the will and wisdom and understanding and knowledge of God's Blessed Will which are called the "brain," and these are the same as the vitality that clothes the Torah and mitzvot. (*Tanya*, *likutei amarim*, chap. 52).

When someone decides that there is more than just the world they see, but that there is a hidden force uniting everything, then there is a revelation that everything is connected, clearly similar to this last stage. In our Survey of Rabbis, this stage was represented by those who stated they find God as a source that "connects everything and everyone to everyone else." Some have stated that because of these feelings, they can have mystical experiences in which there is "an ecstatic communion with the sparks of God."

It should be noted that not everyone achieves each stage, and the stages do not always occur in the specified age range, nor does each successive stage equate to a "higher" Jewish understanding of religion or God. For example, some may achieve a sense of mystical union with God in their twenties or thirties, while some might get stuck in the synthetic-conventional stage or even the mythic-literal stage. Hopefully, with continued education and experience, people move through the stages to find perspectives that are most valuable to them.

Overall, these developmental stages also parallel the Jewish approach to learning into adulthood. Initially students are expected to simply memorize a tremendous amount of material when they are younger. However, as students become older, they can begin to synthesize that memorized information appropriately and understand the more advanced level of learning. The prior larger unkempt "mass" of neuronal connections has begun its pruning process. Thus, the advance to the Talmud becomes almost a natural one that helps students understand the basis for much of the previously learned material. Additionally, for a long time, students of Jewish mysticism felt that Kabbalah was only to be studied once someone had mastered all of the prior Jewish texts of the Bible, Mishnah, and Talmud. This may also be due to a similar educational concept that larger spiritual elements cannot be fully grasped without an appropriate framework to build off of. Only when the brain is more fully developed can it help the individual engage these more complex spiritual and religious concepts, enabling the person not only to become part of the Jewish community, but also to find his or her own identity within it.

Judaism and Psychology

Certainly, entire books can be written regarding the interrelationship between Judaism and psychology. However, we focus here on a few of the psychological concepts that might relate more directly to understanding neurotheology from the Jewish perspective. Specifically, we will briefly explore how some of the Jewish psychological concepts might relate to brain function.

Judaism has typically been relatively accepting of scientific pursuits, especially with regard to psychology. One reason for this is an emphasis in Judaism on morality and goodness in the present-day world. Therefore, processes such as science, psychology, and aesthetics can all lead to a greater goodness among people. If psychology can help an individual or a group of individuals toward becoming better people, then it is a worthy endeavor. Psychological principles typically do not interfere with the central tenets of Judaism. In fact, Judaism generally considers the human person as being in a continual state of spiritual and psychological development. Thus, psychological approaches toward helping people improve their overall mental state can also potentially help aid them in their search for a meaningful relationship with God. And we might ponder which brain areas are most likely associated with fostering such a meaningful relationship. Are areas of the brain that support love and compassion, altruism and charity, or adherence to religious precepts the most important? We can also consider whether certain brain processes, especially those that underlie psychological disorders, might foster immoral behaviors. We must also distinguish immoral behaviors that are truly immoral rather than related to mental illness. Those behaviors that are attributable to a mental illness are considered to be abnormal whereas those that are counter to the basic Jewish tenets are immoral.

Often, groups that are more insular may have a negative perspective regarding mental illness. Among some Jewish groups, mental illness in the past was seen as a blemish on the family and something that could be destructive regarding the potential for arranged marriages. As such, the emphasis was not on open discourse and appropriate treatment, and instead this approach resulted in hiding significant issues in the community. Additionally, early psychologists were often openly anti-religious, which made promulgation and discussion of mental health in religious communities a difficult task. However, substantial progress has been made on this problem, with more and more organizations and individuals speaking out, as well as psychologists specializing in mental illness in the religious community. This particular problem is not as prevalent in the Reform and

Conservative Jewish ideologies, as their theology focuses more on practical day-to-day issues, and therefore mental illness is accepted as a potential problem that can be overcome with appropriate therapeutic interventions. It is also interesting to note that brain-related abnormalities are generally regarded as medical problems, compared to psychological abnormalities that do not have clear brain pathology. In fact, showing the brain-related processes that underlie various psychological disorders might help clarify which behaviors and mental processes are considered abnormal versus immoral. There could be some important theological implications that arise from research exploring such questions.

Another psychologically oriented problem is that the basic impulses and neuroses described by Sigmund Freud and elucidated in the fields of cognitive behavioral therapy and psychoanalysis can also masquerade as religious motivation. For example, the Talmud (*Sotah* 4b) states that a person who is conceited is considered as if he has worshipped idols. Many of the commentators understand this line to mean that if he thinks he is the most important thing in this universe, and that nothing else matters, it is as if he is now worshipping himself. Someone with that sort of personality might be described as having narcissistic personality disorder. Often those with true neurosis and/or significant psychiatric conditions can end up incorporating religious themes into their schema when in fact it is their mental illness that uses religious ritual and/or myth as a shield to hide behind.

One person may need to wash her hands repeatedly before eating bread instead of just one time, or she can never be sure she truly "cleaned" her hands from the bathroom. Another can live in fear that he is sinning constantly and God will smite him at any time, or he may feel that he can never live up to the standards of what his religion has taught him and is therefore a failure and feels depressed. These feelings, while partial interpretations of one's religious standing and spiritual well-being, must be interpreted within the context of the individual's mental struggle. A truly skilled psychologist should be able to divide the religious principles and emotions from the psychiatric, and thus enable the individual to live a more complete and meaningful life in both the intellectual and religious realms.

Repentance is also a great example of how Judaism views the mental life of every individual to be in a constantly developing state. While ordinarily we might think our spiritual state to be static, the Jewish view of repentance allows one to rapidly reorient the self as a newly "born-again" person, not the same as the one who came just before.[26] This new person, dedicated to a new life, does not bear the sins the other one had and is able to develop new relationships and spiritual perspectives not simply by creating a brand-new relationship with God, but by transforming the "prior" sins into "merits." This new person may have had the prior experiences of sinning, but as he or she has not gone through the process of *teshuvah* (repentance), these components now make up part of a new personality dedicated to God. This formulation of repentance allows for significant psychological insights and changes even at an older age and ensures that the above-mentioned models of brain development can be maintained even within the system and structure of Judaism.

Conversion is another example of Judaism's view that one can change one's personality and mental structure in a way that others might say is impossible. "One who converts is like a child born" (Talmud, *Yevamot* 22a). A person can be radically reborn with a conversion process as he or she accepts all the tenets and obligations of Judaism and Jewish life. However, the change in the person's mental life is not assumed to be willy-nilly and easy to do. Conversion also requires a significant amount of study and commitment prior to the actual process. The Talmud (*Yevamot* 47a) also states, "You should know that before you came to convert, if you ate *chelev* fat there was no punishment of *kareit* [lit. "cut off," most likely meaning death via heaven]. If you transgressed Shabbat you would not be punished with stoning. However once you convert, eating *chelev* fat is punishable with *kareit*, and profaning Shabbat is punished with stoning." So, converts are substantially warned about of the gravity of the commitment they are undertaking, as well as actively dissuaded by the members of the conversion court in order to ensure that conversion is the right path for the individual. If despite all this hardship, a person still wants to join Judaism, he or she is given the opportunity to do so. These two examples—repentance and conversion—may give insight into the neurotheological mental underpinnings behind Jewish practices. There may be a Jewish psychological perspective that believes in the transformative potential of the human mind and does not take a solely negative view of human nature, though it may require significant effort to overcome. In addition, we might wonder from a neurotheological perspective what changes in the brain occur during repentance or conversion. Are there actual changes that occur in the emotional or social centers of the brain? In fact, would it even be possible to ascertain to what extent a person's brain is reacting to a state of repentance or to the enthusiasm one has for conversion?

While much work remains in defining an exact "Jewish psychological theory," the fact that much religious thought and writing have already been dedicated toward introspective analysis and explanations of human motivation and thought should not be ignored. Only by plunging into the resources of religious thought, as well as bringing in the neuroscientific perspective, can there be developed a fully cogent Jewish psychology. Based on what has been discussed up until now, while we cannot posit a full psychological theory, there do appear to be basic principles of Jewish thought that would contribute to a better understanding of Jewish psychology. Judaism has basic principles that guide it religiously. However, in addition to those principles, which mainly manifest in differing laws and Jewish practices, there is an underlying psychological "mental status" that the Jew is meant to have and cultivate. Ultimately, this has been a substantial part of the discussion surrounding the different camps of Jews throughout history and across cultures: how exactly should Jews be thinking not only about Judaism, but about themselves? Judaism does not view itself as simply a religion of the weekend. One who is intent to commit to its practice, in whatever denomination, must be willing to incorporate its values and worldview into the entire week. It becomes a living and breathing part of a person and, as such, triggers significant questions regarding the general mental mind-set a person should have.

Rabbi Joseph B. Soloveitchik, in *Halakhic Morality: Essays on Ethics and Mesorah*, has argued that one of the unique aspects of the religious Jewish individual is emphasis on Torah study as a religiously uplifting principle: "We exist in God and through Him. Without access to Him, man is doomed and his existence forfeits its very core and logico-ethical ground."[27] Both the Mitnagdim discussed above and the Hasidim agreed to this principle.

> Therefore the vitality light and existence of all the world are right only when we are properly engaged in [Torah], for the Holy One, blessed be He, the Torah, and Israel are one. (Rav Chaim of Volozhin [the quintessential Mitnagid], *Nefesh HaChaim*, chap. 11)[28]

> Therefore, when a person knows and comprehends with his intellect such a verdict in accordance with the law as it is set out in the Mishnah, Talmud, or later authorities, he has thus comprehended, grasped, and encompassed with his intellect the will and wisdom of the Holy One, blessed be He, whom no thought can grasp, not His will and wisdom, except when they are clothed in the *halakhot* that are set out before us, and also His intellect is clothed in them. (Rav Schneur Zalman, *Tanya*, chap. 5)[29]

Thus, both the spiritualistic and intensely intellectual camps admit to the basic premise of study of Jewish text and intellectual endeavors to be a founding principle. Despite the fact that one may not be able to achieve spiritual connectedness every time, the fact that there is a tradition one can tap into intellectually should always be in one's mind as a place to return to in times of need.

This leads back to the notion of continual growth, intellectually, psychologically, and spiritually. Is it possible that the Jewish concept of such personal growth helped inform one of the most influential Jewish psychologists, Abraham Maslow? Maslow's work focused on the concept of self-actualization, which guides how human beings work in general. He created a "hierarchy of needs," a theory of psychological health predicated on fulfilling innate human needs in priority, culminating in self-actualization. This was defined by humanistic psychologists as the notion that every person has a strong desire to realize their full potential, to reach a level of "self-actualization." The goal of this approach emphasized the positive potential of human beings and differed from the work of Freud: "It is as if Freud supplied us the sick half of psychology and we must now fill it out with the healthy half."[30] Maslow also helped toward developing the fields of transpersonal psychology and positive psychology. Maslow envisioned moments of extraordinary experience, known as peak experiences, which are profound moments of love, understanding, happiness, or rapture, during which a person feels more whole, alive, self-sufficient. He or she feels an intimate part of the world, more aware of truth, justice, harmony, goodness, and so on. Self-actualizing people have many such peak experiences. In other words, these "peak experiences," or states of flow, are the reflections of the realization of one's human potential and represent the height of personality development.

Additionally, Judaism recognizes the importance of speaking out one's concerns and worries, even if they are not able to be easily remedied. Simply having someone to speak to is often helpful, and so Jewish prayers are often replete with requests to God for help—help with personal health, another's health, happiness, bringing redemption, or even to smite evildoers. Whatever it may be, Judaism is clear that such prayers seem to be asking God for something mundane that God should have given already if God felt it was needed. How can this be? Rabbi Soloveitchik writes:

> On the one hand, Judaism realized the transcendence, omnipotence, infinity, and omniscience which saw God as exalted in His inapproachability and incomprehensibility, distant from and exalted above all forms of creation. On the other hand, it recognized man's helplessness, wretchedness, absurdity, and nihility. . . . What Judaism aimed at with introducing supplication-prayer is just the opposite of suggesting some kind of action to God—namely the manifestation of absolute dependence on Him.[31]

By telling people to pray, Judaism is making them recognize that they need help and are willing to admit they cannot manage on their own. However, despite this, Judaism also emphasizes that human beings are inherently good. Maimonides believes that when God states, before the sin of Adam and Eve, "Behold it was very good" (Genesis 1:31), God is referring to the inherent state of humanity's ethical nature at that point. Human nature in its primordial state was established on the ethical principles of truth and falsehood, and there was no need for any subjective "good or bad" to define how human beings were meant to live. Rabbi Soloveitchik believes that when one reaches a state of spiritual ecstasy, that feeling of connectedness eliminates that dichotomy of the "good and bad, as opposed to the true and false" and one "attains full redemption in his harmonious purified being."[32] When one is able to look past the difficult times to a better future, this can bring great consolation and hope. However, this perspective is often difficult to maintain. At times, one will need to fall back on a more basic human ability, and Judaism chose the ability to think and act using the higher cognitive areas of the brain.

Even those who are not spiritually inclined have emotional needs and mental health requirements. Judaism says that the basic needs of a community should be met with an attempt to embrace people at their religious understanding and let them gain meaning from engaging with the tradition. No matter one's background or religious underpinnings, in order to understand a true Jewish psychology, the cultural components of Judaism become of great importance. Many Jews do not have a strong connection to the religion, but that does not mean that they have not had significant life relationships to other Jews and figures they associate with Judaism. Jews who survived the Holocaust and decided to dedicate their life to the rabbinate may not have had the largest Jewish experience and background prior to that experience, but something as earth-shattering as the Holocaust will leave an impact on them, their children, and all congregants for potentially years to come.

The perspectives of Judaism are only as strong as those who espouse them and adhere to them, and as such a cogent Jewish psychology also needs to take into account the mental health of the spiritual leaders of Judaism as well. Often quoted is the maxim "As a hammer strikes a rock" (Jeremiah 23:29), meaning "Just as this hammer sends off many sparks, so too a verse can be expounded to have many reasons" (Talmud, *Sanhedrin* 34a). This saying needs to be taken to heart particularly regarding mental health and psychology. Judaism has guidelines and laws, restrictions and motivational pieces. However, ultimately it relies on human beings to carry out the practice faithfully and with good intention. Jewish psychology is no different. One must take into account the person's nuances and personal influences. One's education, experiences, religious education, and ideas about the world will all influence how Jews perceive their religion as well as their place in the Jewish community. Only with a thorough understanding of the person and his or her perspective on Judaism will we be able to truly understand the Jewish psyche. "One who holds up [supports] one soul of Israel, the verse considers them to have upheld an entire world" (Mishnah *Sanhedrin* 4:5). Every person is a world of nuance to themselves, and a proper Jewish psychology would balance general principles of mental health and religion within the psyche of the person.

How Neurotheology Relates to Jewish Thought

We had never had any interactions before, but the meeting was a fascinating one for both of us. David was still an undergraduate at Yeshiva College in New York as well as a rabbinical student at RIETS (Rabbi Isaac Elchanan Theological Seminary), thinking of applying to medical school, and his father, a physician at Jefferson, had set up a meeting with someone in whom David might find common ground for scholarly research. Andrew was the director of research at the Marcus Institute of Integrative Health at Jefferson and had an expressed deep interest in religion, and David was fascinated by the topic of neurotheology. When we first met, we discussed a possible work on analyzing the specific concept of the oneness of God in kabbalistic thought and Jewish philosophy. However, neither of us pursued the project further after our initial meeting, since we both were busy with other things. Two years later, Andrew lectured to the first-year medical students during their neurology block. He spoke about the human mind and consciousness, and his appearance and lecture reminded David of the conversations they had had two years prior. So we decided to meet again, now with David's newly acquired medical knowledge and Andrew's further developments in neurotheology. Over the course of the conversation, it became clear that we were both excited about the possibility of exploring the notion of a Jewish neurotheology, recognizing the value of the combination of rabbinical training and neuroscience. The summer project turned into a much larger and longer conversation, and ended with *The Rabbi's Brain* as you see it here now.

—David Halpern and Andrew Newberg

How do different groups view neurotheology, including different Jewish sects, Jewish fundamentalists, scientists, health-care providers, theologians and philosophers, and even atheists/

materialists? This is an important question when considering whether neurotheology should encourage science and religion to be brought together in the first place. Of course, there are no formal statements about neurotheology, but from what we know of these different groups, we can infer how neurotheology might be regarded in terms of both its pros and cons. There are other religious and theological implications as well, such as whether neurotheology might provide a new perspective on which aspects of religion are the most important, how to optimize the quality of theological and scientific research, and the importance of carefully drawing conclusions from neurotheological research.

Neurotheology should be a highly complex, multidisciplinary analysis of the structure and function of religious and spiritual phenomena. Whole departments and seminary schools are devoted to the full understanding of specific religions such as Judaism. It is certainly not possible to cover all of the elements of Judaism as they would reflect on neurotheology. This full understanding could take many years, and most would argue a lifetime, or as in Judaism, many lifetimes. In spite of these daunting prospects, we can at least begin to explore neurotheology as it relates to Jewish thought.

We can also focus on how Judaism, as well as several other traditions, views the relationship between the mind and religious beliefs in general and what this might mean for neurotheology in particular. As we have stated previously, there are important questions with regard to the mind, the brain, the soul, and the other components of our lives that come together to make us who we are and also define our religious beliefs.

The Human Person: Spiritual or Biological?

As we have previously described, neurotheology takes a scientific perspective on the human person, but neurotheology is also a two-way street. Thus, we cannot simply use science to understand everything about ourselves, including our religions. We have to engage both perspectives with an equality and reciprocity deserving of both. We cannot reject science or religion out of hand until we more deeply understand their relationship. This is the fundamental approach in neurotheology. But this approach in and of itself can sometimes lead to consternation on the part of both scientists and religious people.

Those who adhere strongly to one side or the other might have difficulty with the apparent openness of the neurotheological approach. A religious individual might be concerned that science is treading into areas that are better left to the Torah. And scientists might feel that religious beliefs are too unscientific to even be considered. Neurotheology asks us to be patient with both sides, to acknowledge the importance of both sides, and to make conclusions only when they can be considered carefully. It is entirely possible that neurotheology will yield information that favors one side or the other, but if we are to regard the pursuit of truth as the ultimate arbiter of things, then we should all be willing to accept what is true if we are able to fully determine what is true. We will consider this issue of neuroepistemology in the last chapter. But it is important here to realize that for neurotheology to work, it must be carefully developed from both sides—the scientific and the religious.

From the scientific perspective, the human person is essentially a pile of elements and molecules working in an incredibly complex and cohesive way. We understand that the brain and body are composed of billions of cells that perform different functions. The billions of brain cells ultimately are associated with our various thoughts, feelings, and experiences. This is a point that requires great care in terms of how causality is assigned. The hard cognitive neuroscientist usually concludes that our mental processes arise from or are caused by the neurophysiological workings of the brain. This is based in large part on the cognitive neuroscientific studies utilizing brain scans in which different parts of the brain become activated during tasks that involve specific processes such as memory, emotions, or language. There are always one or more parts of the brain that seem connected to a particular mental process. Lesion studies in which a person has damage to a specific part of the brain have also contributed to the notion that the brain causes our thoughts and emotions. If there is damage to the emotional centers or the language centers, we lose those abilities. The implication is that the damaged brain areas are *responsible* for those abilities.

Even from a strong neuroscientific perspective, there is one very challenging question about the brain and mind: where in all of the neurophysiology does the mind actually arise? What we mean by this is that there are so many different physiological processes going on all the time that it might be difficult, if not impossible, to discover which ones specifically cause a thought or a feeling. For example, when a neuron fires, sodium and potassium atoms rush across the cell membrane to cause it to depolarize. Once the signal gets to the end of the neuron, there is the release of a neurotransmitter such as serotonin or dopamine. That neurotransmitter affects the next neuron, causing it to turn on or off. Each neuron is connected to thousands of other neurons also all firing and transmitting. And then there are changes in the electrical activity and blood flow associated with each of these processes. So where in all of that—the sodium and potassium, the neurotransmitters, the neurons, the network of neurons, the electrical activity—does the thought actually occur? Does it occur with only one of those processes or all of them?

There is even a new neurophysiological model based on the concept of "emergence," which means that our thoughts and feelings *emerge* from the physiological processes. Emergence implies, however, that the thoughts and feelings are also distinct from all of those physiological processes. A common example of this has to do with the properties of water. We would never say that a single water molecule is wet, but trillions of water molecules can be wet. Wetness emerges from the interacting properties of trillions of water molecules with their oxygen and hydrogen and the ways the molecules interact with each other. But we can ask whether wetness is caused by the molecular processes of water or whether wetness arises out of those processes. Similarly, perhaps the mind emerges out of the full complexity of the brain's processes. Thus, we can see that even a scientific perspective on the mind and brain is not straightforward.

Neurotheology would ask us to utilize these various pieces of information regarding the brain functions and mental processes. But we should also look at religious views regarding the nature of the mind and brain. In this regard, there are a number of oppositional concepts that religions consider that deal directly with the human person. One of the primary dualities that religions contend with is the physical versus the nonphysical. This single duality is recapitulated throughout a number of other dualities that are considered by religious and spiritual beliefs. Religions themselves are usually predicated on a nonphysical aspect of the universe. This nonphysical aspect may be God in the theistic traditions or some type of ultimate reality or universal consciousness in the nontheistic traditions. This nonphysical part of the universe is considered to transcend the physical part and thus has priority. One of the important dualities that arises, then, is the relationship between the body and soul.

Religions must consider if there is a nonphysical part of the human being that can connect to the nonphysical part of the universe. In the Western religions, this "soul" is what ultimately connects us back to God. From a scientific perspective, there is also a related, and fundamental, duality, which is the brain and mind. This duality is of crucial concern to neurotheology, since it is not clear whether the human mind and human consciousness actually exist beyond the biological workings of the human brain. The unidirectional causal relationship of the materialist perspective on the human person is generally in contradistinction to what is described by religious and spiritual traditions as the soul.

Judaism has traditionally believed in the existence of a soul and has attributed the mind of the individual to this soul as well. Despite the fact that the brain has processes that are correlated to the said experiences the individual has, the assumption many Jewish scholars make is that the brain, like the body, is the physical house of the person's "self." A person can only function appropriately if the house is well kept and works well. Jewish thought has focused on the debate regarding just how physical versus spiritual is the soul. Does the soul include the more basic drives of a person, or are those drives relegated to a "lesser soul" or animalistic tendency that must be overcome?[1] And how much is the human being a fundamentally "this-world being," meant to follow God in this world, or a fundamentally "spiritual being," who seeks to engage the spiritual world? From the neurotheological perspective, we might consider several relevant topics with regard to the soul. There may be different brain areas that help us to perceive physical objects versus nonphysical objects or concepts. After all, regardless of our beliefs, we all can at least consider nonphysical objects, such as thoughts and ideas or perhaps angels and God. How does our brain assign existence to these different types of objects? And how might we reconcile physical and nonphysical aspects of the human person on either scientific or religious grounds? Neurotheological research could explore these questions in more detail by bringing in neuroscience to explore unusual states of mind that are part of spiritual traditions.

In recent years, much research has focused on meditation practices, which are relatively easy to study and result in altered states of mind or consciousness. Studying highly proficient practitioners may provide an important window into understanding the mind, since we can find how these practitioners view their practices and experiences. Then we can utilize neuroimaging or other measures to determine how our biology interacts with our mind and soul.

Within Judaism, kabbalistic thought and practice would seem a ripe area for such research. We could study how the brain works when a kabbalist is in deep meditation, specifically when he or she is engaged with particular *sefirot*. The *sefirot* refer to the attributes or emanations of God through which God reveals God's self to the universe and continuously creates both the physical and metaphysical realms. Each of these *sefirot* can be engaged through meditation and achieving specific levels of consciousness in the kabbalistic system. A neuroimaging study of a kabbalist who is in deep meditation via continued recitation of specific Hebrew letters might reveal similarities and differences compared to individuals of other traditions reciting mantras or other texts. To assess the neurophysiology of someone undergoing *hitbodedut*, literally "self-seclusion," would be fascinating. Perhaps we will find that the brain plays a pivotal role in such experiences.

Or, perhaps the most exciting finding would be that when a person has an intense spiritual or mystical experience, nothing changes in the brain. If the brain has no changes when one feels the presence of God, what does that mean for the scientific claim of the need to use our neurological synapses in order to perceive reality? Perhaps through this type of subtraction technique, we might actually find something that is spiritual rather than psychological or neurological.

Another fundamental dualism set up by the human brain is the "self-other" dichotomy, which eventually can be expressed as the "us versus them" concept. We typically regard our self or our group as distinct from the rest of the world and other groups. On one hand, this distinction is important for survival, since we must be able to recognize our self as distinct from other objects in order to find the right food and avoid certain dangers. After all, we would not want to eat our own foot for nourishment. On the other hand, this "us versus them" mentality can produce substantial animosity between people, resulting in aggression and violence. On a larger scale, this can result in racism, hatred, persecution, and war. There are studies of this phenomenon that suggest that human beings have an inherent brain processes that help establish an "us versus them" mentality.[2] When this is coupled with doctrines of ultimate reality and strong emotional responses, the results can be extremely dangerous. But there is also a balance that must be achieved so that we also can dissolve the self-other distinction sufficiently to form a family, to have children, or to connect with God. The ability to experience empathy and express compassion and love is also part of the brain. It is perhaps within this balance that we find the most important perspective on the relationship between the mind and brain, whether these distinctions are biologically driven or placed there by God.

The human perception of time is also an important element of most religious systems. Religions typically describe time in very specific ways and refer to important time points as having religious meaning. In Buddhism, for example, certain times are associated with the ability of the Buddha to come about. In the Bible, there are many time periods that have importance, such as six days to create the world, forty days of rain, seven weeks from Passover to the harvest festival Shavuot, seven years in the *Shemitah*/Sabbatical cycle in the Land of Israel where the land lays fallow, and seventy years in the *Yovel*/Jubilee cycle where slaves go free and debts are canceled (See Exodus 21:2 and *Mechilta* on the verse, as well as Mishnah *Rosh HaShana* 1:2). Judaism believes that mastery of one's time is one of the most defining principles of being a free person. When the Jews were freed from Egypt, the first commandment they were given was the sanctification of the new moon (See Exodus 12:1). For a new people broken by slavery, there was no concept of having one's own time or setting a schedule; everything had been forced upon them. Judaism states that there are two time realities—that of the world that God made, and that of human beings that we constantly remake. Every week the Sabbath comes every seven days whether we like it or not. However, the other Jewish holidays only come at the times of the calendar that is established by two Jews coming and testifying in the court that they have seen the new moon. The world can also shape its own destiny, and people can establish the times that are necessary for them to appropriately connect to God.[3] There is also an important distinction between how human beings and how God measure time. God tends to be conceived of in terms of "eternal." God is not created nor does God die. God understands how human beings perceive time, but this is very different from how God perceives time, since God is really beyond time. When we stand outside of time and see "the larger picture," then the life events that we experience make sense in a different way.

The perception of time itself is a curious thing. To our knowledge, human beings may be the only animal that perceives the passing of time, can assess that time, and perhaps more important for religion, has a notion of the end of time. Several studies have evaluated how the brain perceives time. It seems that the frontal and inferior parietal lobes, along with the insula and putamen, may be particularly important for time perception.[4] Time perception might require a combination of the functions associated with these structures, namely executive functions, attention, and working memory. And interestingly, the same areas of the brain involved in cognitive processing, particularly cognitive effort, are also involved in time perception. The question is, how do these areas function when we consider time perception in a limited manner versus an eternal manner? And how do we distinguish the physical limitations of time with the cognitive understanding of time?

Along similar lines is the notion of mortality as distinct from immortality. Mortality is something that our brain is set up to comprehend in a very strong way. As the Talmud says (*Moed Katan* 28a) "were the Holy One, Blessed be He to say to me: "Go back to that world (the physical world) as you were." I would not want to go, for the fear (of death) is great." The brain must fear death as a motivation to do things to survive. We eat and drink because we want to live, and we avoid predators and cliffs because we don't want to die. This notion of mortality is deeply tied to our brain's ability to interpret and respond to the world and is a primary reason that many cite as the origin of religion. For those trying to explain religion from a materialist perspective, religion is a method that many individuals use to deal with their own mortality. However, for most religious individuals, their own mortality is a small aspect of why they believe. Religious beliefs appear to arise more from an experience of God in the world rather than as purely a way of managing the fear of death. A brain-related study might help distinguish the emotional and cognitive elements of the awareness of death or immortality.

Another duality relating more specifically to our behaviors and thoughts involves ethical beliefs. In particular, we can conceive of thoughts and behaviors as being either good or bad, as ethical or unethical. There is actually a distinction between good and ethical and bad and unethical. Good and bad are more general terms and may relate to more pragmatic aspects of our behaviors. For example, killing someone may not be moral, but in war it might be good, especially if your brain does not want to be killed. What makes certain behaviors moral or immoral, though, is extremely complex and depends on beliefs, doctrines, and culture. However, neurotheology would certainly focus on the mental processes that ultimately help us make moral decisions. How do we decide if robbing a pharmacy to save someone's life, for example, would be good or bad, ethical or unethical? One might say this is a good action since you are saving someone, but unethical since you are violating another's property to do so. Brain imaging studies of morality have focused on a variety of functions, including emotions such as disgust, social context and relationships, and theory of mind processes that help us perceive what another person is thinking.[5] However, it is difficult to delineate how religious beliefs might affect moral decision-making and whether the impact of religion is felt via emotions, cognitions, or both. Results of studies exploring how the human brain processes moral dilemmas might provide important information for understanding morality, religion, and ourselves.

Although there may be some fascinating data that we could use to explore the many aspects of being human, there are of course some important concerns that religion, or more precisely, religious individuals, might raise about the neurotheological perspective of the human person. Does the human person as described in religious scriptures have to be related specifically to a soul? Do religious texts argue for a way to be a person versus a literal interpretation of how humanity came about? How much should we try to shape the religious notion of a person into a biological one?

In the past hundred years, there has been a growing emphasis on the relationship between the human mind and the human spirit. At the turn of the previous century, William James, in his *Varieties of Religious Experience*,[6] raised a number of important issues that are highly relevant to neurotheology. He began with addressing how religion and religious experiences come about and noted that we also have to address the question "Now that it is here, what is its meaning?" James did not feel that religion was diminished by examining it. Simply because we study love or happiness does not mean that we will someday eliminate it as a part of human life. Similarly, a biological basis for spirituality does not necessarily diminish its value in the human mind.

Pierre Teilhard de Chardin, a Jesuit priest and paleontologist, discussed the relationship between the mind and spirituality in his book *Phenomenon of Man*.[7] In this work, he argues for meaning and purpose within the evolutionary process. This is distinguished by more recent evolutionary scholars such as Stephen J. Gould, who argued against a specific "direction" of evolution. Thus, Teilhard's approach is not necessarily viable from a biological evolutionary perspective, but it does represent another influential approach. Teilhard argued that the "superiority" of human beings was related to their ability for reflection, or the ability of consciousness to turn in upon itself. "Hominization" he described as the leap from instinct to thought, but also the progressive spiritualization in human civilization of all the forces in the animal world. Teilhard suggested that the development of the "noosphere," the thinking layer that exists above the biosphere, represents that which differentiates human beings from animals and that this difference is mediated by God.

From a Jewish perspective, Rav Abraham Isaac HaCohen Kook believed that the development of human beings, both evolutionarily and historically, was heading in a direction similar to that described by Teilhard.[8] This development would enable the maximum number of people to appreciate God and serve out their true religious purpose. Again, this does not coincide well with the current understanding of evolutionary theory, but represents a religious perspective on human evolution and the mind as it relates to religion. Evolutionary theory currently recognizes that species evolve to be most adaptive with respect to given environmental forces, and there is not a specific directionality toward which evolution moves. Thus, evolutionary theory would argue that humanity is the most adaptive species at the current time but that there may be other, less intelligent, species that can be more adaptive in the future. If human beings destroy the world through nuclear war or global warming, it might be cockroaches or bees that take over the world.

More recently, several scholars have suggested a kind of hybrid between religion and natural science. This has taken on several different terms, such as religious naturalism and non-reductive physicalism.[9] These approaches focus more on the unique human qualities, such as language, empathy, memory, future orientation, and emotional modulation, that are the properties representing the human soul. In this hybrid perspective, the soul can be described as the unique emergent elements of the neurocognitive processes of the brain and are not independent or do not operate differently from other neurological processes. The soul in this definition does not necessarily have to have a divine or metaphysical nature, though it is still a distinct possibility. This approach is more congruent with current cognitive neuroscience but is somewhat different from more traditional Western religious approaches to the soul.

Judaism and Neurotheology

How do we begin to understand the potential relationship between Judaism and the emerging field of neurotheology? As we have described throughout this book, the religion of the Jewish people is the religion ascribed to the people of the kingdom of Judah, and the Torah is the primary body of divine Jewish teaching, which refers specifically to the Five Books of Moses. Jewish beliefs also come from other parts of the Hebrew Bible; the entire *Tanach* (Torah, *Nevi'im*/Prophets, and *Ketuvim*/Writings) was canonized from around 140 BCE to the second century CE, depending on scholarly opinion. Jewish tradition ascribes the origin of the Torah to Moses writing down God's word (Deuteronomy 33:4). The Torah was accepted by the Children of Israel on Mount Sinai when the Jews, through Moses, entered into a covenant with God (Exodus 24:7). Although this covenant was accepted by the Jews, the basic message of the Torah is technically for all humankind. But does this basic doctrinal understanding of the roots of Judaism relate at all to neurotheology?

From a neurotheological perspective, there are several interesting aspects of Judaism. For one, depending on the value placed on modern science, various Jewish groups might be more or less likely to embrace neurotheology. If we are to go and learn about the Torah, and hence the universe, then science can certainly play a role in that investigation. Neurotheology, as a hybrid of science and religion, might be a particularly exciting intersection for Jewish scholarship. By combining the commentary in the Talmud and other texts, this information can inform us about how the human person, mind, and soul are interrelated from the Jewish religious perspective. Neuroscience can be used to help uncover the value and/or validity of such concepts. Perhaps some ideas about the mind and soul will need to be modified, while other ideas might be confirmed.

The Jewish people have always placed substantial emphasis on education and learning in order to utilize rational thought processes to understand the human being and the human relationship to the world and God. The monotheistic view of Judaism is also important, since the human mind must strive to understand a great unity and oneness of God. There is also a strong emphasis on ethical behavior, and there is some distinction between behavior and thought. In Judaism, behavior has a somewhat higher priority compared to Christianity, in which even sinful thinking is considered unethical. In this regard, Judaism focuses on behaviors rather than thoughts, even though there is a general understanding of how thoughts ultimately affect behaviors. Judaism also has its share of rituals, which include specific ceremonies such as the Brit Milah and the wedding ceremony, as well as the longitudinal telling of stories of religious importance, such as of the Exodus from Egypt, which is retold at the Passover seder each year.

The belief system of Judaism has evolved into several major denominations, which each might have their own perspective regarding neurotheology. In addition, the study of neurotheology has the potential to contribute to our understanding of many of the different beliefs within Judaism and particularly with respect to differences between Jewish denominations and subsections within those denominations. By studying the various correlates of the brain to spiritual phenomena, the possibility for explaining the purpose and potential origin of religious rituals and beliefs particular to Judaism becomes a real driving force for any student of religion.

For the Reform movement, the perspective of science strongly influences knowledge about the nature of human beings and the world. Thus, Reform Judaism would arguably be the most tolerant of neurotheology. The ability to identify brain areas that are associated with religious and spiritual beliefs or practices might be highly regarded by Reform Judaism because such findings help demonstrate the naturalistic impact of the religion. As such, scientific studies could be regarded as lending support to those following Jewish traditions and beliefs. There becomes an added validation through science of the Jewish beliefs. Neurotheology might even consider which brain areas are most active in Reform Jews compared to the other denominations.

Correlations between mental processes and spiritual practices also could help inform individuals looking to work on specific spiritual goals via defined rituals and practices. In Judaism, it would make religious sense to begin with the ethical principles derived from the Torah and determine how they might be related to specific brain processes. As such, the idea of giving charity to those in need, or of being tolerant to others could potentially map onto various brain structures. Once these general ideas are associated with specific areas of the brain, practices could be identified that might strengthen those processes. For example, if sending loving prayerful thoughts to friends and relatives helps make someone more empathetic and compassionate, and this is documented not only subjectively but via changes in brain function, then such a practice might become a new "ritual" for Reform Judaism.

Conservative Judaism also would generally have a positive perspective on neurotheology. Although more strict about following Jewish laws and practices than Reform Jews, Conservative Jews also have a deep respect for the value of scientific discovery. Conservative Judaism could take many of the rituals of Judaism and interpret them as potentially beneficial, as the Orthodox would, but would be able to eliminate any that would be considered too difficult or not potentially in line with the proposed neurotheological framework of the movement's goals and theological objectives. As an extreme example, the blowing of the shofar, used by many to signify the season of repentance around the High Holy Days, might be discontinued if the majority of constituents no longer understood or received the proper training in understanding the meaning behind the shofar being blown. The general weltanschauung of the people would determine what would be considered useful in perpetuating the halacha/Torah and observance in general. By correlating the different rituals to specific neurological processes and networks, the Conservative leaders would be able to adequately explain to their congregants why a specific law was still in use, without the need for the older reasoning used by previous generations of Jews. This additional benefit might, in fact, allow for the continued observance of many more Jewish rituals than is currently the norm for many officially "Conservative Jews."

Our Survey of Rabbis survey showed that Conservative rabbis reported God as a "real" concept more regularly than did Reform or Reconstructionist rabbis and were far more consistent in keeping kosher laws as well. One particular response in our survey is indicative of the possible trends Conservative Judaism might take regarding neurotheological principles: "I place just as high, and sometimes higher [than standard kosher laws], a priority on organically grown food, pasture-raised animals, ethical treatment of workers, animals, and the earth. . . . We teach mindfulness practice for balance as well as to help cultivate a deep ethical connection to self. So, my *kashrut* lived and taught, is perhaps not the average." Perhaps not the average currently, but possibly a workable response to the theological and religious issues many Conservative Jews may grapple with.

Within Orthodox Judaism, there are a plethora of differing opinions regarding the approach one should take toward science and how one should approach science that is not consistent with religious doctrine. As discussed previously, the three main divisions of rational, indifferent, and spiritualistic will all have their own perspective on neurotheology, and even within these groups the perspective may be different depending on the cultural norms of engagement with science in general and how much trust one has of the research reported by the scientific community. For those in the rational group, the studies correlating brain activity to specific religious principles should in theory be welcomed. After all, simply having the experience in one's brain does not mean that the experience is any less real (we will discuss this issue in more detail below). If this is the case, then science can complement Jewish theology and philosophy of mitzvot. When discussing the giving of charity, the shaking of the four species, or prayer, one would be able to point to the specific neurological correlates of the feelings and motivations that one is supposed to develop by engaging in the rituals.

By continued study and correlation, one might even be able to develop an overarching "theory of the Jewish brain," as touched upon previously, where the entire Jew is meant to represent some specific outcome of specifically activated neural networks. The implications for such work are tremendously exciting, but potentially dangerous to some who might shy away from trying to box Judaism into a specific archetype. Within the indifferent group, there are those who would be perfectly happy with the research of neurotheology, in line with many of the rational group discussed above. However, there might be some who would have the same reservations that many of the spiritualists would have. The difference in this case would be that those members of this group would not see the issue as one that would need to be decided at the current moment. They would argue that "the jury is still out" on neurotheology's research and organization. As such, they would caution against drawing any conclusions theologically, ethically, or scientifically from current speculation. For the spiritualists, there might initially be a strong reaction against the idea that the brain is the location of the spiritual experiences that are meant to be "out of body" or associated with the soul. However, among this group, there is a strong tradition, which as we have mentioned is often linked to the works of Rabbi Abraham Isaac HaCohen Kook, that the general natural processes of the world are part of God's spiritual design. According to this perspective (mentioned above in relationship to Teilhard's perspective on the evolution of the world), evolution is actually designed by God to create a better world, and we are still evolving to some extent even if we cannot see how God could rule the world with chance mutations. According to this perspective, the neurons firing in specific ways may not be just an accident, but actually the intention of God in order for the person to have the experience that he or she is meant to have.

The route one takes to neurotheology therefore could come from any of the different potential perspectives Judaism offers toward science. Some, however, who have been identified as "Jewish fundamentalists," meaning that they would not be willing to accept any new change of understanding to the current theological framework, would potentially see neurotheology as a threat. From their perspective, the soul exists entirely independent from the body on a metaphysical plane that one cannot access through science, and the entire study to correlate brain processes to religious feelings and acts is therefore misplaced. Their claim would be along the lines that correlation does not equal causation, and therefore, neurotheology is irrelevant. This explanation allows for the dichotomy of body and soul to continue to exist and states that the phenomenon of neural activity correlating with thought and action is simply a by-product of the way God made the world, to test us or for some other unknown reason. A similar analogy is made by some to explain dinosaur fossils and carbon dating, that God somehow made the world this way to make people think it was older than the literal biblical account states it is. This position is the one that most Jewish scientists would strongly disagree with, especially since this stance leads to the science versus religion antimony that is so well popularized.

Health-care providers might welcome the additional benefits of a neurotheological framework to enable them to engage more with religious patients who are looking for benefits from their religious rituals and meditations. By understanding scientifically how there might be a health benefit gained through various religious practices and acts, patients could develop greater therapeutic alliances with health-care providers and even potentially use religious thought processes as models to help maintain health and continue health-improvement programs such as diet, weight loss, and chronic disease management.

Neurotheology touches on the larger issue of whether science and religion should be brought together in the first place, a debate that has framed much of the general Christian theological discussions. However, Judaism does not need to follow the same path as Christianity in this particular dispute, and as discussed at length in chapter 1, Judaism has been traditionally very accepting of science and religion creating a potential for actual synthesis. The distinctions between groups, however, would still remain in determining just exactly how neurological correlates could be interpreted. Minimally they could show that God is in some way truly experienced by human beings, and maximally they could be used as a framework to interpret the purpose behind certain key Jewish beliefs and practices.

Another question with respect to the neurotheological debate is if Judaism has a position on whether experience/faith is the only important aspect of religion or whether there is something more significant than this that exists beyond one's subjective experience. Based upon what was discussed previously, the perspectives on this question could potentially fall as well along denominational lines. For Reform Judaism, since the idea of religion is constantly understood to be developing and evolving with one's personal relationship to God, then there really could be nothing more significant and important than one's own faith and personal relationship with God. The individual spiritual experience is paramount. While there is still a strong foundational bias of the Jewish people being a people, the history and commandments specific to Jews are meant to be taken as learning points for the individual's relationship to the spiritual. In this regard, areas of the brain that support the personal ego and self-experience may be most prominent. Such areas would include those that support autobiographical memory, sensory experience, and the sense of self.

Conservative Judaism would have a slightly different perspective. The Torah is still seen as something imbued with divine significance, even though it can somewhat change with the times and norms of society. As such, while there is an emphasis on the individual's relationship to the Godhead as well as individual subjective experiences, there is a far larger emphasis on the group's needs and the continued evolution of halacha/ Jewish laws over time. Thus, when there is a debate between the experience of an individual and the group, Conservative Judaism would tend toward the group theologically. The group experience is made up of many individuals, and that experience trumps a single person's spiritual experience. However, that is not to say that the individual messages and lessons one personally receives are not as significant for Conservative as for Reform Judaism; there is simply an additional element of the group structure of the religion that requires some constraint from its members in their imposition of personal values on the larger group. As an individual, however, practitioners may maintain their own personal experiences and would not be questioned as to the legitimacy of their practice. Regarding brain processes, we might speculate that the social areas of the brain would function more prominently compared to those supporting personal experience.

This is not the case with Orthodox Judaism. Orthodoxy believes in the divine origin and continued divine authority of the Torah in its original form and in the authority of the traditional rabbinate to create rules meant on protecting that authority. The complex legal system is one that has been codified by the Mishnah, then Talmud, down to the current books of law, such as *Mishneh Torah* by Maimonides or the *Shulchan Aruch* by Rabbi Joseph Caro with the gloss of Rabbi Moshe Isserles, known as the *Rama*. These codifiers sought to compile their unique perspectives on what the authoritative law was on all aspects of applicable Jewish law. However, the varied traditions of learning and studying the Torah within the traditional framework created a plethora of positions, some of which are more or less authoritative or have inherent contradictions or inconsistencies with other positions. It was the goal of these codifiers to create a work so the average members of Judaism would be able to read that work and understand what was expected of them from Jewish law/the halacha. However, since these codifiers disagreed on what was authoritative at times, the overall result was diverse practices among different groups of Orthodox Jews. Additionally, cultural differences and locations isolated over time gave rise to distinctly unique practices.

From the perspective of the Orthodox Jew then, the individual's religious perspective and experience, through which he or she is able to personally connect to the spiritual world and speak to God, are valuable. However, the individual must also conform to the group's structure and laws deemed to be as authoritative as if God gave them. As such, if one wishes to meditate as a kabbalist in one's prayers, while still remaining within mainstream Orthodoxy, one would need to first make sure to pray at the appropriate times and ideally with a minyan. Only after fulfilling that requirement would one then be able to spend time in intense meditation. This particular struggle between the individual's spiritual needs and the communal requirements has been a common one among Orthodox Jews for many generations.

> As a matter of fact, the great conflict which divided the Jewish people into *Hasidim* and *Mitnagdim* at the end of the eighteenth century never revolved around the way of doing things. The Ba'al Shem Tov and the Vilna Gaon ate the same *matzot*, and they both said the same *Hallel* (Praise to God in Prayers). The *Hasidim* and *Mitnagdim* were careful and meticulous in observance of the dietary laws, the laws of family purity and the Sabbath. Their conflict revolved about *style*, not about *way*. Hasidic performance was ecstatic, in a more democratic and popular manner. They acted not only with their hands, but with their hearts as well. On the other hand, the Gaon and Lithuanian Jewry knew very little of ecstasy. Their motto was not ecstasy, but discipline and surrender. Their respective styles differed and clashed, bringing about the great controversy between *Hasidut* and *Mitnagdut*.[10]

Many others have spoken about the balance between the personal and communal aspects of Judaism and how both sides can be embraced.

> In other words, there is in Judaism a universal, impersonal aspect as well as an individual, personal aspect. Judaism is on the one hand an objective discipline. The *Shulhan Arukh* is the book of an institutionalized, organized, fixed, statutory religion. On the other hand, Judaism is also an adventure, a great romance, a heartwarming and ecstatic experience. How to execute the law belongs under the category of *derekh* (way/path). But how to enjoy observance and to reach out for the Eternal belongs not under the category of *derekh* but of that of *signon*, of style. There is one *derekh*, but there are a variety of styles. Each individual interprets his own performance in a unique singular way.[11]

It is further interesting to consider how each of these paths may be associated with different brain processes. An experience of ecstasy would likely be associated with powerful emotional responses and the release of neurotransmitters that lead to extremely intense experiences. Social interactions require social areas of the brain in the parietal lobe and likely involve mirror neurons that enable us to emulate others in our group. Discipline appears to be associated with strong controls of behavior that may be associated with either an internal source, sometimes referred to as our conscience, or an external source such as the Torah or God. In fact, our research has suggested that specific brain patterns, namely decreased frontal lobe function, are associated with an experience of surrendering one's will.[12]

With respect to neurotheology, it seems that the brain must play an important role in all of these approaches within Judaism. Whether one follows a more objective path that may proceed from the functions of the rational areas of the brain or the more spiritual path that is associated with emotions and spiritual experience, these all contribute to the Jewish tradition and the life of Jewish individuals. Ultimately, the path a given individual follows may depend heavily on the overall patterns and balances within that person's brain. It is possible that neurotheology may provide a foundation upon which to understand the different neurological influences on an individual and how they pertain to the beliefs, Jewish or otherwise, that arise from them.

Christianity and Neurotheology

In rounding out the discussion about how Judaism might interact with neurotheology, it may be helpful to see how other traditions, not just Judaism, might interact with the concepts associated with neurotheology. As with much of our discussion above, this becomes a two-way street in which religions might examine neurotheological concepts and neurotheology might examine religious beliefs. Let us briefly look at the basic mythic concepts in several of the major religious traditions as they relate to neurotheology. Because of the particular emphasis of neurotheology on brain function and the mind, the focus will be on the relevant elements of the specific traditions. Eventually, the list and discussion can be greatly expanded with information regarding almost any belief system, but it is also important to focus on the most common. From there, we can move on to the broader concepts of theology.

As a belief system, Christianity is typically defined as the existential recognition either that "Jesus is Lord" or, in the theological formulations of the Councils of Nicea, Constantinople, and Chalcedon (in 325, 381, and 451 CE respectively), that the Divinity is a unity comprising the Father, Son, and Holy Spirit. Furthermore, Jesus is considered "God incarnate" and thus is both fully divine and fully human. In the New Testament, Paul describes the fundamental aspects of Christianity as the trio of faith, hope, and charity (love) (1 Corinthians 13:13). As we have described throughout this book, many of these religious concepts can be considered from a neuropsychological standpoint. Faith requires various cognitive and emotional processes, as do hope and love. Christianity also deals with the particular problem of the divine trinity that must be maintained and also merged into a single wholeness. Both the quantitative and holistic processes of the brain would be required to help resolve this potential discrepancy and maintain the oneness of God.

While Christianity itself does not place great emphasis on psychology and the human mind, there is no question that the Bible deals directly with thoughts, feelings, and behaviors. The teachings of Christianity focus substantially on concepts such as love, forgiveness, and charity. Such concepts also bear directly on the psychological abilities and limitations of the tradition's followers. After all, it is essential that people have the ability to forgive or be charitable if they are to be held to such behaviors. Similar to Judaism, it is recognized that the requirements of the religion are things that people can do. As it says "One does not decree on the community a decree unless it is something that the majority of people can keep" (Talmud Avodah Zara 36a). Not all Christians need to act like Mother Theresa. Through their own life, behaviors, and attitudes, any Christian is capable of following the path towards spiritual salvation. Of course, even the notion of salvation is interesting from a neurotheological perspective. Is the brain saved or only the soul? And how does salvation actually occur? What are the psychological and even neurophysiological effects?

Just like Jewish individuals, such questions, and many others, may or may not find additional answers through neurotheology. For example, many Christians can appreciate that there is a value in understanding how the brain might help people to be spiritual and to connect to God. On the other hand, fundamentalists, like some of the Jewish Orthodox, are likely to argue that while it might be interesting to understand the relationship between the brain and religion, it is ultimately irrelevant in terms of how we are to behave and believe with respect to God.

Neurotheology also might provide a unique perspective on Christian rituals which play a fundamental role in the tradition from holidays such as Easter or Christmas, to the singing of hymns, to the taking of Communion. Each of these rituals helps to bring the religious doctrines of Christianity to its followers and engrain such concepts in the neuronal connections within the brain. Christianity also uses the other senses such as vision (the image of Jesus on the cross), smell (the burning of incense), and taste (the taking of the wafer and wine as the body and blood of Christ). But each of these ritualistic elements has an impact on the brain. Ultimately neurotheology might seek to understand the similarities and differences between Christianity and other monotheistic traditions, as well as other traditions such s Buddhism and Hinduism.

Islam and Neurotheology

The word *Islam* is Arabic for "submission" (to God). One who submits is a *muslim*, which is from the same root as *Islam*. Muslims consider Islam to be a fulfillment of Judaism and Christianity and the restoration of a fundamental Abrahamic monotheism. The Prophet Muhammad (c. 570–632 CE) proclaimed Islam to the polytheistic Arabs of Mecca between 610 and 632 CE. Muhammad is described in the Qur'an to have received a revelation from God through the archangel Gabriel and from that revelation developed the religion of Islam.

The Qur'an (meaning "recitation") contains the essential teachings of Islam, known as *iman* or "faith." First and most important in Islam is the belief in the divine unity of God. In this way, Islam shares the basic monotheistic belief of Christianity and Judaism. In Islam, the belief toward God is very uniform among the followers, and thus there is a common pattern of worship. The second major doctrine is the belief that Muhammad is the last in a series of prophets that includes Adam, Noah, Abraham, Moses, and Jesus. A third major belief of Islam is that angels are God's messengers. The fourth belief is that the holy scriptures represent the literal word of God as revealed through the prophets. The

Qur'an is the final, definitive scripture according to Islam. The fifth major belief of Islam is in a Last Judgment when each person will stand before God and the righteous will be rewarded with a heavenly afterlife, whereas those who have committed evil will go to hell. From the neurotheological perspective, we can wonder how each of these essential beliefs are forged within the human brain. Believing in the scriptures or the prophets requires certain ideas to be accepted as true while rejecting others. The brain helps everyone determine which beliefs to hold and which to reject through its cognitive, emotional, and experiential processes.

Muslims must follow the five "pillars" which refer to devotional duties: (1) The *Shahada* refers to "witnessing" the unity of God and the messengerhood of the Prophet Muhammad by declaring, "There is no god but God [Allah]; Muhammad is the Messenger of God [Allah]." (2) *Salat* is formal prayer performed five times a day, either in congregation or alone, while facing the holy shrine in Mecca. (3) *Zakat* is annual almsgiving amounting to a set percentage of one's wealth for the benefit of the poor, new converts in need of assistance, and defenders of Islam. (4) *Saum* is fasting during the holy month of Ramadan between sunrise and sunset, when no food or drink may be consumed or smoking or sexual relations engaged in. (5) *Hajj* is the pilgrimage to the holy city of Mecca which is to occur at least once in a Muslim's life, if circumstances and resources permit. The hajj is typically considered the high point in most Muslims' lives.

From a neurotheological perspective, Islam has several unique features, since there is an emphasis on surrendering oneself to God. This is a different conception regarding the relationship between human beings and God than in most other religions. The concept of surrender could relate to the brain in specific ways that are also associated with willfulness and purposeful behaviors. Thus, brain processes involved with willful processes (increased frontal lobe activity) are inversely related to surrender processes (decreased frontal lobe activity).[13] Furthermore, the rituals in Islam are very structured and can be evaluated from a neurophysiological perspective to determine the specific brain processes engaged. Prayer practices in Islam may have similarities and differences compared to prayer practices in other traditions. Also Islam also has substantial dietary rules, particularly fasting during times such as Ramadan, which might result insimilar and/or different brain changes compared to comparable practices in Judaism or Christianity. In addition, it might be helpful to observe how fasting could augment various spiritual states in Islam and compare those to fasting states in other traditions.

In terms of the Muslim perspective on neurotheology, as with Judaism and Christianity, it probably depends on the general perspective one takes toward science. Muslims who believe that the natural world is derived from God would most likely see value in exploring the biological part of human beings as it relates to our ability to be religious. Such individuals might find it fascinating to understand how specific pillars of Islam are manifested within the brain, and which brain areas may be involved in supporting specific Islamic beliefs. On the other hand, those with a fundamentalist view might deem any scientific investigations as not relevant for following the Muslim religion. However, we would argue, as for every other religion, that neurotheological investigations have the potential to contribute to our understanding of how human beings are able to engage religious beliefs and practices.

Buddhism and Neurotheology

The founder of Buddhism was a Hindu, Siddhartha Gautama, a prince who saw four visions: the first was a man weakened with age; the second was a man who was emaciated from hunger and supremely unhappy; the third was a band of individuals crying over the death of a loved one; and the fourth was a recluse, calm and serene.[14] These encounters with human suffering deeply affected Gautama as he struggled to find peace and understanding about the nature of himself and the world around him. He eventually found a path of enlightenment that revealed the four noble truths of Buddhist ideology: the Noble Truth of Suffering—that suffering exists in the world; the Noble Truth of the Origin of Suffering—that suffering arises from the attachment to bodily desires; the Noble Truth of the Cessation of Suffering—that suffering will cease when attachment to these desires ceases; and the Noble Truth of the Way Leading to the Cessation of Suffering—that freedom from suffering is possible through the eightfold path toward wisdom and detachment.

Various sects of Buddhism follow variations on how to eliminate human suffering and ultimately find mystical enlightenment. Tibetan Buddhism uses texts (called Tantras) based on deliberately obscure symbolism along with mantras (specific words of sacred significance) and visualization techniques in their meditation to address the noble truths. Theravada Buddhism suggests that one must withdraw from everyday life, because involvement with others inevitably results in good and bad karma. "Salvation must therefore ultimately be sought in a total renunciation of society and of the world."[15] Mahayana attaches great significance in being a bodhisattva, a person who is capable of reach nirvana but delays it in order to bring help and compassion to others. And Zen Buddhism utilizes meditation, particularly focusing on paradoxical questions called koans. Koans such as "What is the sound of one hand clapping?" help bring about enlightenment by stripping the mind of its usual logic and emotions.

Given Buddhism's focus on suffering, emotions, thoughts, and consciousness, the tradition has a long history of interest in psychology and the mind. Thus, Buddhist ideas would seem to resonate well with neurotheology, which could function as a tool for better understanding the biological substrate of human suffering and help show how practices such as meditation can reduce or eliminate suffering. Emotional areas of the brain (i.e. the limbic system and insula) are affected by a variety of spiritual practices, particularly meditation. It seems reasonable that neurotheology could help establish a neuroscientific source for Buddhist concepts and provide insights into how meditation might be most effective. On the other hand, the state of enlightenment is supposed to bring a person's consciousness to another level that transcends the physical world. In such a state, the existing paradigm of science becomes limited in understanding the universe. Of course, neurotheology would argue for developing a link between our understanding of the brain and consciousness and understanding how the physical and mental worlds interrelate. Buddhist enlightenment likely has important parallels with kabbalistic Judaism with respect to the use of meditation practices and the goal of achieving mystical experiences. Neurotheology must be open to new paradigms for understanding the universe, whether they derive from science, religion, or mystical experience.

Hinduism and Neurotheology

Hinduism is the dominant religious tradition in India and its primary elements include a cyclical notion of time, a pantheism that infuses divinity into the world around us, an intimate relationship between people and God, a priestly class, and a tolerance of diverse paths to the ultimate ("God").

The three central tenets of Hinduism on the transcendent level are dharma, karma, and moksha. Dharma is the general moral force that holds the universe of sentient beings together. Karma is individualized such that a person's present status in life is a consequence of good or evil deeds done in past lives. In addition, the individual's present behaviors hold the key to future existence. Moksha refers to transcending an individual's karma and ending the cycle of births and deaths. The Bhagavad Gita (Divine Song) is a primary text of Hindu thought that shows the path through which an individual finds enlightenment through detachment from the physical world and comes to comprehend a new understanding of the universe and God.

Hindu and Buddhist beliefs share a number of similarities which makes sense historically since Siddhartha Gautama was a Hindu. Similarly, Hinduism, like Buddhism, places a substantial emphasis on the mind and mental processes that lead to human suffering. As with Buddhism, Hinduism should theoretically be quite open to scientific endeavors for understanding the human mind, human emotions, and the effects of practices such as meditation. Such an understanding could provide information on the most effective ways of engaging the spiritual concepts of Hinduism. And while scientific analysis would not be regarded as being able to specifically lead one to enlightenment, it could help provide some direction towards enlightenment. Similarly, such a neurotheological analysis might provide insights into Jewish mystical practices, their methods, and concomitant experiences.

Judaism also believes in a world of mental processes that affects us all and struggles with the larger issue of good and evil in the world. As the original monotheistic religion, Judaism has had to struggle with the idea of theodicy and thus cannot rely on the Hindu or Buddhist assertion that human mental processes are the source of suffering. Rabbi Akiva Tatz asserts:

> Avoiding suffering may be inadequate as an absolute goal. You can see that there are many ways to avoid suffering that are not intrinsically good. By avoiding certain distressing situations we may avoid suffering, but these means may be wrong. The point is this: we need to eradicate suffering at the root, not at the level of a symptom. Therefore the aim must be a perfect world, not primarily the avoidance of suffering. The absence of suffering that is the assured result of human perfection will be the real and good state. . . . I would suggest an analogy from the practice of medicine: the physician's primary goal is not always properly the alleviation of suffering as a primary act. In fact, more than occasionally he must *inflict* some degree of suffering in order to achieve health—a deeper and more lasting freedom from suffering.[16]

Whether suffering from the Jewish or Buddhist perspective, as well as its alleviation, can be differentiated through a neurotheological analysis remains to be seen. Perhaps brain imaging can help distinguish various effective ways of reducing or eliminating suffering. On the other hand, if suffering can only be achieved by enlightenment in Buddhism or by attaining human perfection in Judaism, we might have a very long wait to see which approach works best.

Taoism and Neurotheology

Taoism refers to a religious and philosophical tradition founded in China and based on the philosophy of the mystic Lao Tzu (or Tze), a contemporary of Confucius. An essential concept of Taoism is the Tao, or the Way, which refers to understanding and following the fundamental ways of the universe. The goal of life is to keep within the natural Tao or rhythms of the universe. Maintaining good health and hygiene while properly cultivating the Tao are required for adaptation of one's vital rhythms to those of the universe. The concepts of yin and yang represent the natural order of the universe based upon this duality. This duality is further expressed in such pairs as masculine and feminine, hot and cold, dark and luminous, and heaven and earth. Thus, in Taoism, one can see the initial notion of opposites that are presented by the binary processes of the brain. These opposites are unified by following the natural way of the universe in a holistic manner. Understanding that way, both philosophically and scientifically, would seem consistent with Taoist principles. In fact, science might be able to help clarify the natural ways in which the universe functions in terms of its binary and holistic properties. Neurotheology includes the fundamental opposites of the autonomic nervous system and the antagonism between excitatory and inhibitory neurotransmitters, all coming together in how human beings think about and interact with the world.

An analysis of the Taoist idea of unifying opposites would prove interesting in comparison to Maimonides's review of

Aristotle's concept of the golden mean in his discussion of human character traits. Maimonides highlights the need to find a balance between opposing emotions and natural tendencies of human beings as an important component of character development toward serving God. (See Shemonah Perakim Chapter 4:4, *Mishneh Torah Hilchot Deot* (Law of Human Dispositions) Chapter 2, and his commentary on Pirkei Avot 4:4). However, the concept of a balance of tensions as described by Maimonides is not the only important viewpoint on this issue in Judaism. The fact that opposites are tolerated in tension may be consistent with the rationalist perspective, and the concept of two opposing ideas being integrated is scattered throughout kabbalistic and Hasidic thought. The *sefirot* themselves are often seen as layered versions of conflicting aspects of God's influence, which can merge and be contained together in the next *sefirah*. *Chesed* (Kindness) and its opposite, *Gevurah* (Strength), are unified in *Tiferet*, (Balance/Glory). This combination is often seen as kindness tempering harsh judgment, or judgment giving balance to the world that would otherwise be too expansive to form from the outpouring of *Chesed*. One example of this spiritual contradiction in the Torah is in describing the plague of hail on the land of Egypt: "And there was hail, and fire lighting within the hail strongly, such that had never been in the entire land of Egypt since it had become a nation" (Exodus 9:24). According to Rashi, this verse represents "a miracle within a miracle, the fire and the hail mixed, and hail is water (and it still mixed), and to do the will of their Maker in heaven they made peace between themselves (the fire and water)." Opposites unifying is a concept well known to Jewish thought, and it would be fascinating to explore just how similarly this concept is to the Taoist ideals and ideas.

Other Religious Approaches and Neurotheology

There are many hundreds, if not thousands, of approaches to religion and spirituality that each have their own specific belief system and doctrine. Some have more spiritual components, some are more ritualized, some are more group oriented, and some are more individualized. While it is not possible to list or describe all of them here, suffice it to say that all of them have some mixture of cognitive, emotional, experiential, and behavioral elements. And in that regard, a larger analysis of religious traditions and approaches can inform Jewish neurotheology whether directly or indirectly. And much like Judaism, these other traditions range from very practical to very mystical. Each of these elements of religions can be considered from a cognitive neuroscience perspective leading to neurotheology. Thus, it is important to determine which elements are present, how they relate to each other, and how they ultimately relate to the human mind and brain.

A few other general conceptions of God are worth noting because they offer alternative approaches to the theistic conceptions found in Western religions. *Deism* is the belief that God exists but is not involved in the world that God created. This is the classic notion of the cosmic clockmaker who has wound the world up and now sits back passively and observes its events. *Pantheism* is the belief that God and the universe are more or less coextensive, that is, God is the universe and vice versa. *Polytheism* is the belief in many "gods," which is found in animism and the "folk" forms of Hinduism and Buddhism. *Agnosticism*, from the Greek *gnosis*, meaning "knowledge," is an inquisitive perspective that does not flat out reject or support the notion of God, but rather intensely questions the notion of God. God might or might not exist, but it is not yet known. *Atheism* is the belief that there is no God, but within atheism, there is a tremendous continuum of beliefs. For example, some believe that there is no personal God such as exists in the monotheistic religions of Christianity, Judaism, and Islam. However, some atheists believe that there is a spiritual or metaphysical realm, only it is not an actual being called God. Another atheist perspective is pure material reductionism, in which it is believed that there is nothing beyond the physical world. All that exists can be explained completely by scientific means.

Again, it can be stressed that neurotheology enables one to explore how the brain might function in relation to these different belief systems. In this way, neurotheology tries to answer questions such as the following: What makes someone an atheist? What makes someone an agnostic? What makes someone religious? And are there differences in their biological makeup that result in an individual finding agreement with one or more of these beliefs?

In the end, neurotheology is something that can be applicable to all of these traditions and perspectives. Judaism, because of its emphasis on learning about the world and developing commentary and study of the world around us, would seem to be particularly appropriate for embracing neurotheology. Given the general enthusiasm for scientific discovery among Jewish people, neurotheology would seem a natural fit that can then be incorporated not only into the secular side of Judaism, but into the religious and spiritual as well.

Jewish Theological Implications

Great philosophical systems are never produced in a scientific vacuum but usually follow the formation and completion of a scientific world-perspective.

—Rabbi Joseph B. Soloveitchik, *The Halakhic Mind*

Basic Structural Elements of Religions

One of the most important aspects of neurotheology is to consider the theological implications of understanding how the human brain intersects with religion and spirituality. A systematic approach is to consider how the brain functions and then determine how those functions are expressed in various theological concepts. All religions must begin with a foundational myth that is described in the sacred texts of that religion, such as the Torah and *Tanach*. The foundational myth might also be described through oral and written traditions (e.g., the Talmud) and handed down from generation to generation through the retelling of various religious stories, rituals, and holidays. From the foundational myths are developed a series of theological and philosophical interpretations of the religion that can then be translated into behaviors, morals, and thoughts of the participants.

As we have discussed previously, "myth" is not a synonym for "fantasy" or "fable." Thus, it does not imply falsehood or fabrication. Instead, in its classical definition, the word has an older, deeper meaning. It comes from the Greek word *mythos*, which translates as "word," but one spoken with deep, unquestioned authority. *Mythos* is, in turn, anchored in another Greek term, *musteion*, which, according to religion scholar Karen Armstrong, author of *The History of God*, means "to close the eyes or the mouth," rooting myth, Armstrong says, "in an experience of darkness and silence."[1]

For Joseph Campbell, this darkness and silence lies at the heart of the human soul. The purpose of myth, he says, is to plumb these inner depths, and tell us in metaphor and symbol, of "matters fundamental to ourselves, enduring principles about which it would be good to know if our conscious minds are to be kept in touch with our own most secret, motivating depths."[2]

Myths, Campbell says, show us how to be human. They show us what is most important and what, in terms of the inner life, is most deeply and profoundly true. The power of myth lies beneath its literal interpretations, in the ability of its universal symbols and themes to connect us with the most essential parts of ourselves in ways that logic and reason cannot. By this definition, religions *must* be based in myth if they are to have anything meaningful to say. In this sense, the story of Noah is a myth even if it is literally and historically true. Likewise, even if the extraordinary events that myths chronicle never happened and the beings they portray never walked the earth, the lasting myths of past cultures all contain psychological and spiritual truths that resonate with the psyches and spirits of readers today.

This perspective is mirrored by Maimonides in his *Guide for the Perplexed* (Introduction: Preparatory Remarks), in which he argues that the messages of much of the Bible are equally if not more important than the actual specific details of many of the stories. When there is a conflict between science and the literal biblical story, it is very plausible and religiously acceptable to understand the biblical story as a parable. That is not to say that Maimonides did not believe in the biblical accounts of creation and the Exodus. He simply was stating that if it were shown not to be plausible, then certain stories would not need to be seen as literal truth. The storyline of the creation of the Jewish people, however, he understood to be factually accurate. For if that story was not seen as fact as well as myth, then there would be no basis for the Jews in the first place. Therefore, while the psychological and spiritual truths may resonate, it is still important for much of Jewish tradition to understand the said myths as having actually occurred. This point is one that is significantly debated between the sects of Judaism as well. While Orthodox and Conservative Jews will traditionally subscribe to the position that the Bible was given to them from Moses and that the stories told therein are accurate accounts for the most part, the Reform movement sees the stories as providing subjectively relevant truths that individuals must explore for themselves and is less concerned about the factual events in the biblical narrative.

From a brain perspective, we might consider how various structures help us to take different ideas in a literal or allegorical manner. On one hand, the brain typically works better when it can rely on definitive concepts. It is easier for us to understand that "thou shalt not kill," rather than a long-winded discussion about times when killing might be morally allowable, such as in defense of one's family or community. Neurologically, the brain can function more simply when there are simple options to consider.

However, the world and our interactions with the world are anything but clear. Thus, the brain must contend with a great deal of ambiguity. It is in this regard that myths and stories can often help. But it is also through discourse and abstract thinking that the brain can help us to consider the many complex aspects of what it means to be a human being.

A fundamental problem though is that when there is ambiguity, the brain can elicit an anxiety response, since the ambiguity implies a lack of understanding about the world and hence a potential danger. Some individuals are less reactive to ambiguity and thus may be more inclined to see various problems and challenges, while others are more reactive and appreciate more definitive information about the world. Is it possible that these two types of brains would set people up to be more or less open to complex theological questions? Perhaps neurotheology can help elucidate whether the brains of people from different denominations or different religions are more or less reactive to ambiguous information. Or perhaps it has more to do with the relative contributions of emotions and abstract thought that make a person's brain engage one type of religious belief over another.

Theology Basics

Of course, a special emphasis of neurotheology must be theology itself. As mentioned earlier, theology refers primarily to the analytical developments that arise from a foundational myth of a particular religion. The word "theology" is derived from the two Greek words, *qeos* (*theos*), meaning "God," and *logos*, meaning "word, discourse, or reason." We might then simply define theology as "words about God," or as H. Ray Dunning does, as "disciplined thinking about God."[3] In some sense, it also represents a methodology for interpreting the relationship between God and human beings. This methodology can take many forms but is usually based on a systematic approach (not to be confused with systematic theology) that may contain philosophical, religious, and spiritual ideas.

A number of other definitions have also been offered for theology. While many of these other definitions are not fully comprehensive, they are helpful in providing a better understanding of what theology is and how it works. The nineteenth-century British Methodist theologian W. B. Pope defined theology as "the science of God and of divine things."[4] This is an interesting definition because it suggests that theology is a science, and others have suggested similar definitions. Of course, theology as a science does not refer to a physical science, but more to the realm of social science. With the development of neurotheology, theology itself has the opportunity to add a more physical scientific component to its other characteristics.

A number of other definitions are here given that all include the idea of theology as a "science." What is important to note is that theology has often referred to Christian beliefs, but this certainly need not be the case. Any tradition that has a foundational myth that expounds its basic beliefs can produce a line of inquiry that seeks to better understand the basic beliefs and tries to help incorporate those beliefs into human thought and behavior. Judaism with the Torah and subsequent Midrash and Talmud commentaries clearly fits this description. The Princeton theologian and Presbyterian Charles Hodge sees theology as "the science of the fact of divine revelation, so far as those facts concern the nature of God and our relation to Him, as His creatures, as sinners, and as the subject of redemption."[5] Thus, according to Hodge, theology deals with incontrovertible facts, but from a neurotheological perspective, these facts must be considered beliefs. The Baptist theologian Augustus Hopkins Strong broadens the focus to all of creation when he defines theology as "the science of God and the relation between God and the universe."[6] To the extent that theology uses rational thought to explore beliefs and relate them to behaviors, one can see how neurotheology might help to tie these different elements together in new ways. Karl Barth defined theology as "the reflection on the Word of God, which is spoken by God in revelation, recorded in Scripture, and proclaimed in preaching."[7] In this way, theology is deeply tied to religion and the ways in which religious knowledge is believed to come about and subsequently expressed to the general population. In this sense, theology also has a practical application.

Jewish theology is an ancient concept going at least as far back as the time period of the Talmud (c. 100–600 CE), in which the sages sought to explain certain beliefs about God and the universe as well as the metaphysical and physical reality created by God. These beliefs, however, were not all set out in a rigorous independent work, but were given over in the form of stories and parables scattered throughout the Talmud and Midrash. Later, with the closure of the Talmud and rise of the Geonim (the leaders of the Jewish schools of learning and thought in the Babylonian towns of Sura and Pumbedita from 589 to 1038 CE, who rose in authority and power with the close of the Talmudic era and began to write their own responsa and rulings on issues), there was a first attempt to rigorously organize a Jewish theological treatise. Rav Saadiah Gaon was one of the first to write a full treatise on Jewish theology, in his *Book of Beliefs and Opinions*, in which he discussed the positions of the Greek and Islamic philosophers in relation to Judaism. After the time period of the Geonim, with the rise of the *Rishonim* (eleventh- to fifteenth-century rabbis and commentators on the Talmud and Geonim's works, who sought to elucidate and issue rulings on the Talmud where there was confusion or new situations not previously discussed), the number of works on Jewish thought and theology began to multiply significantly. Some of the more famous authors on many of these works are Maimonides, in his *Guide for the Perplexed*; Nachmanides (Rabbi Moshe ben Nachman, Catalan philosopher, Bible commentator, and mystic); Gersonides (a.k.a. Ralbag, Rabbi Levi son of Gershon, medieval French scholar and theologian); Rabbi Judah HaLevi (medieval Spanish poet, physician, and philosopher), in his book *The Kuzari*; Rashi (Rabbi Shlomo Yitzchaki, medieval French rabbinic scholar, whose commentary on the Bible and Talmud features prominently in many printed Hebrew texts); Rashi's grandson Rabbenu Tam (Joseph son of Meir of the *Tosafists*, commentators on the Talmud; French rabbi and theologian); and Chasidei Kreskas (medieval Spanish rabbi, theologian, and philosopher), to name a few.

These authors began to tackle a systematic interpretation of the prior stories and explanations about Judaism and God in light of the other scientific and theological work of their time. The goal was to create a framework for Jews and others to understand what Judaism had to say about topics in theology such as theodicy, free will, God's omnipotence and omnipresence, and the end of days. These works remain important even as new scientific and philosophical developments required continued redevelopment of some of their ideas along with a return to the initial sources the *Rishonim* utilized by the current *Acharonim* (the Jewish leaders and sages since the sixteenth century, many of whom we have cited and discussed throughout this book).

It is our hope that neurotheology might provide yet an added perspective to the overall understanding of Jewish thought, rituals, and laws. Neurotheology represents the latest in scientific developments bringing in the most recent use of neuroimaging and neuroscientific studies, qualitative research, medical and genetic research, anthropology, and psychology. Each of these developments might contribute, as we will describe, to the ongoing debate and analysis of Jewish thought.

One important point, especially with regard to neurotheology, is required here to differentiate theology from religious studies. Theology typically has the role of analyzing religious doctrine from within the religion itself. Thus, theology is typically considered to be under the auspices of a religious authority. This is clearly distinguished from "religious studies," which arose primarily in the academic setting and seeks to understand and analyze religions from a secular perspective. There are obvious benefits of both, depending on the goals of the particular study. Neurotheology can be considered a bit of a hybrid in this regard. On one hand, neurotheology can be approached from within a particular religion, as we are doing in the current book, in order to enhance overall theological knowledge of that tradition. However, neurotheology can also provide an important component to a secular, religious studies based approach to religions in general. Thus, neurotheology has the potential ability to work solely within a religious tradition, but also extend beyond any given tradition to observe religions in a secular manner.

Theological Methods and Neurotheology

There are a number of theological methods that we will need to consider with regard to both theology and neurotheology.

Theological methods have been categorized in a number of different ways, and any given approach to theology should generally be considered not to be competing with other approaches, but complementary. Biblical theology is the study of the contents of sacred scripture, systematically arranged, and arrived at through exegesis or interpretation. This approach mirrors that of many of the biblical commentators such as Rashi, Nachmanides, Ibn Ezra, and many modern rabbinic commentators. Through analysis of scripture, they attempt to determine what the *peshat*, or most clear meaning, is of the verses. Many scholars have contributed additional layers to the interpretation of the text, including legal and metaphysical interpretations, that have become part of the Jewish tradition, *masora*. However, these interpretations would not be part of this first theological component. Historical theology seeks to evaluate how various doctrines and theological ideas have arisen over time and the impact of cultural and historical influences. This provides a longitudinal perspective on how a given religion evolves over time to maintain its divine message. A similar approach to this would be the Jewish system of halacha and the analysis of the development of the law as it is applied to new situations and circumstances throughout the ages. Even if the principles and systems created remain the same, novel technologies such as electricity or computers, as well as modern circumstances such as instant communication and global travel, create new challenges to the traditional system that requires extrapolation from the given laws and precedent.

Systematic theology is based on a strong logical and deductive approach and is currently an important approach to theology. Systematic theology makes more use of philosophy, apologetics, and ethics than do other disciplines. Systematic theology in some ways incorporates both biblical and historical approaches to theology. Practical theology seeks to make religious knowledge applicable to everyday life and the ministry of the religion, similar to the halacha and Rabbinic weekly sermons. Process theology was originated by Alfred North Whitehead and was based on the notion that the world is dynamic and always in motion—always in process. Underlying this dynamic process is a permanent background of order that is mediated by God. Process theology also observes God's causality in the world as influence and persuasion rather than direct causal intervention. Whether Judaism would be able to incorporate process theology would likely depend on what denomination of Judaism was asked about it.

In our survey, one Conservative rabbi wrote that "God is a dynamic force for positive change and growth in the world," and a Reform rabbi wrote, "God as described at the burning bush *Eheyeh asher eheyeh* [lit. "I will be what I will be"]. A God who is continually transforming and adapting. Process theology speaks to me." One Reconstructionist rabbi wrote, "God is a process and a presence; God is beyond description in words." Two rabbis, previously Reform and Conservative respectively, each wrote also about process theology speaking to them as a description of God. This perspective would be a difficult one for an Orthodox rabbi to maintain, as it assumes a dynamic, changing God, which would undermine the authority of tradition and the more literal and stringent view of the Bible and Talmud.

Regardless of the approach one takes toward theological method, it is important to observe the cognitive and emotional elements involved. For example, all theologies are based on a primary faith system. Thus, belief is first and foremost at the foundation of any theology. However, the analytical component can have an emphasis on thought, feelings, experiences, practical behaviors, or other elements that eventually can be related back to various functions of the human mind and brain. Thus, any method of theology can theoretically be evaluated from a neurotheological perspective in addition to its more traditional approach.

For example, to initiate any theological method, one must first read the sacred text, such as the Torah. Reading requires the ability to understand syntax, grammar, meaning, and also various inflections and symbolism. Once these are accomplished, there is a second level of analysis that most likely incorporates emotions and cognitions as ways of further analyzing what was read. This second level of analysis might also utilize memory and personal experience. Some individuals might be drawn to the practical implications of theology and hence focus on how the sacred text has meaning in everyday life behaviors. Another individual might have a brain with more analytical predispositions and prefer process theology as a way of employing causality and other cognitive elements into theology.

With regard to theology in Judaism, it will be of great importance to separate those theologies that place primary emphasis on thought and intellectual comprehension compared to those that focus on emotion compared to those that focus on spiritual experiences and connection to God. We will explore how the various brain processes might lead to specific theological concepts later in this chapter.

Theological Concepts in Neurotheology

In addition to general theological methods, there are a number of theological topics or questions that are important to consider and determine their relevance to neurotheology. The following questions are at the center of much theological inquiry:

1. Can the existence of God be proved?
2. What is the nature of God?
3. What is the nature of good and evil?
4. What is the nature of spiritual revelation?

Theology attempts to make logical arguments that address these and other issues related to God and God's relationship to the world. To some extent, proof of the existence of God is not completely necessary from a theological perspective, since the foundational element of religion, namely that there is a God, is taken on faith. A number of "arguments" regarding proof of God's existence have been offered throughout the history of theological development, which include the cosmological argument—that since the world exists and since the world cannot come from nowhere, there must be an original or first cause, which is God; the teleological argument—that there is a purpose and intelligent design in the universe, which must arise from God; the moral argument—that God is what must have provided human beings with their sense of morality; and the ontological argument—that "if [we] could conceive of a Perfect God who does not exist, then [we] could conceive of something greater than God, which is impossible. Therefore God exists."[8] What is interesting about each of these, and the many other arguments put forth to prove or at least support God's existence, is that they each depend on various functions of the human brain. If our brain did not perceive causality in the world, then we would not conceive of a cosmological argument. If the brain could not perceive meaning in things, then we would not conceive of a teleological argument. Thus, the sense, or lack of sense, that these arguments provide is highly dependent on the brain functions that conceive of them.

Neurotheology may play a prominent role in the discussion regarding the existence of God. The reason for this is that if God does exist, then neurotheology can provide information about how human beings relate to God. The beliefs, behaviors, and rituals associated with any religion all can be correlated with various brain processes. In this way, we can help establish what emotions and cognitions are utilized in a given religion. This might lead to understanding how some religions are more emotionally based, while others are more cognitively based. Of course, all religions require the full repertoire of human brain function. We have frequently argued that given the richness and complexity of religions, it is not one part of the brain, but virtually every part of the brain, that becomes involved with the elaboration of religions and religious beliefs, practices, and experiences.

There is also the possibility that neurotheological studies might determine which ways of relating to God are "better" than others. Although fascinating, this is a potentially dangerous proposition, since the implication is that various religious groups could utilize such information to proselytize, criticize, oppress, or attack other groups. However, it would seem unlikely that any neurophysiological study could provide the kind of evidence that would support which beliefs are more accurate, but results from such studies might help individuals determine what works best for them. There is probably too much variability in normal human brain function to clearly differentiate the effectiveness and accuracy of certain beliefs or practices. Nonetheless, neurotheology has the potential to be thrust into the middle of many different kinds of conflicts, and anyone seeking to be a scholar in this field should maintain a very cautious position regarding results and interpretations of such studies. Additionally, neurotheology might be able to help describe the benefits of specific religious practices and ideas for religious groups who have not found satisfactory answers to questions regarding their practices previously.

If God can be proved not to exist, then the only other alternative is that God is a manifestation of the human mind. If the human brain creates the concept of God and its elaboration through religion, then neurotheology could provide critical information as to how this might happen. A neurotheological perspective would also have to take into consideration the possibility of religion being a cultural construct as well as other factors such as philosophy, nationalism, racism, myth, and ritual. However, neurotheology must also constantly remind scholars of the limitations imposed on human beings in discerning reality. For example, a brain scan that demonstrates changes in certain structures when a nun experiences being in God's presence could indicate that the brain changes created the experience or that the brain was responding to the actual experience. The scan itself should not be construed as proving the existence or nonexistence of God in this context. Neurotheology should continue to encourage research of brain function during religious experience and seek to determine if a study design might be possible that could more specifically address the proof-of-God question. The methodological challenges of such a study are clearly very substantial, but it is important to stress the need for careful planning and even more careful interpretation of results.

Another interesting theological concept is the determination of the attributes of God. God's attributes are sometimes divided into those that cannot be shared with human beings (incommunicable attributes) and those that can be shared (communicable attributes). Here there is a clear distinction between what the human brain can and cannot perceive. Incommunicable attributes are those related to God being considered omnipotent, eternal, infinite, omniscient, and omnipresent. These are things that the human brain can help us contemplate, but we cannot directly experience these attributes. On the other hand, communicable attributes are related to those things that human beings can experience directly such as mercy, justice, wrath, and love. After all, everyone feels these emotions at one time or another. And through personal experience, people are able to understand God's attributes. However, we must consider how accurately any human thought or experience is at capturing what God might be.

A related question is whether these attributes can be ascribed to God in any real way at all or if they are simply the only way for human beings to relate to God through a construct that we can comprehend. Often, then, attributes are seen as a way for human beings to strive for their own moral perfection, through imitation.

> It is impossible to do *hesed* when its essence is the absence of moral necessity. The commandment is worded in an entirely different fashion: "You shall walk in His ways" (Deut. 28:9), and just as He does *hesed* so shall you do *hesed*. In other words, the imperative refers to *imitatio Dei*: we are not obligated to do *hesed*; our obligation focuses on the imitation of God, one of whose attributes is *hesed*.[9]

The command of *imitatio Dei* has a central place in the moral fabric of Judaism. The book *Tomer Devorah* (*The Palm Tree of*

Deborah) is a kabbalistic work by Rabbi Moses Cordovero on the thirteen attributes of God and how to emulate God via *imitatio Dei*. He states right away that "it is proper for man to emulate his Creator, for then he will attain the secret of the Supernal Form in both image (*tzelem*) and likeness (*demut*)" (chap. 1). The goal for the kabbalist in emulating God is not only to be an ethical being, but to achieve spiritual heights and a metaphysical/mystical connection to God otherwise unattainable.

Imitation as a cognitive construct is an interesting field of study in cognitive neuroscience. The human brain is able to imitate other's behaviors. In fact, there has been much attention paid to "mirror neurons," which actually reflect what another person is doing. Thus, if we see someone smile, our mirror neurons will make us smile, at least internally. Although this is an overly simplistic example, the point is that there are parts of our brain, including specific neurons, that help us to reflect internally what we see someone else doing. In that way, our brain mimics what it sees. We have the opportunity to allow or reject the actual act of imitation, but our brain appears primed to do this. Research also suggests that if we witness a person being kind or smiling, then our brain responds with similar attitudes and behaviors—we try to help someone or we smile. And if we see someone become angry, it is more likely to make us angry. The notion of trying to emulate an ideal being such as God would likely take advantage of these social imitation areas of the brain. Furthermore, the more we use any given neural connections, the stronger they become. If we practice being compassionate, then we strengthen the neurons that support compassionate thoughts and behaviors, and we are more likely to be compassionate in the future. These physiological processes seem to support the idea of *imitatio Dei* as a fundamental way in which our brain works.

Reflecting on God's attributes either through prayer or meditation is an important way of understanding God and the relationship between God and human beings. For example, there are traditionally thirteen attributes of God used in the prayers leading up to the High Holy Day prayers derived from Exodus 34:6–7 ("Adonai, Adonai, the compassionate and gracious God, slow to anger, abounding in love and faithfulness, showing kindness to thousands, and forgiving wickedness, rebellion, and sin, and cleansing"). Celebrating these attributes helps the person understand God and why the Holy Days are so important. These attributes also enable the person to embrace the importance of forgiveness and repentance. From a brain perspective, understanding how a person forgives or repents might provide a unique perspective on important human thoughts and feelings. Are there more effective ways of repenting or forgiving? Is it better to forgive another person by recognizing that he or she is human and fallible, or is it better to forgive the person knowing that God might decide the best course of action for dealing with another person's mistakes? Does one approach to forgiveness or another lead a person to feel more at peace with the forgiveness process? And could we help to measure this in some way?

The kabbalistic approach to the attributes of God also uses verses from the prophet Micah:

> Who is a God like You, who pardons sin and forgives the transgression of the remnant of his inheritance? You do not stay angry forever but delight to show mercy. You will again have compassion on us; you will tread our sins underfoot and hurl all our iniquities into the depths of the sea. You will be faithful to Jacob, and show love to Abraham, as You pledged on oath to our ancestors in days long ago. (Micah 7:18–20)

These verses, recited in Orthodox synagogues on the afternoon of Yom Kippur, are meant to indicate the thirteen "higher" levels of mercy that God can show and that should be emulated by people as best as possible. Again, we see the importance of imitating the ideal and morals of God. Rabbi Cordovero first discussed how these higher attributes mean one must be tolerant, particularly of others who do not follow their path to higher worship, and should be willing to work to undo others' sins, since that is similar to God forgiving sins. The implication is that one should love and feel pain with every other Jew, and ideally with all of God's creations. One should remember the good another has done instead of the bad, strive to have faith in humanity, and go beyond the letter of the law to help people.

Cordovero also discusses how people should model their behavior after the ten *sefirot*, or emanations in the Kabbalah of the light of God used to create the world through many metaphysical descents until the physical world (chap. 2). Remember that these ten *sefirot* are *Keter* (God's "crown"), *Chochmah* (Wisdom), *Binah* (Understanding), *Chesed* (Kindness), *Gevurah* (Strength) or *Tiferet* (Glory, often integrating *Chesed* and *Gevurah*), *Netzach* (Victory), *Hod* (Splendor), *Yesod* (Foundation, often integrating *Netzach* and *Hod*), and *Malchut* (Kingship). These ten *sefirot* are systematically utilized repeatedly throughout kabbalistic literature as building blocks for both theological and ethical principles of how people should act in the world. Cordovero goes through all ten and explains how they should determine how one should act and then concludes in his final chapter, "He must act at the right time—that is, by knowing which *sefirah* dominates at a particular time, he can bind himself to it and carry out the adjustment associated with the ruling attribute" (chap. 10).

This interpretation implies that the attributes used by a person are based on a metaphysical or kabbalistic understanding of what *sefirot* are in control at specific points in a person's day and life. This attitude runs in clear distinction from the more intellectual approach to the concept of *imitatio Dei*, whereby God's commands and principles are meant to be simply ethical lampposts for people to emulate. The rational perspective argues that emulating God is not based on a mystical reality, but is done because it is what is best for humanity and the world. According to this theory, the Jewish legal system and the ethical system are not to be seen as entirely separate. There is in fact much debate between Jewish theologians about there being an ethic independent of the halacha. Rabbi Aharon Lichtenstein is famously quoted as the defender of the position that the halacha is equivalent at its higher levels to an ethical imperative, as he states:

> Halakhic commitment orients a Jew's whole being around his relation to God. It is not content with the realization of a number of specific goals but demands personal dedication—and not only dedication, but consecration. To the achievement of this end, supralegal conduct is indispensable. Integration of the whole self within a halakhic framework becomes substantive rather than semantic insofar as it is reflected in the full range of personal activity. Reciprocally, however, that conduct is itself stimulated by fundamental halakhic commitment.[10]

According to this approach, when someone is being ethical, the person is following halacha by definition, and supralegal conduct is also somehow subsumed in the larger legal ethos. However, Rabbi Yehuda Amital is often quoted as saying, "According to this point of view [that the halacha is what is ethical], which zealously tries to defend the honor of the Torah, there is no connection between God, Creator of man, and God, Giver of the Torah, as if that which God implanted in man's heart does not belong to God,"[11] and this is a position he is unwilling to accept.

The nature of good and evil, particularly in relation to God, has great importance for theology because it helps to establish a sense of morals, but also tries to explain the many good and bad things that happen in the world. One of the pressing concerns most individuals have is why bad things happen to apparently good people. We might question our faith if we feel that in spite of doing everything we are supposed to, bad things continue to happen. We might feel that our religious beliefs are not helping and thus ultimately reject them. In the Bible, the story of Job plays a pivotal place in considering this issue. Theology itself strives to address such questions, but the most common answer is that the mind of human beings cannot grasp the full divine plan. Human beings can have a basic moral understanding of how to act in the world but are limited in their ability to determine what is ultimately right and wrong. Of course, such a position is often criticized as a convenient inability to understand God. How can people understand certain aspects of God and not others? Since the brain is continually trying to understand the world, it makes sense that we should try to understand things, and when we do not, we can either admit our lack of knowledge or project that onto God by stating that God knows what we don't. This can also fall into a "God of the gaps" concept, which is usually difficult to maintain as human knowledge continually increases. We used to think thunder was produced by God, but now we understand that thunder occurs as the result of alterations in electrical discharges in storm clouds, creating lightning, which compresses the air and produces thunder. As things in the physical universe become answerable through science, the God of the gaps concept loses steam. However, one might argue that certain things like eternity or omniscience are things we will never fully understand. But we do not yet know the limit of what our brain can know or comprehend. In addition, kabbalists often reflect on understanding issues such as why bad things happen to good people from a mystical perspective. Thus, there may be a variety of ways of comprehending how good and evil interact in the world and affect human beings.

Spiritual revelation in the context of neurotheology is akin to the ability of the human brain to receive God in some sort of spiritual manner. Revelation addresses the issue of how human beings come to have any understanding that God exists and that God wants us to do certain things. Revelation is more religiously, rather than neurologically, oriented. However, there is much that can be considered from a neurotheological perspective. For example, how are human beings limited in what can be revealed? Often the explanation given for many of the more esoteric revelations in scripture by the prophets is that the human mind is limited in its scope and understanding. Moses is said to have seen the closest "truth" of God's existence "through a clear mirror, while all other prophets saw through a cloudy mirror" (Talmud,

Yevamot 49b), and still it says in Exodus 33:20, "You cannot see My face, for no one may see Me and live." And later 33:23: "Then I will remove My hand and you will see My back; but My face cannot be seen." Even the greatest prophets cannot breach the final limits of human understanding. If human beings can have access only to communicable aspects of God, then there may be specific limitations that are placed on the ability to perceive and understand God. On the other hand, perhaps there are ways to connect parts of the brain that previously were not connected at all, through prayer and meditation, thus giving rise to new experiences and perceptions that may be an aspect of what is understood to be prophecy or divine inspiration.

In Judaism, revelation is considered in two general ways.[12] The first is what in rabbinical language is called *gilluy shechinah*, a manifestation of God by some wondrous act. These acts, such as the parting of the Red Sea or the presentation of the commandments on Sinai, are so overwhelming that they instill in human beings a profound sense of awe in which they have an intense experience of God's presence. In general, the *gilluy shechinah* was also often associated with an actual cloud symbolizing God's presence dwelling in a certain place. It was most often used with the symbolic resting of God's presence on the Tabernacle in the desert and on the *Beit HaMikdash*, the Temple in Jerusalem. The second major form of revelation is via oracular words, signs, or laws. After the five books of the Torah, there is a clear drop in revelations related to *gilluy shechinah* and a significant increase in revelation via personal prophecy. The notion of revelation is somewhat different in Judaism compared to Christianity in that manifestations of God's presence through wondrous acts and miracles is far less typical in the literature after the First Temple period. However, in the Bible, there are many examples of revelations that occur on a one-to-one basis with God, often involving a vision such as the burning bush with Moses (See Exodus Ch. 3) or the vision had by Samuel in his dream (Samuel I: Chapter 3). In fact, God frequently comes to prophets through dream states.

From a neurotheological perspective, the above-described types of revelation in Judaism would have different effects on the brain. Experiences are perceived through our senses, such as vision. Thus, something very remarkable like the parting of the Red Sea would be experienced by the visual system of the brain. A question to be posed is why certain visions have such a significant impact. There are many wondrous experiences, such as witnessing a total solar eclipse. Such an experience could have been attributed to God, although we now understand the celestial effects of the moon passing in front of the sun (an action that can still be attributed to God, but can also be explained by physics). Our brain can be easily fooled visually, as anyone attending a proficient magician can attest. So how does our brain know what things to believe when we see them and what things not to believe? The data is not clear, since our brain can respond in a similar way when we actually see something versus when we imagine something. It would be difficult if not impossible to use brain imaging to prove which experiences of the world are real and which are not. Thus, it would be difficult to use brain imaging to ascertain whether Moses actually saw a burning bush or imagined it.

Similarly, when a person has a dream, no matter how real it feels, when we awake, we recognize that the dream represents an inferior version of reality. However, there is no clear neurophysiological basis for making such a determination. Thus, it is difficult to know, when an individual perceives a revelation from God, how that is determined by the individual and how we, as outsiders who have not had the same experience, should regard that revelation experience. One would hope that cross-referencing experiences with others provides a better way of determining whether a revelation has occurred or not.

At Sinai, revelation occurred to an entire group of people, a point often used by many arguing the veracity of Judaism. Of course, there are examples of "mass hysteria," and if the brain of each person can resonate or mirror another's, then it is entirely possible that a revelation experience to a group of people could still reflect a brain experience. Ultimately, regardless of the veracity of an experience, the brain must make sense of the revelation in terms of an understanding about God or the universe and about how to utilize the information that is revealed. In other words, the person must come to understand what new behaviors to perform or beliefs to hold. Thus, after Sinai, the Jews understood that they needed to follow the commandments and obey the Sabbath. These are processes that ultimately arise from the brain's abstract thought, memory, and behavior areas.

An important part of revelation as conceived of in Judaism is that it is not the product of human thought, but rather human beings are the "instrument upon which a superhuman force exerts its power."[13] In discussing the revelation at Sinai, it is important to note that the tradition recognizes the superhuman feat of that revelation and its impact on the individuals who experienced it.

> And Rabbi Yehoshua ben Levi said: From each and every utterance that emerged from the mouth of the Holy One, blessed be God, the souls of the Jewish people left their bodies, as it is stated: "My soul departed when he spoke" (Song of Songs 5:6). And since their souls left their bodies from the first utterance, how did they receive the second utterance? Rather, God rained the dew upon them that, in the future, will revive the dead, and God revived them, as it is stated: "You, God, poured down a bountiful rain; when Your inheritance was weary You sustained it" (Psalm 68:10). And Rabbi Yehoshua ben Levi said: "With each and every utterance that emerged from the mouth of the Holy One, blessed be God, the Jewish people retreated in fear twelve *mil*, and the ministering angels walked them back toward the mountain. (Talmud, *Shabbat* 88b).

If we were to invoke neurotheology here, we might argue that the parts of the brain that perceive the body were somehow disengaged from that body. The experience sounds much like an out-of-body experience that is typically described with respect to religious experiences. The transcending of the body is part of how we identify an experience as spiritual. It is interesting that once the people left their bodies, there was a recognition that they could no longer perceive God through the normal senses. On one hand, such an experience could elicit fear, but it can also feel deeply powerful and spiritual. The tradition has it that after experiencing this level of revelation for the first two commandments, the people opted for Moses to deliver the rest (see Talmud, *Makkot* 23b, and the language change in the verses of the Ten Commandments between the first two and the rest). Thus, it would appear that there are limits to the human mind of an "average" person that indicate that some are better able to receive the spiritual experience of divine communication than others. And if such an experience altered the cognitive processes of the brain, those people who received it and could tolerate it might experience a profound sense of wisdom and understanding.

In addition, it might be argued that the revelation also is experienced by people as a continuum through their actions. This is similar to the more modern argument by theologian Abraham Heschel—that leading a life of holiness is the way to apprehend God. Thus, this additional mode of connecting with God through our acts could still be consistent with the manner in which the brain helps us to enact our behaviors, ultimately resulting in the emotional experience of revelation and "radical amazement" he talks about. This may also be similar to the general understanding of Maimonides (See *Guide for the Perplexed* Part 2:Chapter 36), that prophecy is a gift from God but also requires intense preparation and constant work to be eligible to receive said prophecy.

Further, "the more of itself the divine mind imparts to the susceptible human mind, the higher will be the degree of the revealed truth."[14] Might there be a way of observing such intense changes in the human mind and brain? It is an intriguing possibility. Still further, Jews have argued that they have produced a long line of prophets and thus might actually be "endowed with peculiar religious powers that fitted it for the divine revelation."

Whether or not that is truly the case, the ability to reflect on the unique characteristics of the Jewish brain might be helpful in trying to better understand this. Are there certain brain processes or certain neurotransmitters that might make a person more or less likely to receive revelation? The interesting work by Dr. Dean Hamer suggests such a possibility. People with genes that express a specific receptor in the brain that regulates serotonin and dopamine appear to have higher levels of self-transcendence. This might also help explain that in descriptions of prophecy and how one achieves it, it was a common occurrence for prophets of old to surround themselves with peaceful environments and music, increasing their feelings of happiness and contentedness prior to being able to receive prophecy. A dopaminergic or serotonergic increase would certainly help facilitate the spiritual process. In fact, drugs such as LSD that produce intense "spiritual-like" experiences are known to work by greatly stimulating the serotonin system. Thus, it might be argued that there are certain brain types that are more able to receive something from God or at least perceive that they are receiving something. On the other hand, the atheist might argue that these brain changes simply make it easier for a person's brain to "make up" some divine experience even though it is not real. Neurotheology offers a perspective that might help delineate such differences by exploring both the brain and religious experience in an integrated manner.

A final point about revelation is that in Judaism, the Torah is considered a powerful source, since it is regarded to be completely written by God via Moses. The rabbinic phrase *Torah min hashamayim* (lit. "the Torah is from heaven") implies that the Torah is a God-given gift from heaven. As the *Jewish Encyclopedia* states, "Revelation, in the sense of a manifestation of the will of the Deity, is identical with '*debar Yhwh* .'"[15] This means that revelation is viewed as a direct word from the Deity, and the Torah is seen as the physical manifestation of that word. However, receiving revelation through reading and understanding the Torah is a lifelong process that involves cognitive and emotional processes to take in the teachings and incorporate them into a person's life. Of course, this is done to different extents in Orthodox, Conservative, and Reform Judaism.

It should also be clearly stated that whatever limitations the human brain places on our ability to conceive or receive God, this has no impact on whatever is the true nature of reality. If the human brain could not perceive causality in the world, then God could not be understood as the first cause. The inability to understand God as the first cause has no bearing on whether God actually is the first cause. Furthermore, one has to be very careful interpreting neurotheology as being able to comment on whether God does exist and whether the brain creates God or God creates the brain. This is an extremely complex question that often is approached with substantial biases from both believers and nonbelievers. The most appropriate neurotheological perspective is to carefully evaluate all ways of understanding God, including an absence of God, in order to best determine what the brain can know about reality (see chapter 13).

However, the very notion that theology pertains more to the human understanding of God is commensurate with the goals of neurotheology. Neurotheology necessarily must explore how the brain can think, feel, and perceive the concept (or the actual reality) of God. More specific theological analyses can be developed depending on the focus of a particular approach. What is important in terms of neurotheology is to observe how the various developments in theology pertain to human perceptions, feelings, cognitions, and behaviors. Any time the focus turns to one of these aspects of theology, a neuropsychological perspective can be added that deepens the understanding of these concepts.

Brain Processes and Theology

Up to now, we have explored Jewish theology while integrating certain aspects of brain function. However, we can now turn to an analysis of Jewish theology constructs from the "inside out" by starting with brain functions and trying to determine what theological ideas might arise. In particular, theological concepts can now be interpreted in relation to specific brain functions that might be manifesting themselves either singly or in combination with others. On an "everyday" basis, many of the cognitive and emotional processes of the brain are brought to bear on all daily experiences. Therefore, human beings assess various relationships, problems, and issues utilizing the various cognitive functions of the brain. Furthermore, certain functions may be more appropriate for certain situations. Causal thinking might be necessary to solve a particular problem with regard to fixing a car, while emotional processing might be necessary to help deal with an interpersonal relationship.

Each of the many brain processes may also have a role with regard to theological principles and concepts. Ideas related to causal thinking, holistic thinking, and emotions all can have a different impact on theological development. Theologian Paul Tillich stated that religion is at its essence involved with that which human beings consider as "ultimate concern." Thus, it certainly seems appropriate that theology should involve some power being that is considered to be the ultimate cause of the universe (if derived from the causal process of the brain) or the ultimate unifying force of the universe (if derived from the holistic process of the brain). However, it could be argued from a neurotheological perspective that the driving force behind this desire to seek out ultimate things might be based partly on the brain striving to understand the ultimate questions of the universe and partly on an experience representing this ultimate level of the universe.

Such an experience may be associated with the functioning of various cognitive processes on reality. In fact, it might be possible to consider major theological or philosophical principles from the perspective of various brain processes acting on reality in specific ways. Several possible neuropsychological mechanisms might be postulated that could have a direct impact on theological conceptualization.

For example, it may be possible that the total experience of reality is "filtered" through one particular brain process. Whether it is possible to physiologically view all information through one brain process has not been fully established, but it is certainly possible for a significant amount of that information to be handled by one particular brain process such that the individual has the experience that everything is perceived in that manner.

The filtering of all information regarding reality through one brain process might be referred to as the "total" functioning of that process. In these instances, one particular brain process, and hence one approach to the experience of reality, supersedes all other functions. The person becomes convinced that the universe is fully understandable from the perspective of that particular brain process. In the case of the causal process of the brain, its total functioning would be experienced as having the entire world analyzed according to various causes and effects. This has formed the basis of much scientific and philosophical analysis, from Aristotle's four causes to Newtonian physics. From a theological perspective, the notion that would arise would be that of God causing all things to occur. The idea is that everything can be considered to have been caused by something and causes something else to occur. And when things do not appear to follow a clear causal pattern, those things are sometimes perceived as unreal.

We have previously considered how another type of brain function, namely deafferentation, is associated with the blocking of input into various brain structures. In the case of deafferentation, we have argued that the deafferented structure underlying a particular brain process continues to perform that process although there is no input upon which to work. The result is an experience in which the process itself represents a fundamental law or nature of the universe. We refer to this experience as the "absolute" functioning of a particular brain process. It is absolute because it represents the process in and of itself without necessarily operating on any particular input. It is the pure or absolute function of some cognitive process. In the case of the causal process of the brain, it is not so much that everything can be understood through the idea of cause and effect; rather, causality itself is perceived as the fundamental property of the universe. From a theological perspective, the result would be the notion that God is the fundamental cause of all things—the ground cause of being.

It should be mentioned that the "total" and "absolute" functioning of any cognitive process also may occur together. This would result in all things being filtered through some fundamental principle of reality. Both of these types of functioning of specific brain functions may be more of a theoretical construct, although there is sufficient experiential as well as growing neuroscientific evidence that such functioning might occur. Detailed analysis of how certain critical brain processes may function in an absolute manner and what such a functioning results in as far as the experience itself will be the focus of the rest of this chapter.

How such an absolute functioning of a cognitive process might occur can depend on a number of factors similar to those described for the spiritual experiences. An example of how such a sequence of events might occur is the following: The philosopher or theologian thinks very intensely in a particular way. Perhaps he or she is thinking, almost meditating, about how things are caused. The intensity of this use of the causal process of the brain eventually may produce total deafferentation. Suddenly, our thinker experiences a profound sense of all of reality as cause and effect. This is not yet a philosophical concept. It is infinitely more powerful than a concept. It is the profound sense that one has had a glimpse into the ultimate and that, in this case, it is cause and effect. After this philosopher's or theologian's flash of insight (when he or she recovers from the total deafferentation of the causal function, in this case), a philosophical or theological concept is derived from the experience. The philosopher or theologian then goes about constructing a logical system in the firm certainty that he or she has fundamentally comprehended what is "real." Such an experience can theoretically happen with any cognitive process, generating diverse ultimate realities and diverse philosophies and theologies.

One can see such an experience arising from deep study of the Jewish tradition. Often, Jewish study of halacha or exegesis requires intense concentration and focus. Some students have reported at times a profound experience in which they simply "get" the concept being taught or understand what question the commentator is asking. This may also invoke a refocusing of attention as the person deafferents all of the other connections and solely focuses on the intellectual inquiry. This may also be the explanation for how much of the Chabad philosophy as well as the *Nefesh HaChaim* (4:7) written by Rabbi Chaim of Volozhin (a student of Kabbalah but strong opponent of the Hasidim) praise the study of Torah as a way to achieve *devekut*, connection with God. If one achieves this level of focus, one might be able to feel the Divine Presence while focusing so intently on one's studies.

If theology is a rational deduction from a foundational myth, neurotheology can offer a great deal about how this rational deduction arises. Rather than selecting specific theological concepts and describing them from a neurophysiological perspective, we are going to examine several specific brain processes to determine how theological concepts in general might be derived. Further, we will see how the derivation of various theological concepts might arise from either a total or an absolute functioning of these brain processes. It is important to emphasize that the use of the term "cognitive process" is not specifically biological, although there are certainly biological correlates, as will be described below. The terms we will be using are based on how the brain works and also how various philosophical and theological principles have been described. Thus, just like the term "neurotheology," the terms "total" and "absolute" functioning refer to an integration of biological and mental processes. In terms of overall theological constructs, there likely needs to be some degree of absolute functioning of a cognitive process to generate the profound sense of the underlying nature of reality and consequent ideas about the ultimate. However, many "lesser" theological derivations may require lesser degrees of deafferentation (i.e., partial) of a given cognitive process. Lesser theological concepts may also result from the combination of the function of two or more cognitive processes. Either way, these cognitive processes must eventually be brought to bear on the foundational myth for theology to arise. With this basic overview, several specific brain processes now can be considered.

The Holistic Process

The holistic process, as with all processes of the brain, can be conceived of as being experienced in a "total" and "absolute" manner. In general, the holistic process of the brain most certainly is associated with the experience of the oneness of God—the fundamental idea of Judaism. In the brain, this process likely resides in the upper part of the parietal lobe and, by its function, forces the theologian to account for God's omnipresence, omniscience, and ability to bind and maintain the entire universe. Thus, the total working of the holistic function necessitates, at least, considering the expansion of any foundational myth to apply to all of reality, including other people, other cultures, other animals, and even other planets and galaxies. In fact, as human knowledge of the extent of the universe has evolved, the notion of God has evolved to incorporate this expanding sense of the totality of the universe. The holistic function requires that however far we expand our reach into the universe, God must be there. No matter how small a quantum particle might be, God must be there as well.

The human ability to comprehend such an infinite God with our brain is necessarily limited in our cognitive understanding of infinity. We can easily make the statement that "God is infinite," but we cannot cognitively comprehend what this means. The religious literature of all traditions acknowledges that God cannot be described cognitively. Even Maimonides, the ultimate Jewish rationalist, states that God can only be described in negatives—what God is not. Perhaps only through mystical experiences has any person even tried to describe what the perception of an infinite God might be like. However, the true mystic will maintain that it is impossible to humanly experience this state (especially since in these states, there theoretically is no discrete existence that allows for our human experience). Even though we may attain a mystical state through meditation or prayer, the experience is so ineffable as to defy any cognitive understanding. For the theologian, this mystical notion of God must be incorporated and maintained within the foundational myth if the foundational myth is to be experienced as valid.

In Judaism, the unity of God has been a basic tenet of almost all components of the faith since their inception. The only group that actually differs from this holistic perspective is one group of kabbalists, who took the ideas of the *sefirot*, God's emanations that are divided into ten parts with unique traits and abilities of God, and said that they are not simply *or ein sof*, "the light of the infinite," but a piece of the infinite as well. This perspective has traditionally been rejected by almost all Jews as against the basic tenet of the religion, presumably in line with the neurotheological holistic principle.[16]

The absolute function of the holistic process leads to a different perspective on the unity of God, but a perspective that is very much the central tenet of Judaism. Absolute functioning of the holistic process would lead to the perception not of God as one, but of the oneness of God as the fundamental starting point of the universe. All things in the universe are essentially derived from the oneness of God. Put another way, God's oneness is the essence of the universe. In Jewish tradition, this would be best expressed by the *Tanya* in the section titled "*Sha'ar Hayichud*" (The Gate of Oneness), which states that God is a unique oneness that fills the world, if such a thing were possible. *Ein od milvado*, "there is none beside [God]" (Deuteronomy 4:35), can be interpreted literally to mean that God encompasses everything and everything is God in some sense. This can be interpreted in a literal sense (like some kabbalists have, with much pushback from the larger religious community), or it can be metaphysical (like many Hasidic interpretations assume) as well as largely philosophical. Maimonides describes this holistic feeling to be about truth, that "there is no true being beside God, like God" (Maimonides, *Mishneh Torah*, Laws of Foundations of the Torah 1:4).

There is one other important point to be made regarding the holistic process and our neurophysiological approach to mystical and theological concepts. It is interesting to note that many religions seem to exclude the possibility of other religions. One may wonder why this should be the case. One might consider the argument that if God is truly infinite, then God should have infinite manifestations. Why, then, should any particular version of God be set completely apart and exclusive of any other version? While it may be more evident in terms of religious ritual leading to the development of a group cohesiveness that excludes others not in the group, the question remains as to whether religious ideologies should be exclusive at all levels of religious experience. A state of absolute unity, in which all things are one, cannot have exclusivity because of its infinite and undifferentiated nature.

The question then is, can slightly lesser unitary states be exclusive? A neuropsychological analysis of this question would suggest that the highest unitary states short of absolute unity can, in fact, be exclusionary. When one considers the notion of the absolute functioning of the holistic process, this would theoretically be associated with a total deafferentation of the parietal lobe regions supporting this process and the concomitant experience of spacelessness and wholeness. If an individual does not achieve absolute functioning of the holistic process, but comes very close, then one would expect that there is incomplete deafferentation of the parietal lobe even though the person may feel totally absorbed into the given idea or focus (e.g., God, Christ). This is a state of total absorption into that object, but it is not a state of universal oneness. Thus, the entire universe is perceived to be that object to the exclusion of all other things, since anything other than that object must be a part of that object or must not exist in reality. If something were to exist in reality outside of the object of focus, this would present an irreconcilable paradox. One resolution of that paradox is that the abnormal object is really part of the object of focus even though it does not seem so. Therefore, any notion of God, Christ, Brahman, or Allah that results in a total absorption into that object necessarily excludes all other interpretations.

However, if the person's brain were able to go one step further to the absolute functioning of the holistic process, then there can be no exclusivity and all things must be considered to be inclusive. Certainly, the problem of exclusivity is prevalent throughout theologies. All religions, including Judaism, must somehow come to terms with the fact that other religions exist. This neuropsychological approach may help show a method by which the problem can be resolved or at least explained. This perspective, in fact, perfectly parallels that of Rav Kook in his work *Orot HaKodesh* (3rd *Maamar HaAchdul Ha'kolelet*—3rd set: 20-*HaHitachdut Ha'Elyonah*):

> The more the world develops and becomes whole, the more its parts are united and its organic quality becomes more apparent. The highest unification is the unification of the human's intellect and will with all existence in general and in its particulars. The unification with the Divine Matter (*haInyan haElohi*) makes this action complete, and there is no wonder that the will of the righteous, those who cleave to God, acts in the world and their prayer bears fruit. And so all humankind is destined [to unite], certainly as the prior special and senior quality of Israel will be revealed and spread to everything. "And God's glory will fill all the earth" (Psalms 2:19).

The implication is that by continuing to advance, humanity will eventually realize the holistic universalization of itself, in a literal-yet-metaphysical rendering of God's glory filling the entire world.

The Reductionist Process

The antithesis of the holistic process is the reductionist process, which is likely associated with the lower or inferior part of the parietal lobe. Total functioning of the reductionist process would theoretically lead an individual to observe all of reality by breaking things down to their component parts. This, in many ways, is the approach of science—to apply more and more precise measurements to establish the basis of every object in the universe. In terms of a scientific methodology, this reductionist approach clearly has its benefits for providing an accurate analysis of material reality. For science to occur, it is important that a small, clearly defined object is studied to eliminate the possibility of confounding factors. Thus, science often strives to isolate a given part of reality. This is where the reductionist function operates at its best—on a small piece of reality that is irreducible. However, recent scientific thought has realized the limitations of such an approach and has sought new avenues for understanding wholes rather than parts. The fields of cosmology and ecology have been particularly interested in the holistic understanding of the universe, thus invoking the holistic process as much or more than the reductionist process. Newer disciplines such as the study of chaos theory allow for the development of models that take into consideration the whole rather than the parts. More practically, holistic medicine has become popular as a total approach to health of the mind and body. Where this exploration of the parts and the wholes will end is difficult to determine. Certainly, science has recently made great strides in combining the holistic and reductionist functions. But science is just beginning to realize the importance of looking at the whole in addition to the parts.

The absolute functioning of the reductionist process on all of reality would theoretically lead to a primary intuition and existential sense that the universe at its essence is composed of the smallest objects. The Greek atomists would hold such a notion. This concept might develop into the theological or philosophical concept that the whole is composed of, and only of, the sum of the parts. When applied to the foundational myths of the monotheistic religions, the result is the notion that God is actually composed of the totality of all of the parts of the universe. This is akin to the concept of pantheism in which God is considered to be the universe.

It seems that the absolute function of the reductionist process would not lead to the notion of a transcendent God—a God that extends beyond the individual parts of the universe. The absolute working of the reductionist process necessarily contradicts the transcendent notion of God derived from the holistic process. Interestingly, however, it seems that the absolute functioning of the holistic process usually takes precedence, perhaps both biologically and philosophically, over the absolute functioning of any other cognitive process such that all of the parts previously perceived as being discrete are now considered to be one. Thus, in its absolute functioning, the holistic process actually combines both the reductionist and the holistic processes in a uniquely mystical experience.

The Quantitative Process

The quantitative process appears to be associated with function in the upper part of the temporal lobe and helps us to quantify various objects and ideas in the world. If this process is applied to the totality of objects, the result might be the notion that mathematics can be used to explain or understand all things. Similar to the reductionist process, the quantitative process both underlies and supports science and the scientific method. Science essentially is based on a mathematical description of the various objects and forces in the universe. Thus, Isaac Newton developed calculus to help us understand how gravity and other scientific processes work.

The absolute function of the quantitative process in the brain should lead us to the notion that mathematics is the essence of the universe. In other words, everything that exists is mathematical at its core. Pythagoras was one of the first to consider mathematics representing the true nature of the universe. In terms of philosophical and theological implications, the quantitative process appears to have heavily influenced the ideas of Spinoza, who often used mathematical concepts such as geometry to help explain the nature of God and the universe.[17] For thinkers such as Spinoza, and perhaps Albert Einstein as well, it might be said that God *is* mathematics. For Judaism, while God and the world may not be mathematics per se, the numerical correlates of existence are a large component of ancient kabbalistic texts. In Hebrew, the letters themselves are numbers and correlate to specific amounts. The fact that there are numerical values that have religious significance is encountered repeatedly in Hasidic literature and utilized to connect biblical verses and concepts together. Everything may relate back to not only the numerical value of the letter, but to the letter's shape and what it is supposed to represent. Kabbalistic names of God are often used as meditative mediums, with significant use of the phonetic spelling of each letter in the name. The concept of the gematria, where letters are used to equate to numbers and the amount interpreted to mean something spiritually, is often utilized in sermons and homiletic teachings on much of Jewish text.

The quantitative process provides a very important element to religions, and especially Judaism. If numbers are fundamental to the universe, then they should carry some meaning that is greater than simply the fact that they represent an amount of objects. In fact, we begin to imbue certain numbers as being important or magical. Throughout history Judaism has placed strong emphasis on certain numbers. Numbers abound in the Bible and lend their significance in terms of time, people, and places. There is a reason that there are ten commandments, ten plagues, and ten *sefirot*. Various numerologies in the folk practices of Christianity and Islam, as well as the gematria in Judaism, all bear witness to the powerful force of the quantitative process of the brain to impress the mind with the "mysticism" of numbers. For example, the number eighteen in Hebrew represents the symbol for life, and the number seven and multiples of seven are used often in Kabbalah and the Jewish calendar as a means of creating meaningful relationships between different spiritual religious practices (See Zohar 3:96b Parshat Emor).[18]

An interesting question from a neurotheological perspective is why certain numbers have meaning while others do not. Is it purely random or are some nu mbers more universally recognized as important? Simplistically speaking, it might make sense that 10 is a special number. We have 10 fingers and toes. And the use of base 10 in numbers is particularly easy to use for mathematical problems. The reason is that the number 10 exists in every base system and reflects the next addition of values in that system. Thus, base 8 goes from 1 to 7 and then "10" actually represents the number 8. One can see that using "10" to represent the number 10 makes the most sense. Other numbers are more difficult to explain. Why is 40 an important number? Certainly 40 is a round number and is a multiple of many other numbers, but so is 24. And 18 makes sense because it "accidentally" is represented by the letters *chet* and *yod*, which together spells *chai*, "life."

It seems that the quantitative process has played a major role in Judaism. However, when applied to God and broader theological concepts, the quantitative process appears to have its shortcomings. The quantitative process appears to be too narrowly focused to allow for the more holistic and, certainly, the more emotional aspects of God. This is evident in the thinking of some scholars who have been concerned that too much dependence on mathematics tends to lead one away from how God is immediately manifested in the universe. Thus, numbers seem to be useful religious symbols as a means for connecting to

God but in their own do not typically represent the true nature of God.

The Binary Process

We have already considered how the binary process appears to have a crucial role in the formation of mythic structure in all religions. This function is clearly important for theology as well. The opposites that are set by the binary process of the brain allow for the concepts of good and evil, justice and injustice, and humanity and God, among many more. Many of these polarities are encountered throughout the Bible. And Jewish thought and theology ultimately strive to solve the psychological and existential problems created by these opposites. Theology, then, must evaluate the myth structures and determine where the opposites are and how well the problems presented by these opposites are solved by the myth structure, specifically how the power of God brings together the problematic opposites. If the binary process were to function in an absolute manner, the result might be the notion that the universe itself is derived from oppositional forces. Perhaps the best-known example of such a notion of the universe is the yin and yang of Chinese philosophies. The notion is that you cannot have a back without a front, and hence everything in the universe is essentially dualistic, but holistic at the same time.

Thoughts and problems arising from the binary process of the brain are significantly discussed and debated between the different camps within Jewish theology as well. The spiritualists will see the traditional understanding of thesis, antithesis, and synthesis to be a true explanation of how the world functions. According to the famous kabbalist the Ari (Rabbi Isaac Luria), ultimately all will be seen to be from God and all will be well, but humankind is first tasked with raising the "sparks" of holiness from the antithesis, the *sitra achra*, the "other side," as is referred to by the kabbalistic literature. Only by bringing the holiness from all things that appear on the surface to be separate from God will this synthesis come into being. However, this perspective has its own dangers. During the uprising and messianic movement of Shabbetai Tzvi, after he converted to Islam many Jews continued to believe he was the Messiah by using the concept of "bringing sparks from the darkness." This was immediately attacked and shunned by the larger Jewish community as antithetical to their way of life, but was a possible consequence of stretching the limits of this theology given the historical circumstances. For a rationalist, however, the thesis and antithesis of good and evil, or any dynamic of I-Thou with humanity and God, does not necessarily need to be resolved at all. Rav Joseph B. Soloveitchik famously states in his work *Lonely Man of Faith*, "If one would inquire of me about the teleology of the Halakha, I would tell him that it manifests itself exactly in the paradoxical yet magnificent dialectic which underlies the halakhic gesture."[19] This dialectic is valued and championed as the true intention of man's existence, not to be a great unifier, but to live within that tension and balance both sides simultaneously, that of the imminent God with that of the transcendent one. Thus, God still unifies all things even though humankind must persist in living with the dualities of the world.

Another excellent example of the binary process pervading Jewish theology comes from the work of Martin Buber in his, *I and Thou.* In it, he discusses the importance of various types of oppositional relationships. The I and thou relationship is a personal one between two beings and can be in the form of two people or even a person and God. The holistic union of these two beings is ultimately reflected in how a person is engaged at all times with the world and this ultimately leads to the actualization of that person's existence. Thus, the opposite of I and thou is reflected in the opposite of the person and the world. And in the end, it is the holistic process that helps to resolve these opposites. Buber does discuss that the I and thou relationship with God is fundamental in Judaism and occurs when we allow God to come to us. At such a point, this would be, according to Buber, the ideal life which is lived in the unconditional presence of God, and there is no separation between religious experience and one's daily habits, rituals, and activities.[20]

It is clear that Buber draws heavily on much of the Hasidic tradition in creating this approach. The desire to have God come and enter into ones life as an "other" that one simply enjoys basking in the presence of *ziv hashechinah,* the glow of the Divine Presence, is a concept throughout the Kabbalistic and Hasidic literature. The ideal union is one of man and God, becoming entirely embraced by the "*or ein sof*" and becoming nullified within God's neverendingness. This may be contrasted by the "dialectical tension" model often ascribed to Rabbi Joseph B. Soloveitchik discussed above, where man is intended to live with the struggle of defining themselves and the other simultaneously, but never unifying.

The Causal Process

The causal process of the brain is also crucial to Jewish theology. The causal process of the brain, located near the junction of the temporal and parietal lobes, tries to find the cause for any given strip of reality. The brain has this fundamental ability to seek out causality in the world and try to understand causes and effects. When applied in its total function, the causal process requires us to find what caused each thing in the universe to happen. As with quantitation and reductionism, the causal process is essential for scientific applications. Scientists strive to understand what causes all biological, chemical, and physical phenomena to occur. When the causal process functions in an absolute manner, causality is perceived to be the root of the universe. Put another way, the entire universe is derived from causality itself applied either with or without the need for God. If without the need for God, the cosmologist might suggest that the cause of the universe resides in quantum forces that cause reality to come into existence and to continue its existence. If God is needed, then God provides that fundamental first cause of the universe. In fact, God continually causes the universe to exist.

The Midrash (*Genesis Rabbah* 10:6) states that there is no blade of grass that does not have its own *mazal* above that strikes it and says, "Grow!" *Mazal* in this sense is interpreted by Rabbi Zalman of Liadi, in his *Tanya*, to mean that God's influence taps the potential within the grass and causes it to grow (see *Tanya*, section "Holy Letter" and "Gate of [God's] Unity and Faith"). God is perpetually influencing the world through the "light" that emanates from God and goes through an infinite number of contractions in order to create the physical world. After this connection is made, according to the Hasidic and kabbalistic masters, it is constantly flowing forward. If the connection is cut off for a minute, then the world would cease to exist. While this position is not universally held (see our earlier discussion about there being a natural order out of God's immediate actions), it does speak strongly to the casual process of the brain that aids toward the development of religious and theological ideas.

Theologically, the causal process of the brain might help lead one to conceive of the notion of an uncaused first cause. This was the classic argument from Aristotle and Saint Thomas Aquinas, who carefully followed their causal process in the brain to a conclusion that there must be a first cause of all things. But why must this be so? Is it because there is an inherent need in the universe for a first cause, or is it because our brain leads us to consider a first cause? If we had a different brain, would it be possible that we would not seek to understand a first cause? For Judaism, along with other monotheistic religions, the foundational myth posits that God is the cause of all things (i.e., is the uncaused first cause). However, this very question of how something can be uncaused is a most perplexing problem for human thought.

The Abstractive Process

The abstractive process is tied into the brain structures that underlie language and conceptual thought near the junction of the temporal and parietal lobes, particularly in the left hemisphere. The abstractive process of the brain creates general concepts from a larger group of objects. Thus, dog, cat, horse, and tiger are all animals and also all mammals. In some senses, this process distills the basic characteristics of whatever objects it is working on and helps to define and categorize it in some abstract manner. In some ways, this brain process presents us with a sense of "thingness" or "being," since it generates the basic components of any object and declares that object as a certain thing. If the abstractive process engaged the world in a total manner, we would find ourselves naming, categorizing, and conceptualizing everything. In philosophy, this process might be regarded as establishing the metaphysical realm—the realm of ideas and forms rather than physical objects. The function of the abstractive process might have been prominent in the book of Genesis where Adam is asked by God to name each of the animals on the earth. "And out of the ground the Eternal God formed every beast of the field, and every fowl of the air; and brought them unto Adam to see what he would call them: and whatsoever Adam called every living creature, that was its name" (Genesis 2:19). Arguably, Adam's abstractive process was operating in a total manner by providing the name and category of every animal.

If the abstractive process functioned in an absolute manner, then the perception would be that the universe was composed in its essence of "thingness" or perhaps existence. Beingness or existence would be the essence of the universe. In Judaism, again, that essential existence rests with God. The absolute functioning of the abstractive process gives a profound sense that reality is fundamentally Pure Being, having the same relationship to physical matter as the pure concept "animal" has to the billions of living things in the world.

From this profound sense of being arises the philosophical/theological concepts such as Plato's "the Good," Aristotle's "hylemorphism," Aquinas's "essences," Tillich's "ground of being," Maimonides's "Foundation of all foundations" (*Mishneh Torah*, Foundations of Torah 1:1), and the *Ein Sof* of Kabbalah as a description of God. Certainly, the foundational myths of Western religions imply that God not only is the creator of all things in the universe, but continues to give substance and existence to all things all the time (an assumption maintained by some Jewish theologians as well). Theology must then be forced to explain how God can be the ground substance of all being while performing other roles stipulated in the foundational myths. Issues as to whether God constantly supports existence or simply created the universe and then let it run its course is an important theological controversy. However, it seems that the notion of God as the ultimate being who supports all of existence would be a natural consequence of the absolute operating of the abstractive process of the brain.

The Emotional Value Process

The emotional value process of the brain lies primarily within the limbic system structures (and insula) and pertains to the various emotions, both positive and negative, with which we evaluate everything that happens to us. This process imparts emotional values on whatever is presented within our experience. If this emotional process operates in a total manner, then we would strive to emotionally evaluate everything that happens in the universe. We would try to determine whether everything that happens to us or in the universe leads to a positive or negative emotional response. Although the notion of emotional responses may seem odd in the context of physical objects such as the sun or a galaxy, traditions such as Buddhism and Hinduism do imbue all objects in the universe with a sense of consciousness. Whether the sun is "happy" or whether we simply feel happy when the sun shines on us would be interesting philosophical and theological questions from the perspective of our emotional processes.

The absolute working of the emotional processes of the brain would lead to the view that emotions are the essence of the universe. Again, this might sound like a strange perspective to take, but from a religious point of view, one might argue that *God's love* is what suffuses the existence of the universe. In this case, an emotion is the primary force that creates and guides the universe. What is interesting about the emotional processes of the brain is that there are many types of emotions supported within our brain. There are basic positive and negative emotions, such as happiness or sadness. And there are more subtle emotions, such as pride or jealousy, that have generally positive or negative elements. When considering how the emotional processes might function in an absolute manner, there appear to be three possible types of emotional experiences that would group all of one type of emotion into one. Thus, a person might experience the positive, negative, or neutral emotions in an absolute manner. There may be a fourth option in which all emotions are perceived together. The result could be an overall neutral emotional response, since everything cancels out. Alternatively, it might be that every emotion is felt simultaneously. This is a rarely described experience, but possible at least from a brain and neurotheological perspective.

If there is absolute functioning of the positive emotions, then the result is that the entire universe appears to be an overwhelmingly beautiful, blissful, and loving place. Such a state might be similar to what Abraham Heschel was talking about when he described the notion of "radical amazement" in the context of the personal awareness, or even mystical experience, of God. For Heschel, the problem with human beings is that they have been separated from the world and from God. To reconnect required letting God back in to an individual's life which leads to a powerful feeling of astonishment. But it is a notion that this astonishment is a fundamental force in the universe that helps bind the person to God. The astonishment is not just of God, but of the ability to perceive God, and the of the fact that the self is able to perceive God. When applied to mystical states, the experience appears to be perceived (after the fact) in a personal manner such that the individual has become one with God or some form of a divine being—connected through the ultimate positive emotion of love. When applied to the concepts of theology, God is the primary driver for this overwhelmingly positive affect that pervades the universe. This being the case, God is essentially pure love and benevolence. This is characteristic of the kabbalistic mystical experience of God. The individual essentially feels at one with God and is privy to the unique force of love with which God governs the world. This exposure may be from the side of *Din*, which refers to judgment and harshness, but most often is perceived as recognizing God's trait of *Chesed*, or absolute kindness and overwhelming benevolence.

However, this positive experience of the world immediately presents high theological problems, since the pain and suffering that exist in the world must somehow be explained in light of the overwhelming love of God. In other words, we are left with the enduring question "If God is ultimate love, then how can God allow all the suffering that occurs?" This clearly has been a very difficult question for all theistic religions to address, and is likely why much of Jewish thought has focused not on the above-mentioned mystical experience, but on the fact that we do not have the answers for why bad things sometimes happen to people, and that God's knowledge of the world is beyond our understanding at times. However, if the absolute working of the emotional value process truly represents all possible emotional responses happening simultaneously, then there would be room for negative emotions even if the overall perception is positive.

If the absolute working of the emotional value function is perceived as neutral, then the universe is considered to be impersonal. In terms of mystical experiences, this neutral affect likely is associated with void consciousness or nirvana, in which there is an empty, impersonal consciousness that lies at the foundation of the universe. When the neutral affect is applied to theological or philosophical concepts, it may underlie the notions elaborated on in existentialism. Existentialism is based on the fact that all we can do is exist and "feel" our way through.

Other than our emotional sense, we can get at no other understanding of that reality. From a theological perspective, the conclusions drawn from the neutral interpretation suggest that God is impersonal or perhaps that there is no God at all and everything simply is without purpose or even meaning. This existential approach is antithetical to most theistic religions, and certainly Judaism; however, theology must contend with the possibility of an existential universe.

The third possible interpretation of the absolute working of the emotional value processes is a negative one. The result is that the entire universe is viewed as intrinsically evil and horrible. There are very few examples of absolute negative emotions in the mystical literature. The absolute unitary state has rarely, if ever, been associated with negative emotions. Indeed, anecdotal reports have suggested that such a state is impossible to attain while maintaining normal life functions. This suggests that a negative absolute unitary state may actually be incompatible with life. While there is no solid documentation of this bizarre notion, there are occasional rumors and anecdotal reports of mystical sects that try to achieve such a state. Whether they truly exist remains unknown. Of course, the negative interpretation applied to theology may be responsible for the notion of hell in which all of existence becomes horrible and terrifying. In Judeo-Christian theology, though, it becomes challenging to explain how such a negative existence can be maintained alongside the generally positive image of God.

The Willful Process

One part of the brain that is very relevant to spiritual practices such as prayer or meditation, as well as to theological principles in general, is the frontal lobe. The frontal lobe is activated whenever we purposefully perform a task such as driving a car, solving a math problem, or trying to read a book and is sometimes regarded as the seat of the will. If this willful process is applied to all things in the universe, the total functioning of the willfulness process, then we might see purposefulness in everything. We can understand why certain things happen and who or what is the purpose behind them. This process might be applied to our own life as we strive to find meaning and purpose in our job or our relationships. If the willful process functions in an absolute manner on all of reality, it would likely generate a sense that everything is derived from intention or will. This is similar to the concepts elaborated on by the philosopher Arthur Schopenhauer in *The World as Will and Idea*.[21] In this work, he presents the notion of will as the striving from which all things are derived. However, Schopenhauer goes further in describing that with increased will, there is increased suffering. This ultimately explains why there is so much human suffering, because human beings are endowed with an enormous amount of will.

The notion that the entire universe is derived from will also leads to several Jewish theological implications. Certainly, the "will of God" has a prominent place in Judaism. God used will to create the universe and to carry out all divine actions. The entire beginning of Genesis is God "speaking" and things coming into being simply due to God's command. Thus, it could be argued that will is the driving force of the universe, only in religion it is God's will. That human beings have a part of the universal will also leads to an analysis of free will. Schopenhauer actually addresses this issue, arguing that will does act with freedom. However, it is only the conscious analysis of human actions that leads to the conclusion *a posteriori* that there is no free will. The point of this description of the absolute functioning of the willfulness process of the brain is to indicate that it leads to certain concepts that are an integral part of theology, particularly when considering the will of God and free will. In the next chapter, we will discuss more about Judaism's take on free will. However, it is clear that Judaism has traditionally espoused a view that free will is a necessary prerequisite both for God and for human beings.

Final Thoughts on Theological Implications

Whatever area of the brain one is using and whatever neurotheological questions are considered, the larger question still remains: to what end? Will the pursuit of knowledge about the neurophysiological activities of the brain during religious and spiritual activities help us gain a better knowledge of both the physical and mental existence of the religious individual? We believe the answer is definitively yes. Even in these few scenarios discussed above, the potential for a greater "truth" to be found exists, one that encompasses not only science but religion and theology as well. Pursuing such questions using our various brain processes, particularly the holistic centers, we may truly be able one day to discover the unifying theory of everything. Perhaps the best way to uncover such a theory is not by simply allowing the areas of the brain to operate in independent silos, but to integrate them in ways that have been thus far unexplored. When the Jewish people sent spies to report on the Land of Israel's inhabitants, they came back with terrifying stories of giants and heavily fortified cities. The people were fearful they would never be able to overcome this great barrier and were about to give up. However, two of the spies, Joshua son of Nun and Caleb son of Yefunah, stood up for what they could see beyond the initial fear of the unknown. "We shall surely conquer it!" they said, and showed the great promise of the Holy Land, large fruits and great bounties (Numbers 13:17–30). Sometimes the truth is there, but it is scary and hard to acquire. With enough confidence and continued research, the field of neurotheology can be significantly expanded and give definitive insights into the stories and actions of Judaism and other religions across the world.

Free Will and the Brain

The room was dark and cold, and the queen stood, about to enter into the king's antechamber unannounced. Everyone knew that this was punishable by death if the king did not decide the intrusion was allowed, and yet, the queen felt a need to enter. "Who knows if for a time like this you ascended to the monarchy?" Mordechai tells Queen Esther, during the Purim story (Esther 4:14). Her rise to the the throne in such a short period right when the Jewish people needed help is seen as a sign of *hashgacha*, God's influence working behind the scenes. However, it is clear that without Esther deciding to act, she would not have been the one to engineer the salvation of the Jews of the time period. The same verse even states, "If you remain silent, relief and salvation will be given to the Jews from another place." Mordechai was convinced the Jews would be saved, but human action was still required. So how free was Esther to act? From the verses it appears entirely so. But what about the fact that the Jews would be saved either way? Free will may be an inherent contradiction in terms, but it may also be the only way humans have to create a meaningful experience for themselves. Jewish theology has long grappled with this issue, and a full overview of Jewish theology would seek to explain how these struggles may in fact correlate with much of what we have been discussing until now.

—David Halpern

The Nature of Free Will

Free will is something that has been hotly debated by philosophers, scientists, and theologians for centuries. The basis for the arguments grow out of the functioning of the binary process that sets up the opposites of good and bad and right and wrong. If operating in its totality, this binary right and wrong can be applied to virtually everything, or at least every human activity. Every thought we have and every behavior we enact can be evaluated according to these opposites. Of course, one of the problems with invoking the binary process is that many times there are a lot of gray areas. Actions can be partially good and partially bad or mostly good and just a little bad. And there is the added issue as to how much ends justify means, a notion that arises in part out of the causal process of the brain. Eventually, free will, or at least the feeling of having free will, appears to derive from the willful areas of the brain, particularly in the frontal lobes. The frontal lobes turn on when we perform purposeful activities and when we make decisions. But is that really free will? Are we fully conscious when we make our choices, and do we need conscious choice or can free will also be based on unconscious processes that are still our own? By exploring these various issues, along with the overall perspectives that Jewish thought has on morality, we can discover a fruitful area of neurotheological scholarship.

From a purely causal perspective, if everything human beings do is broken down into its prior causes, then the complex neural interactions of the brain can be said to "determine" the choices, thoughts, and thoughts about choices we make. This position is described as the "determinist" position and is based on the assumption of materialism and of cause and effect. There are four basic ways one can respond to the argument that everything is predetermined: acceptance, rejection, denial, or synthesis.

A number of scholars currently argue that there is no free will. They accept that there are so many forces acting on us that we never consciously make a free decision. They further argue that there is no conscious part of our self that sits within our brain and makes decisions *de novo*. Noted neuroscientist and atheist Sam Harris argues, "I generally start each day with a cup of coffee or tea. This morning, it was coffee. Why not tea? I am in no position to know. I wanted coffee more than I wanted tea today, and I was free to have what I wanted. Did I consciously choose coffee over tea? No. The choice was made for me by events in my brain that I, as the conscious witness of my thoughts and actions, could not inspect or influence." However, many scholars argue that we must have free will, both because we seem to feel like we do and for theological reasons. How can God hold us accountable for our actions if we have no free will?

The incompatibilist rejects the deterministic assumption and believes that it is still possible to have free will despite, or even because of, recent scientific discoveries. Others will deny the determinist position entirely, often by rejecting the scientific theory and method as untrustworthy and insignificant, and thus it can have no say in the debate in general. Lastly, there are the compatibilists, who argue that one can be a determinist and still believe in free will. However, the arguments are complex in that free will must be redefined within the position of physical determinism. That is why many indeterminists argue that for one to have free will in the strong sense of the word, it is not possible to uphold a thesis of determinism.

Robert Kane and John Searle have defended the incompatibalist position that there is free will and therefore not absolute determinism. Kane argues that there is clearly a physical relationship between the neuronal firings of the human brain and thought, emotion, and decision-making. However, the effort of will creates indeterminate processes in the brain that allow for the firing connections between neurons to change without another external input.[1] Thus, an entirely physical system is one that allows for free will, relying on the quantum understanding of small changes that work together to allow for larger-scale changes on the neural network.

Much of the determinist argument stems from the scientific work of Benjamin Libet, who performed an experiment where subjects were told to be aware of the time on the clock when they thought about flexing their finger.[2] The neurological signals were then correlated with the thought to flex and the flexion of the finger itself. Afterward, subjects reported the time they were aware of the desire to flex on the clock. The main issue that Libet discovered was that subjects' muscles flexed two hundred milliseconds after they became aware of the stimulus, but five hundred milliseconds after the action potential in the brain associated with the action had gone off. This would seem to imply that subjects begin to do an action before even recognizing that they intend to do so. However, Alfred Mele, in discussing this issue, recognizes that there may still be an intention to act that is created prior to the initial reading, which would allow for free will to still play a significant role in creating action.[3] Thus, the debate is left open again, with significant implications for either position. Neurotheology would argue that we need much more data to truly understand how conscious decisions are made within the brain. Libet's study, while fascinating, was performed on only a few subjects, and subsequent studies have suggested that there are a number of other factors, such as general monitoring of the environment, that might help explain some of the results.

Neurotheology reminds us to be careful about our conclusions and make sure we have sufficient data to make them. Neurotheology also encourages us to integrate a number of scientific and theological approaches to the question of free will. First, it is critical to define what is meant by free will and whether it requires conscious choice alone or whether unconscious elements of the brain and mind still qualify as free will. Thus, we need to define "free," and we need to define "will." These definitions might require a number of perspectives, including neuroscientific, psychological, consciousness, philosophical, and theological views. Once we can agree on some initial definition of free will, we must again turn to a combination of scientific and theological concepts to help understand how and when free will might occur and what problems arise if we do or do not have free will. And in an ideal world, all of this would lead to future neurotheological studies that could be used to support or refute our understanding of free will.

Jewish Perspectives on Free Will and the Brain

Judaism has also struggled with the interrelationship of free will and deterministic principles both from God's omniscience and from scientific physical determinism. The differing perspectives on this issue once again reveal the underlying perspectives different groups of thinkers in Judaism may have had about one's relationship to God and one's relationship to the physical universe. Holistic thinkers, particularly those of the kabbalistic/Hasidic heritage, often believe that if God really does ultimately run everything, then the world as we see it is in fact an illusion. This is most famously attributed to the Hasidic thinker Rav Mordechai Yoseph of Izhbitza.[4] However, most theologians even among the Hasidim assume, as does most of mainstream Judaism, that free will is a necessity. An argument more in line with a limited approach to free will is that stated by Rabbi Eliyahu Dessler, who believes that people have a *nekudat habechira*, a "free-will point," that can move up and down on the ethical/moral scale depending on our actions. If one is accustomed to steal, then that may not be a free choice for the person at that time, but that person is responsible for other ethical choices, and when a choice is made, the person gains more freedom at higher ethical levels and can continue to progress.

This is somewhat similar to the position described by Sam Harris in *The Moral Landscape*, in which he states that a person's individual predilection for moral or immoral behavior cannot be undone.[5] We must act according to who we are at that particular moment in time. Hence, according to Harris, we don't have free will, since we are forever influenced by our upbringing, brain, and genetics as these intrinsic factors interact with the external environment. From a brain perspective, we might wonder whether some people have frontal lobes and limbic systems that make them more prone to moral behavior than others. If so, for what actions can we be held responsible? And can we change the brain's functions in certain ways to make us more or less moral?

The Talmudic line "Everything is in the hand of heaven except for the fear of heaven" (*Berachot* 33b) is explained by the Ramchal (Rabbi Moshe Chaim Luzzato), a famous kabbalist and Jewish philosopher (in *Maamar Halkarim* 6), to refer only to things one does related to sinning and doing mitzvot/divine commands. Everything else one does is not actually up to one's own free will, but determined by God. Additionally, the phrase in the Bible "The heart of the king is in the hand of God" (Proverbs 21:1) has been traditionally interpreted to mean that political leaders of countries do not in fact have free will regarding the decisions made on behalf of their countries. Thus, God is actually behind the larger national events. If this were the case, could it be argued that certain brain processes are literally influenced by God? Or if not, could we argue that brain processes occur separate from the mental processes with which we make moral decisions?

This leads to the larger problematic issue separating the mind and brain. As some have tried to argue in courts of law, "My brain made me do it." The argument is that the brain functions in such a way that is beyond our control. If our brain has experienced emotional or physical traumas, or if it functions poorly due to genetic or developmental errors, perhaps our thoughts and behaviors are not our own. This area of exploration and research is important to neurotheology, which can hopefully better delineate what areas of the brain are involved in our thoughts and feelings, and the direction of the causative arrow, especially with regard to behaviors that are moral or immoral.

There are other theological positions within Orthodox Judaism regarding morality and free will. Maimonides famously interprets the Talmudic saying "Everything is in the hand of heaven except for the fear of heaven" to mean the fear of heaven includes all of a person's actions, as they all show whether the person has a true fear of heaven or not. However, even Maimonides himself must deal with the issue of free will being removed, as it seems to be in the Exodus story where the biblical verses clearly state that God hardens the Pharaoh's heart to the point where he does not allow the Jews to leave Egypt (Exodus 7:1–5). How in a system of freedom was God allowed to limit a human being's free will? Maimonides explains that free will, like everything else in the world, is a gift from God, and as such it can be removed as a punishment as well. The Pharaoh had consistently ignored the warnings from Moses and as such was punished not only with the plagues, but by the loss of his ability to stop them (*Mishneh Torah*, Laws of Repentance 6). Others, however, argue that this is not a viable option in allowing free will. On that same story, the Sforno (Obadiah ben Jacob Sforno, Italian rabbi and philosopher) states that "hardening Pharaoh's heart" simply meant that he was given the mental fortitude to withstand the physical and emotional burden of the plagues if he so chose to remain obstinate, which he then did. According to this position, free will is absolute, and there is nothing, not even grave sin, that allows God to remove it.[6]

Judaism clearly also struggles with the notion of God's wholeness and omniscience in the context of human free will. Rabbi Akiva of the Talmud famously states, "Everything is foreseen, yet free will is given" (*Ethics of the Fathers* 3:15). The statement admits to God's omnipotence and omniscience but also simultaneously refuses to deny the free will that is given to human beings. But is this just a statement or can this be argued logically on the basis of the causal processes of the brain? The issue revolves around who or what is causing things and whether any given person is responsible for his or her actions. If the cause of a sequence of reality lies beyond that sequence, then a person within that sequence cannot be responsible. Whether causality exists within a given sequence of a person's reality is what determines if that person has free will. If God knows everything that is going to happen, is it possible for us to have free will? To some extent, the only way to answer this is to argue that while God knows what is going to happen, God does not cause those things to happen. However, if we argue that God causes everything to happen as well, the ability to preserve free will appears to become more problematic. Regardless of the perspective that one takes, neurotheology would argue that there is great value in integrating science and theology in addressing the question of free will.

Eastern traditions have a slightly different perspective in terms of causality. The Buddhist and Hindu ideologies concede true causality only for the realm of the absolute unitary states (i.e., only via the absolute function of the holistic process). The individual ego and material reality are seen more or less as an illusion, with mystical states of absolute unity being the true reality. Causality, as well as free will, only exists on the level of the absolute unitary state and does not apply to material reality or the human ego. However, this still presents a problem with the issue of practical ethics and the accountability of individuals. These traditions suggest that once the absolute unitary state is attained, there is a natural flow of right behavior that derives from it, and this type of behavior is what constitutes ethics. In such a system, the only way to gain a true understanding of right and wrong, free will and determinism, is by attaining the absolute unitary state.

This may be similar to the Hasidic conception of the *tzadik*, the righteous individual. The *tzadik*, as identified in the *Tanya* and many other Hasidic writings, does not operate on the same plane of existence that the ordinary person does. In describing the *beinoni*, the "average person," the *Tanya* states that this identity was assumed by many of the great Talmudic sages. If this is true, it must mean that a *beinoni* is in fact someone who has never even sinned but feels the temptation constantly (See Tanya: *Likkutei Amarim* Ch 12). A true *tzadik*, however, does not have this internal struggle and is a true conduit of the holy to this world and a way for the masses to relate more easily to God. According to the theory of the *tzadik*, this person's natural moral behavior stems from his or her direct connection to the Godhead, and the struggles are of a much more sublime nature. This person does not need to feel pushed and pulled between forces but can simply understand what the right thing to do is. This entire theory is based on the kabbalistic doctrine of God being *soveiv kol olamim*, "encircling all the worlds." If God is the origin of all things, then once one unifies the false divisions of this world by whatever spiritual means, then one can achieve *devekut*, or union with God in some spiritual sense. After this, the spiritual understanding of right and wrong would be obvious, as it was prior to Adam and Eve eating from the "tree of knowledge," which according to some traditions was a tree that made the one who ate of it lose the objective sense of truth and pursue physical pleasure instead (see Maimonides, *Guide for the Perplexed* 1:2). Achieving a state of pre-Eden closeness to God is the mission of the kabbalistic system, and it would thus make sense for that theology to have an underlying holistic causality.

Other modern thinkers have also grappled with the issues of free will and humanity's place in nature in general. Rabbi Joseph B. Soloveitchik, in his *Emergence of Ethical Man*, describes the human distinction from animals as one of degree, not kind.[7] As human beings, we must fight our inner animalistic urges and come to terms with nature in general. This is what Rabbi Soloveitchik understands as our executing free will. While many Christian theologians argue that our freedom is a supernatural ability given by reaching for the higher power of God, Soloveitchik argues that we are given an ethical nature when we are created by God, and as such free will is simply our natural ability to choose between the ethical and the hedonistic animal urges within. This understanding of free will places it entirely within the realm of scientific inquiry, as everything described lies in the physical world. However, it does not openly explain how exactly free will operates on the level of pure physical existence in light of the significant support for determinism in other scientific studies. Again, this is a concept that is ripe for analysis through the lens of neurotheology.

Another perspective is to consider free will as an ability that can be developed or lost like intelligence. Thus, free will is not something that everyone simply has. "Free will is not an organ with which we are born. . . . It is a capacity to respond to the world and its challenges in a certain way. If it is not nurtured properly, it will never develop or it will wither away."[8] By utilizing creativity, we are able to maintain our freedom and continue to develop both morally and religiously. The notion that free will can be augmented or lost, like intelligence, should theoretically be observable through scientific study. Hence, perspectives such as this can be useful as hypotheses to drive future studies to eventually better delineate how free will might work, if it works at all.

Sin and Free Will

Nachmanides on the verse in Deuteronomy 30:6, "And God will circumcise your heart," states that humankind is ultimately destined to return to a pre-sin state of no evil inclination to do bad, and then there will be no more reward and punishment. This perspective indicates that *yetzer*, or desire itself, is what is required for a choice to be made that allows human beings to earn merit. Therefore, it would seem according to this perspective that it is only by subduing that desire that people exercise free will, an idea not dissimilar to the Buddhist concept of removing attachments as a way of ending suffering. It appears that there is an experience bereft of "free will" defined as choosing between things that are good and bad with an internal drive, where one still maintains the possibility of choice. This "pre-Eden" state for Nachmanides is the ultimate goal of man. The ultimate spiritualist goal, then, is not to simply subdue our nature, but to transform it. Once that happens, there is no more free will, but that is not a bad thing. It seems as if this is ultimately arguing for a transformation of the human brain. The brain no longer has to contend with various desires but almost automatically does good. Of course, one might ask whether free will itself is transformed to a different type of decision process, perhaps not so much related to morality but related to spiritual development.

Someone thinking along a more rationalist camp, however, would disagree significantly with this thesis. The purpose of

God's commands are not simply to reach a state where they will not be enforced, but to reach a place where one can truly appreciate God and God's abilities. That is why Maimonides interpreted the position of the Talmud stating, "There is no difference between now and the end of days except for the oppression of the kingdoms" (*Sanhedrin* 99a; *Berachot* 34b), to mean that people would still have free will to serve God, but it would just be in more idyllic conditions. This makes sense if the religious thinker and disciple are not looking for a complete spiritual "oneness" experience, but an experience that is grounded in the concrete components and dialectical tension discussed earlier. Refinement of human beings is the main goal, but this is done with precision and study as well as following God's commands. Sin and free will then become intertwined, in a way that states that people must have the freedom to choose in order to be capable of sin. While free will may lead to much evil in the world, it is also the foundation of morality and our ability to shape the world.

The Brain and Morality

Several research studies have explored the areas of the brain involved with making moral decisions. Such research can further inform Jewish thinking about the process and nature of morality. For example, some early studies suggested that the social areas of the brain are involved in making moral decisions. When people had to think through a moral dilemma involving other people, the social areas of our brain became more active.[9] Thus, these brain scan studies show that our evaluation of the social environment and interactions plays an important role in moral decisions. These areas likely are involved in theory of mind processing that helps us understand what another person is thinking. By understanding the pain and suffering of another, we can better make moral judgments.

Another important aspect of brain function relevant to ethics is emotions. Several neuroimaging studies have shown that our emotional brain areas are activated during moral decision-making. Any ethical decision necessarily requires an ability to place emotional value on various elements. The value placed on each element of an ethical decision is likely determined by our emotional perspective. We might feel disgust, anger, fear, frustration, or a number of other emotions that help us to identify if some act is moral or not.[10] The importance of emotions in ethics leads to an interesting theological and philosophical issue as to whether there are universal ethics based on rationality or whether morals are more relative and based on various emotional values.

Finally, moral reasoning likely requires many of the same brain areas discussed in the prior chapter, such as causal reasoning and the binary and holistic processes. Causal reasoning is essential for delineating what events are caused by what actions. It is the causal process that is challenged most with respect to the deterministic arguments. If we cannot cause our own actions, how can we be responsible for them? And hence, how can we act morally or immorally? The binary process helps us to distinguish moral from immoral by setting up these opposites. And the holistic functions of the brain might help find a way to resolve why there is both good and evil in the world, especially if God is wholly good. Neurotheology will have to continue to explore such questions, adding scientific data to the existing theological and philosophical debate.

Forgiveness and Judaism

Another important element to the free will discussion is its impact on how one views the concept of forgiveness. The Jewish tradition has long held that God gave the people a supreme gift, the ability to do *teshuvah*, "repentance," and to turn away from wrongdoing. The question then becomes, based on our free will and choices, why should we be able to be forgiven? How exactly does forgiveness work? Judaism does not view forgiveness and mercy as divine gifts as Christianity does, given freely for all of humanity to simply accept. Daniel 9:9 states, "To Adonai our God belong mercy and forgiveness, for we have rebelled against God," indicating that forgiveness comes from God. John 1:9 says, "If we confess our sins, God is faithful and just to forgive us our sins and to cleanse us from all unrighteousness." In Christianity, forgiveness comes from God when a person turns to God and confesses.

In Judaism, work is required to achieve full repentance. According to Maimonides (*Mishneh Torah*, Laws of Repentance 2:1–2), people must confess their sins and, if faced with a similar situation, must show that they have changed their response. Forgiveness, however, even for the rationalist Jew, overrides all rational perspectives. It is fundamentally an act of mercy, and in this manner, this area has the most overlap between the two camps of thought we have been discussing. How exactly can the day of Yom Kippur, with its serious introspection and fasting, allow someone to be forgiven for all the things he or she has done during the year? In addition, there is even a concept of the day of Yom Kippur itself atoning for people's sins, without them having to do anything (Talmud, *Shavuot* 13a). The commentaries explain that this is indicative of God wiping the slate clean in an act of mercy. God allows someone to start fresh with a new year. However, Judaism also believes that past actions are not meant to simply be wiped away. Everything one does is integrated into the self somehow, and the most complete repentance is one that transforms the prior actions.

The Talmud states (*Yoma* 86b), in the name of Reish Lakish, that repentance is so great, it can turn intentional sins into merits. If one had previously sinned and then did repent, the spiritual weight of the sin is somehow transformed through repentance not only to be lifted off of the person's burden, but to actually become something uplifting, like a merit. The commentators explain that when people do a complete repentance, they are in effect stating: I am not the person I was previously. They have fundamentally changed and cannot be judged as they were prior. The prior experiences one has are then not simply sins of another;

they are building blocks of experience for this new penitent person. This perspective allows continued development of a person as well as radical change and would be interesting to understand from a neurotheological perspective, given the brain processes happening at the point one is "remaking" oneself. If a person is truly transformed, then we would expect the brain to be functioning in a new way. Perhaps on a brain scan we would find areas previously inactive to become more active or vice versa. Perhaps the cognitive areas of the brain become more active, while the areas supporting negative emotions such as jealousy become less active.

This also underscores the importance free will plays in the traditional Jewish understanding of repentance. Radical choice that fundamentally alters a person's character is possible according to this model. While difficult, this change has been the cornerstone of Jewish character development and the basis of much of Jewish learning and discovery. This process may itself be why Jews have been able to remake themselves so readily in so many different cultures and circumstances throughout the ages.

Forgiveness itself can be attributed to various brain processes. In prior work we have considered the various elements of forgiveness.[11] For example, a person must first be able to experience an injury that can be attributable to another person or sometimes to God. This step, in and of itself, requires memory, the emotional or cognitive sense of injury, and the causal process to establish where that injury came from.

Breaking this down further, assessing an injury requires our emotions and our autonomic nervous system, which are triggered when we have experienced some type of injury. Whether that injury is physical or psychological, our brain responds very strongly so that we feel negative emotions and also feel discomfort throughout our body. This is why a bad relationship ends in a "broken heart." Interestingly, brain imaging research has been able to show that pain from personal or emotional injury is experienced in the same areas of the brain as physical pain.[12] An injury also requires a sense of our overall interpersonal relationship with the other person. A boss is allowed to be insulting if we have done something incorrectly. However, we might be far less tolerant of a friend insulting us. On the other hand, a boss cannot be abusive. Thus, there is a certain social balance that we have to be aware of and through which we assess when we have been injured. Once an injury occurs, we must be able to identify its source, using the causal processes of the brain, and maintain a memory of that injury. The memory is established primarily through our emotional centers in the limbic system, specifically the hippocampus. The hippocampus is one of the key areas that writes memories into our brain. It makes sense that the emotional centers of our brain would also be responsible for our memories, because the brain wants to make sure the things that are emotionally important are the things that we continue to remember. This is true for both good and bad emotions.

In many ways, it is more important to remember something bad so that we can avoid it in the future. If we eat food that upsets our stomach and makes us very sick, the negative emotions associated with that feeling are remembered so that we never eat that problematic food again. The brain's causal process is also required, because we have to be able to determine exactly where the injury came from. Sometimes this is straightforward, such as when a person directly insults us. Other times it is more difficult to make that determination, such as when someone speaks badly of us behind our back. Some people also become angry with God, blaming God for something bad happening in their lives. We see this most commonly with the death of a loved one such as a spouse or child. There are many circumstances where the person feels that God is to blame for the death of the loved one. In these circumstances it is sometimes very difficult for a person to forgive God. This can create substantial emotional and spiritual conflicts for the individual. As we discussed in chapter 9, several therapeutic approaches have been developed that try to address the combination of emotional and spiritual problems.

Once the injury has been fully perceived, an important binary decision has to be made. The person can try to exact revenge, thereby rectifying the social imbalance, or the person can turn to forgiveness. Revenge is certainly a common approach and is widely described in the Bible. The *lex talionis*—an eye for an eye—states how any injury can be dealt with from the revenge perspective. There is an important downside with revenge, which has to do with the degree of the revenge behavior. For most of us, we feel our own pain far greater than we feel another's. For this reason, when we exact revenge, it usually results in a greater injury than the initial one. This can escalate into feuds and even wars, depending on the magnitude of the reciprocal injuries. This may be another reason why the Jewish tradition views the *lex talionis* to be entirely figurative. The rabbis state that "eye for an eye" (Exodus 21:24) actually means to pay the one injured the value they would have lost as a workman or slave with an eye versus without (See Rashi on the verse as well as Talmud *Bava Kama* 84a). It is assumed that Judaism would never condone pure revenge in such a gross physical process. The fact that the tradition takes a verse seemingly literal and says it cannot be so is telling of the Jewish ethos of response to sin and injury.

Forgiveness is an alternative response to an injury that can help dissipate escalating negative emotions and behaviors. In that sense, forgiveness can be an adaptive process for a society by helping to maintain its internal cohesiveness among the members of the group. It is also far easier to forgive someone who is a member of your group than it is to forgive someone outside of the group. This relates to the holistic process of the brain that helps determine who we feel connected to and to what extent. The more connected we feel to someone (e.g., a child or spouse), usually the more willing we are to forgive.

Once the decision to enact forgiveness is made, there are usually a variety of cognitive and emotional processes that are engaged. The manner by which someone comes to forgive another can take various forms. For example, one might come to the understanding that the other person is beset with the same imperfections and flaws as everyone else and thus makes mistakes that can lead to an injury. The recognition of another person's imperfections can lead to the ability to forgive someone who commits an injury against us. Another, more religious approach may be to understand that all of us are created by God, and hence we can forgive another being who is created by God as an appropriate response. Yet another type of religious forgiveness may have to do with relying on God to make the final determination as to the outcome of a particular injury. In this way, a person forgives someone because it is the appropriate mitzvah and God will help to complete the process of forgiveness in God's own way. This is how many commentators explain the inherent issues with Jewish law regarding punishing murder and other capital offenses. Technically, in order to be tried for murder, one must have been warned prior to the action by two witnesses and informed of the consequences of one's action. This is very rarely possible, and as such, a Jewish court putting someone to death would be exceedingly rare. As the Mishnah (*Makkot* 1:10) states "a high court that kills once in seven years is called 'destructive'" (Talmud *Makkot* 7a). There are practical contingencies when societal needs demand greater action, but the fundamental principle assumes that punishing the murderer is often not in humankind's jurisdiction, but that of heaven. Judaism inherently views God as the ultimate arbiter of human sins, and while people may need to judge each other for societal reasons, it is up to God ultimately to determine what is forgiven and what is not.

There are many other approaches to the cognitive and emotional restructuring that is required for forgiveness. The eventual result of the forgiveness process is a reestablishment of the social balance that existed before the injury. We recognize the new relationship between ourselves and the person causing the injury in such a way that we can continue to interact with that person and the rest of the world effectively.

After forgiveness occurs, a number of positive emotional responses may also arise. The forgiver feels a positive emotion toward the forgiven in the form of love and compassion. It is also interesting to note that similar positive emotions may be felt by the forgiven person, who might realize that the injured person is virtuous and not deserving of future injury. In addition, other people in society can recognize the process of forgiveness as having value and bring sympathy and compassion as well to the forgiver. In fact, in the Christian tradition, forgiving their persecutors was something that produced great sympathy and gained many converts for the early Christians. Judaism also recognizes the importance of forgiveness and the recognition that other human beings have the same flaws and suffer the same pains. The gesture of removing a drop of wine at the Passover seder for each plague set upon the Egyptians is a reminder that revenge of any kind is something that must be considered carefully. The reason a full *Hallel* (psalms of praise) is not said on the last day of Passover is often linked to the midrash quoted in the Talmud (*Megillah* 10b; *Sanhedrin* 39b) that states, "The work of My [God's] hands is drowning in the sea, and you [the angels] want to sing *shira* [song of praise to God]!" Even at times of great personal rejoicing, Judaism recognizes that suffering, even of one's enemy, should not be ignored.

We can see how forgiveness can be intertwined with a variety of brain processes. Forgiveness is a fascinating human response, but there may even be the rudiments of forgiveness in animal species.[13] Many animals display various types of conciliatory behaviors that might have laid the foundation for human forgiveness. Whether forgiveness in humans is divine or whether it evolved over millions of years, the underlying brain mechanisms can tell us a great deal about how forgiveness works and may even help us find more effective ways of fostering forgiveness in people. Perhaps this latter application of neurotheology might yield an important global impact, helping to reduce revenge behaviors and encouraging greater compassion and peace among human beings, especially those of different religions.

Future Directions for Neurotheology and Free Will

Clearly the free will debate is one that has gone on for centuries, and it does not look like it will be resolved quickly. However, new developments in science and philosophy have made theological inquiry return to the issue with new insights. Neurotheology in this context would be able to appropriately apply current research about how the brain should react to various moral dilemmas depending on the concepts we have been developing in previous chapters. This research might potentially map how different perspectives and positions regarding free will and human autonomy, in the face of both scientific determinism and God's omnipotence, might underlie different neurochemical and physiological differences.

Is there something unique to a Jewish brain after years of textual study that allows for a dichotomous understanding of free will and omnipotence more than in those of other religious backgrounds? Is there some specific neurological correlate for believing in free will, and is there another that allows for, at the same time, believing that God has given that free will to human beings? When one has a spiritual experience and is entirely enveloped in the "oneness," would that create lasting neurobiological changes in the brain to forever change that individual's perspective on these fundamental issues as well? These questions and more are the cutting edge of neurotheology and will need to be addressed more fully to allow for a greater understanding and framework.

Perhaps neuroscientist Michael Gazzaniga expresses it best:

> I believe, therefore, that we should look not for a universal ethics comprising hard-and-fast truths, but for the universal ethics that arises from human beings, which is clearly contextual, emotion-influenced, and designed to increase our survival. That is why it is hard to arrive at absolute rules to live by it we can all agree on. But knowing that morals are contextual and social, and based on neural mechanisms, can help us determine

> certain ways to deal with ethical issues. This is the mandate for narrow ethics: to use our understanding that the brain reacts to things on the basis of its hard-wiring to contextualize and debate at the instincts that serve the greatest good—or the most logical solutions—given specific contexts.[14]

And yet, Judaism may also have an independent ethic of its own, as Rabbi Aharon Lichtenstein has stated:

> Any supposed traditional rejection of *lex naturalis* cannot mean, therefore, that apart from Halakhah or, to put it in broader perspective, in the absence of divine commandment, man and the world are amoral. . . . It goes without saying that Judaism rejects contextualism as a self-sufficient ethic. Nevertheless, we should recognize that it has embraced it as the *modus operandi* of large tracts of human experience. These lie in the realm of *lifnim mi-shurat ha-din* [lit. "beyond the letter of the law"]. In this area, the halakhic norm is itself situational. It speaks in broad terms: "And thou shalt do the right and the good"; "And thou shalt walk in His ways."[15]

This is the potential contribution of neurotheology to the field of neuroethics—not only to help determine the biological underpinnings of moral behavior as it pertains to religion, but also to help associate the context within which an ethical system develops. A combination of a neurotheological and neuroethical approach might provide our best understanding of human morality.

How Does the Brain Know Reality?

The king's stargazer saw that the grain harvested that year was tainted. Anyone who would eat from it would became insane. "What can we do?" said the king. "It is not possible to destroy the crop for we do not have enough grain stored to feed the entire population." "Perhaps," said the star gazer, "we should set aside enough grain for ourselves. At least that way we could maintain our sanity." The king replied, "If we do that, *we'll* be considered crazy. If everyone behaves one way and we behave differently, we'll be considered the not normal ones. Rather," said the king, "I suggest that we too eat from the crop, like everyone else. However, to remind ourselves that we are not normal, we will make a mark on our foreheads. Even if we are insane, whenever we look at each other, we will remember that we are insane!"

—*Rabbi Nachman's Stories*, translation by Rabbi Aryeh Kaplan

A Neurotheological Perspective of Jewish Epistemology

So what is true, and when can we know if someone's religious ideas and motivations are truly insane? A study of Jewish epistemology would be a necessity before any recommendation or basic Jewish neurotheological principles could be suggested. As a first attempt at understanding this issue of truth and reality, we asked the rabbis in our survey to respond to the following statement: "I think other religions are correct even though they differ from mine." As a group, 38 percent definitely agreed with this statement, 41 percent tended to agree, 12 percent tended to disagree, and 9 percent definitely disagreed (these numbers were similar regardless of gender). Thus, 79 percent of rabbis surveyed thought that other religions could be reasonably accurate with respect to understanding the world or God. Not surprisingly, Orthodox rabbis represented about half of those who felt fairly certain that no other religions were correct. However, one or two rabbis in each group responded that they definitely disagreed with this statement.

We decided to ask a second, similar question, but from a slightly different angle. We asked if they thought that Judaism was the only correct religion with respect to God. Amazingly, only three rabbis (all Orthodox) answered yes to this question. Most rabbis stated that Judaism was not the only correct religion with respect to God, with a number citing that everybody finds the truth that works best for them and connects them to God. A number of rabbis further indicated that other religions have important points regarding how human beings should behave. These religious beliefs are not fully true from their perspective, but nonetheless provide potentially useful insights.

Hopefully, by now, you might question whether the manner in which we posed these questions was valid. After all, what does it mean to say that a religion is "correct"? Does that imply "correct" with respect to its description of God, its description of humanity, or its description of moral behavior? And should we equate correct with truth? It is important for us to understand how every person, Jewish or not, rabbi or not, views reality. This is the ultimate question, and the question that lies at the heart of neurotheology: what is the true nature of reality?

This question also lies at the heart of Judaism. Judaism's fundamental prayer, the *Shema*, which is said at both morning and evening prayers, as well as before bed, frequently has the last word of the prayer, *Eloheichem*, linked with the first word of the next prayer in the service, *emet* (truth). Thus, we have *Adonai Eloheichem emet*, which is often chanted aloud at the end of the *Shema*. This three-word phrase indicates that there is no separation between

God and truth. In Judaism, then, God is truth, and God is the true reality. It is further interesting to add from a neurotheological perspective that the repetition of the combination of the three words, *Adonai Eloheichem emet*, at the end of the three sections of the *Shema*, creates a total of 248 words, which corresponds to the ancient understanding of the number of organs and bones in the human body (Mishnah *Oholot* 1:8). Thus, Judaism may have begun the idea of neurotheology over a thousand years ago by recognizing the importance of combining the body and brain in understanding the nature of God and the truth of reality.

Modern neurotheology provides an important, and different, view from the purely materialistic perspective or the purely theological perspective regarding what represents ultimate truth or ultimate reality. A neurotheological approach seeks to explore the neurocognitive components of the human experience of reality within the context of both science and religion. In order to accomplish this, neurotheology must necessarily include an analysis of the "everyday" experience of reality as well as the religious, spiritual, or mystical experience of reality. This requires that the ability to "know" what reality actually is depends on certain neurocognitive and spiritual states.

Primary Knowing States

To begin with, a neurotheological approach would acknowledge that the only way in which we come to know what is real is by what comes in through the various senses and the brain's response to that sensory input. The brain takes all of the sensory input we receive and puts together a rendition of the world within which we can interact. Outward actions or behaviors then have consequences in the world that are perceived and contribute to our ongoing internal assessment of reality. This is critical. We have an internal set of processes that enable us to think and reflect on the nature of reality. Reality consists of both the external existence and the internal experience of reality. The objective world is what actually is real regardless of what our perceptions are—what exists in external reality. Our consciousness or subjective awareness makes up our inner sense or experience of that external reality. It would seem almost impossible to get at what is ultimately objectively real because any information or sense that is received from this objective reality necessarily must come through and be processed by the human brain.

Judaism has often focused not on what is "real" but rather on what we are expected to do, through study of the Torah and Jewish law. However, that is not to say it does not recognize that there is certainly an area of existence that may often be "beyond what we can see." Rashi cites the Midrash (*Bereishit Rabbah* 65:5,10), which states that when Isaac was bound and almost offered on the altar to God, the angels cried tears, and those tears went into Isaac's eyes, making them "dimmed" (Genesis 27:1). Though he could not see physically, according to some, this apparently enhanced his vision for things not in the physical plane. His positive reception of and desire to bless Esau, despite him being a "man of the field" who shot arrows of contempt at his brother, was due to Isaac's special abilities to recognize the ultimate spiritual potential that Esau had (from the higher spiritual world of Tohu).[1]

Other Jewish scholars argue that the only "reality" one can know is the experience of the truth. The Kutzker Rebbe, a unique Hasidic master, spent long stretches of time on his own trying to understand existence. Maimonides himself believed that knowledge and the study of philosophy were the only way to uncover what was true and "real" in the sense of how the world operated (*Moreh Nevuchim* Part I Ch 34). On a more basic level, Judaism does operate with the assumption that our sense determines what is real. When we see three stars at the end of the Sabbath, that indicates the Sabbath has ended. When we smell nice things, we can say blessings over them, or if we smell something terrible, we cannot pray in its presence. Our senses impact our religious lives and spiritual experience immediately and constantly.

It might be argued, then, that what constitutes something being real is a very strong experiential sense that it is real. After all, why do we think anything is real? We perceive a car or a person as real because there is something about them that simply *feels* real to us. We don't perceive a gremlin as real because it just doesn't seem real to us. In fact, this is part of the problem when atheists and religious people get into an argument. They have such different *feelings* of reality that they do not comprehend how the other person can perceive the realness of God or the lack of realness. In spite of the potential problems with this perspective, such as relativism or solipsism, this statement should not be lightly considered because neurophysiologically human beings have nothing better to go on to help determine what is real.

One might call the sense of reality a *primary knowing state* (or primary epistemic state) of the brain. It should be mentioned that such a state is to some extent a brain state and to some extent a mental state. Therefore, it is the brain that enables that experiential or mental state that subsequently enables an individual to perceive an experience as real in his or her mind or consciousness.

What makes these states define reality for a particular person is the individual's sense, when in one, that what is being experienced is fundamentally or ultimately real. Any other perception of reality is considered to be an illusion, deception, or misperception. The most common example that we all know is the dream state. When we are in a dream state, we act as if it is completely real. It is a primary knowing state. If we see a lion running after us in a dream, we run in the dream. We don't stop to wonder whether the lion is real. On the other hand, as soon as we wake up, we immediately relegate that dream to an inferior experience of reality. We say, "Oh, that was just a dream." We are no longer worried that a lion is chasing us. We are in a new primary knowing state that we call our "everyday reality" state.

So how does this happen? In order to determine what is really real and the characteristics of these primary knowing states, the following model regarding the neuropsychological nature of these states might be useful. Other neuropsychological models might be possible, but for any neurotheological approach, such a model will typically include several important elements. These elements are determined primarily by how we sense and make sense of reality. This requires sensory elements, cognitive elements, and emotional elements. In fact, it might be helpful to break down the primary knowing states into three basic parameters: (1) the sense perceptions of objects or beings that can be experienced as either multiple discrete things or as a holistic union of all things (a unitary reality); (2) relationships between objects or things that are either "regular" or "irregular"; and (3) emotional responses to the objects or things that are either positive, negative, or neutral. These aspects of any given primary knowing state are based on basic brain processes. Our quantitative and holistic processes enable us to determine if there is only one thing in the universe or multiple things. Our causal process and to some extent our abstract process help us assess the way in which different objects relate to each other. When they relate in a causal, rational-appearing way, we say that they have regular relationships. If on the other hand, objects appear to interact in strange ways, we call these relationships irregular. And, of course, we can apply our emotional responses, typically originating within the limbic system, in either a positive, negative, or neutral way. The emotional responses do not refer to the usual feelings of happiness, sadness, etc., but to the overall emotional approach of the person to reality—that is, whether all of reality is viewed as positive or negative.

It is also important to mention that each of these parameters is most likely set along a continuum. We may experience reality in a way that is based primarily on having multiple discrete objects but may also have substantial holistic attributes. Similarly, there may be some regular and some irregular relationships between objects. However, this notation allows for an overall perspective from which more specific elements of primary knowing states can be elaborated. Based on these general parameters there would appear to be nine possible primary knowing states that are internally consistent and readily described by people who have been in these states. Again, though, these nine knowing states should actually be considered as a continuum of knowing states with those mentioned below as "nodal" points along the continuum. The nine knowing states are as follows:[2]

Number of Objects *(Quantitative and Holistic Functions)*	**Object Interactions** *(Causal Function)*	**Emotional Status** *(Emotional Value Function)*
Multiple Discrete Objects	Regular Relationships	Neutral
Multiple Discrete Objects	Regular Relationships	Positive
Multiple Discrete Objects	Regular Relationships	Negative
Multiple Discrete Objects	Irregular Relationships	Neutral
Multiple Discrete Objects	Irregular Relationships	Positive
Multiple Discrete Objects	Irregular Relationships	Negative
Unitary reality	—	Neutral
Unitary reality	—	Positive
Unitary reality	—	Negative

Before we consider each state separately, there are a few general points about this list. One can see in this list that the categories of unitary reality perceived as having either regular or irregular relationships have been omitted. Relationships can only be considered to exist between discrete, independent things. In a unitary experience of reality, there are no discrete, independent things that can be related to each other, so there cannot be any relationships (regular or irregular). In fact, it should be stressed that the unitary reality referred to here is meant to represent an absolute unitary state. As mentioned, there may be many other states that have a significant degree of unitary experience and interconnectedness even though *everything* is not considered to be unified. Unitary states other than absolute unity most likely represent a number of spiritual or mystical states. The absolute unitary state referred to in this model represents a state described in many religious and philosophical perspectives. Thus, nirvana, absolute reality, absolute unitary being, and a number of others are all terms that might refer to this complete unitary experience of the universe.

The first six primary knowing states all could be considered to represent an experience of reality with multiple discrete objects. The person in one of these states perceives individual and independent objects in that reality. These objects can be related to other objects in terms of time, space, and cause. The first three primary knowing states further refer to the perception that there are regular relationships between things. Thus, these relationships are logical and have a logical ordering. It may be said that these regular relationships are predictable and allow for a consistent understanding of reality based on the causal process of the brain. For example, this regularity is what allows science to work in helping to understand what is typically called "physical reality" or "baseline reality." Baseline reality refers to the first state listed above, in which there are discrete objects with generally regular relationships and an overall neutral emotional stance. This is the primary knowing state that most people are in most of the time. For example, most people are quite certain of the reality of the houses, cars, and people around them. Furthermore, few, if any, individuals would question the fundamental reality (or the sense of that reality) of that state.

In Judaism, this state of being would be the one in which most Torah scholars would engage in learning and discourse about Jewish law and study the intricate details and systematic arguments. Even the basic assumption behind prayer—that one should simply pray for what one needs—follows this model. If one prays, then the regular relationship assumed in the Jewish religion is that God will answer (even if the answer is no). In either of these religious acts, learning or prayer, the orientation and procedures are laid out and the reality is already clear to the individual when beginning the practice. It is precisely because this state appears certain to represent the true objective reality, while in that state, that it can be called a primary knowing state. In fact, most people would consider this state to be the true reality and that there is nothing beyond this reality. However, there are eight other primary knowing states. Two of these are very similar to what is experienced as baseline reality and consist of the same discrete objects and the same relationships between these objects. The difference is in the emotional approach we take toward this experience of reality.

The second primary knowing state is one in which there are multiple objects with regular relationships but viewed with an overwhelmingly positive emotional state. It is an experience associated with an elated sense of being and joy in which the universe is perceived to be fundamentally good. There is a sense of purposefulness to all things and to humankind's place within the universe. This purposefulness is not derived logically; it is simply intuited because of the positive emotional state. The onset of this state is frequently sudden and is often described as a conversion experience, especially in religious thought. In psychiatric literature, this state has been called cosmic consciousness by Richard Bucke [3] and is characterized by a state of overwhelming happiness, universal understanding, and love. Although this state may have a sudden onset, it can last for many years and sometimes even for the person's entire life.

A relevant example of this phenomenon in Jewish literature is that of the Talmud's Nachum Ish Gamzu (Nachum man of *Gam zu*). Rabbi Nachum is called this in the Talmud because he always stated about anything that happened to him, *Gam zu letova*, "This is also for good" (Talmud, *Ta'anit* 21a). As such, he believed that everything that happens in the universe might be understood to be for the best because everything ultimately is part of God's creation. Though the emotion usually identified is that of a steadfast belief that ultimately all will be well, Rabbi Nachum's perspective can also be interpreted to mean that everything truly is good, no matter what, even if something seems bad or evil. In a similar manner, when praying in the morning before the recital of the *Shema*, many Jews recite the line that God *oseh shalom uvorei et hakol*, "God makes peace and created all." The original line was changed to *hakol*, "all," from the word *ra*, which means "bad" (Isaiah 45:7). This perspective has God creating both good and bad, and that is somehow still seen as all "good." This may be the perspective of Nachum Ish Gamzu and many Jewish kabbalists later on.

This state experienced as an overwhelmingly positive reality, or cosmic consciousness, is a primary knowing state, since the person senses this positive perception of the universe as fundamentally real (it is not an illusion) and sometimes will look with a sense of pity at those who have only the baseline perception of reality. It is important to note that people in this state are not psychotic, nor do they have any emotional or mental disorder. They perceive the objects and relationships between objects in the universe in the same way as those in baseline reality. They simply have a different emotional vantage point of this perception.

The third primary knowing state is experienced as being composed of discrete objects with regular relationships but is associated with a profoundly negative emotional perspective. It is a state of profound sadness and futility and a sense of the incredible smallness of humankind within a universe of suffering inherent in the human condition. In this state, the universe may be understood as one vast pointless machine without purpose or meaning. This perspective is most clearly described by Job in his call out to God, "I cry out to You but You do not answer me; I stood, and You ponder me. You became cruel to me; with the strength of Your hand You manifest hatred to me" (30:20–21), and in Ecclesiastes "Meaningless, meaningless, says Kohelet, utterly meaningless, all is meaningless" (1:2). Perhaps the closest commonly known psychological state similar to this is major depression. In such a state, people often seek psychiatric help because of their severely depressed emotions even though they perceive this state to be fundamentally real. They are hoping to be restored to a more neutral or positive perspective of reality. At worst, they are hoping to be taught to think in an "illusory," nondepressed way so that they can survive. As with any primary knowing state, this overly negative state can last many years or be very short-lived.

The next three states are associated with the perception of a reality filled with discrete objects but related to each other in irregular ways. Time, space, and normal causal relationships between objects appear distorted, bizarre, and erratic. Examples of this type of state include dreams, drug-induced states, and schizophrenia. In a dream state, you might find yourself at home, but as soon as you step outside, you find yourself in another country, and then suddenly your dog appears. In the dream state, we treat all of these elements as if they are real. We don't typically question how bizarre these things are in the midst of our dream. Drug-induced states are also famous for their wild sensory experiences, with all kinds of colors and distortions of objects. The causal processes of the brain appear to be all mixed up in these states, and we therefore experience strange connections and interactions. Of course, an important question in these states is whether it is the brain or reality that is acting weird. This issue contributes to the general problem that people, including scientists, have with the field of quantum mechanics, because at that level of objective reality, things appear to act in "spooky" ways. This conflict between the irregular interactions in quantum reality and the causal processes of our brain trying to find regular interactions led one of the greatest scientific minds, Albert Einstein, to reject quantum mechanics. Although he tried to find mathematical and scientific evidence to disprove quantum theory, his basic rejection of it was probably more related to the failure of his causal processes in the brain to be able to adequately acknowledge how strange reality can be.

The states of irregular relationships between objects can still have negative, positive, or neutral emotions. For example, the "bad trip" that one can have on LSD can be profoundly disturbing. On the other hand, such states can be quite euphoric and even spiritual. Research by Dr. Roland Griffiths at Johns Hopkins University has shown that drug-induced states, such as those that arise from psilocybin, are frequently reported as the most intense spiritual experiences a person can have.[4] And in our survey of intense spiritual experiences, those occurring while under the influence of psychedelic drugs are every bit as profound and powerful as more "naturally" occurring experiences.[5] Schizophrenia is a pathological condition in which people harbor delusions or hallucinations that also can be associated with negative, positive, or neutral emotions. This results in schizoaffective disorder, in which both mood and perceptions of reality are altered.

The important point regarding all of the primary knowing states with discrete objects is that they all are perceived as real while the person is in that state. As we mentioned above with dreams, once you wake up, you recognize the prior state as an illusion, delusion, or hallucination. This judgment is consistent with the general nature of primary knowing states. As long as you are in a particular state, it is perceived as being reality. And once you move into a different primary state, the first state is perceived to be unreal, an illusion, or a distortion.

While all primary knowing states represent important experiences of reality, the dream state is particularly associated with prophecy as described in the Bible. In fact, prophecies are frequently given to people either through a dream or a vision. A vision is sometimes described as a waking dream. One of the most famous dreams is that of Jacob and the ladder to heaven:

> And he dreamed, and behold, there was a ladder set up on the earth, and the top of it reached to heaven. And behold,

> the angels of God were ascending and descending on it! And behold, Adonai stood above it and said, "I am Adonai, the God of Abraham your father and the God of Isaac. The land on which you lie I will give to you and to your offspring. Your offspring shall be like the dust of the earth, and you shall spread abroad to the west and to the east and to the north and to the south, and in you and your offspring shall all the families of the earth be blessed. Behold, I am with you and will keep you wherever you go, and will bring you back to this land. For I will not leave you until I have done what I have promised you." Then Jacob awoke from his sleep and said, "Surely Adonai is in this place, and I did not know it." (Genesis 28:12–16)

Again, what is essential about these experiences is that while they are recognized as a dream/prophetic state, the realness of what is revealed is undeniable. Most important, the person having the dream usually comes to understand that God is providing information about the world through the dream. In Jacob's case, he came to understand that his descendants would spread throughout the Land of Israel and be blessed by God. The dream also provided proof of God's existence for Jacob. In this way, these specific visionary dreams are not treated like typical primary knowing states, but have the unusual feature of being recognized as reflecting something that is real even when the person is no longer in that state. The realness of the meaning of these experiences endures beyond the primary knowing state, which is part of what denotes the dream as having religious meaning. This is a unique aspect because on one hand, the experience is recognized as a dream and distinct from baseline reality, but on the other hand, there is meaning that carries over into baseline reality.

Rabbi Samson Raphael Hirsch states in his *Commentary on Genesis* (18:2–3):

> Many confuse prophecy—*Jewish* prophecy—with delirium and divination, ecstasy and clairvoyance. As a result, ecstasy is thought to lead to prophecy, and prophecy is considered

> merely a higher state of ecstasy. Even Jewish philosophers are not free of the notion that prophecy requires *hitbodedut*—spatial and spiritual abstraction, physical and mental isolation. Yet a vast gulf separates all these from true prophecy. What leads to God's nearness is not an abstract contemplation, but, rather a life of vitality, flowing from the Source of life. Jewish prophecy is not the product of a morbid imagination, of an agitated abnormal condition; rather, it is part of a healthy life, a product of wakefulness and joyful creativity. As our sages say, "Prophecy does not come when there is sorrow or sloth, frivolity or levity, chatter or foolishness; it comes as a result of Joy in the performance of a mitzvah" (*Shabbos* 30b).

The final three states involve a unitary experience of reality. As mentioned previously, there is no point in referring to regular or irregular relationships in the unitary reality state since there are no discrete objects that can be related to each other. In the state of unitary reality, there is no sense of individual objects, and there is no self-other dichotomy. Everything is perceived as an undifferentiated, unified oneness. Thus, the unitary reality state can only be divided into three possible emotional states, which are positive, negative, or neutral.

If the experience of unitary reality is associated with positive affect, it is perceived as an overwhelmingly positive undifferentiated oneness. It differs from the cosmic consciousness in which the universe is perceived positively, but there are distinct objects, usually interrelated in a positive network of interactions. However, unitary reality with positive affect results in the direct apprehension of absolute oneness—the person actually *becomes* the oneness. This might sound strange, but there are many accounts throughout the world's traditions attesting to this state. This emotionally positive sense of unitary reality is most often described after having the experience as "God" or the "union with God." The positive emotion appears to provide a sense of an ultimate being that suffuses the oneness.

Importantly, this type of experience of reality may be the basis for the Jewish belief in a single God. A person with the profound sense of oneness of an ultimate or supreme being would eliminate any notion of lesser or individual gods. Within this oneness, there can be only one God. And the direct perception of God becomes the goal. This state is the one that many Jewish kabbalists train for years in meditation, prayer, and rituals to achieve.

The emotionally neutral state of unitary reality is associated with the same sense of oneness but typically on a completely impersonal level. Unitary reality is not viewed as good or bad or anything—it just is. The universe is understood on a very existential level. One of the female Reconstructionist rabbis in our survey may have been getting at this type of experience when she describes God as follows: "God is all of existence. Not 'a' universal being, but Universal Being." Everything is because it is, and things happen because they happen. In this context, God might represent "beingness" itself. There is no specific purpose, no good, and no bad. This may be the state that some kabbalistic rituals help achieve. However, Judaism does not believe that there is a void without moral meaning behind it. Thus, this state would be seen as a false one by most Jews. This might be how some interpret the state of *tohu* from the first verse of Genesis with the creation of the world, "and the earth was *tohu* and *bohu*" (Genesis 1:2), which can mean a formlessness prior to the world being created (See *Bereishit Rabbah* 1:5). This formlessness is clearly considered inferior in the Midrash, lacking in the substance of this world and having no purpose that God intended. Some people experiencing this state may go even further, since they understand no particular purpose and may actually interpret such a state as an undifferentiated nothingness or emptiness instead of a oneness. This notion of nothingness is frequently referred to in the Buddhist tradition. For example, the Dalai Lama refers to emptiness as "the true nature of things and events," in his book on the Heart Sutra.[6] However, the notion of emptiness does not mean nonexistence, but rather that the universe and the things in it do not exist as they appear to us.

Thus, we might consider that the unitary state with neutral emotion is the basis for the notion of the void or infinite nothingness frequently described in traditions such as Buddhism.

It is interesting to note that to date there are no clear references to a perception of a unitary reality with an intense negative emotion. It may be that such a state simply is not possible from a physiological or psychological perspective. Perhaps it cannot come about because to psychologically experience all things as an undifferentiated oneness is so powerfully positive and integrative that it cannot be perceived in a negative way. It may even be argued that such an emotionally negative state of unitary reality is physiologically incompatible with life, the brain, or the mind. Thus, until actual evidence can be brought forward to demonstrate the existence of this theoretical state, it must be assumed that it is just that, theoretical.

One additional, although important, point is that the experience of absolute unitary reality should theoretically combine all of the emotional responses into one, just as everything else is absorbed into this oneness. There should not even be different emotional components. Hence, the last three states actually collapse into one, which simply is absolute unitary reality. However, because of the emotional responses a person has after the fact (when back in everyday reality), it may still be worthwhile to consider these unitary states as having positive, neutral, or negative aspects.

When it comes to the neurotheological question regarding the actual nature of these different primary knowing states, we might ponder whether different religious or mystical states are truly different from each other (from a neurophysiological level as well as phenomenological perspective) or whether they are all very similar, if not the same, and are only described differently. The answer to this problem could have profound theological implications, since it could be argued that an "absolute unitary state" in which everything is experienced as a completely undifferentiated oneness is, by definition, the same for everyone. The implication is that the neurophysiology would also be the same. How any given unitary experience of reality becomes incorporated into a person's life could depend on many other variables, including the person's initial beliefs, cultural background, and genetics. On the other hand, if the experiences are fundamentally different from each other, then one can conceive of a typology and a way of relating these experiences to each other based on a combination of neurophysiological and phenomenological elements.

The most important aspect of the primary knowing state of unitary reality is that unlike other primary states, when an individual "comes out of it," that person does not perceive it or the memory of it as an illusion or unreal. Once a person has been in the state of unitary reality, he or she understands it to exist even though the person may not be in those states at some later time. Thus, the state of unitary reality appears to violate the rule of primary knowing states, that they are real when in them and are perceived as not real when in another. When unitary reality is experienced, the person holds this state to be fundamentally real regardless of which other state he or she is in. In fact, the sense of reality is so strong in unitary reality, that when a person comes out of unitary reality and enters into another primary knowing state, the new state is perceived as a mere reflection or distortion of the unitary reality. Everyday reality is perceived to be derived from this unitary state rather than the other way around. This is an important neurotheological point, since many scientists and atheists consider the brain to *produce* or cause our experiences of reality. But those individuals who have experienced a unitary reality see it just the opposite—the brain and everything in our everyday reality comes from the unitary reality. Thus, unitary reality is perceived as real beyond all other primary states even when a person is in those other states.

Jewish Concepts Related to Primary Knowing States

Before we explore the neurophysiological processes associated with these knowing states, it might be helpful to determine how the Jewish perspective can provide additional descriptions that relate to these states. These primary knowing states might link most closely, although in a slightly different manner, to the kabbalistic and Hasidic soul divisions of *Nefesh*, *Ruach*, *Neshama*, *Chaya*, and *Yechida*, discussed in the beginning of the book. The first level of *Nefesh* is understood to be entirely physical and would correlate with the experience of objects with either regular or irregular relationships. The second component of *Ruach* is understood to be where emotional intelligence lies. With this added component, the first six knowing states have been achieved with positive, neutral, or negative emotions. This correlates with what the Kabbalah refers to as the *sitra achra*, things from the "other side," which stem from badness, or negative associations, and *kelipat noga*, which is a middle ground husk of existence that has potential to be used for good or bad. The purely good emotion is associated with anything to do with God and God's worship.

Interestingly, Chabad Hasidim place such a strong emphasis on intellectual understanding as a bridge to connect with God, they are clearest in their description of the next level of one's soul, that of *Neshama*. This knowing state is where intellect rules and understanding everything is key. For Hasidim, this knowledge is not simply understanding, but it is the basis of recognizing that God continuously sustains everything. One might, at this point, enter into the unitary knowing state that is suffused with neutral emotion. The next level of the soul (now only used by kabbalists) of *Chaya* is entirely spiritual, where existence is unified in an unrealized state, which would be likened to the primary knowing state of absolute unity with positive emotion. The final level of *Yechida* is a God consciousness that exists outside of time and space and is never separated from God. At this level, the soul becomes one with the Divine.[7] This *Yechida* level might represent a different interpretation of the absolute unitary state, or it might represent a state related to the soul rather than the mind or brain. It is considered to be the experience in which the soul cleaves to and reflects the original infinite light (*or ein sof*).

This last state also raises an important neurotheological point that relates to the limitations of scientific inquiry. Neuroscience can only measure the brain processes as they relate to mental states. Neuroscience, at least at this point, has no way of identifying processes related to the soul itself. We can determine what happens in the brain when a person experiences some aspect of the soul, but we cannot measure the soul itself. It is not unlike trying to figure out the type of boat from measuring its wake. We get some idea about it, but we are unable to fully understand it. Similarly, if there are true spiritual states that go beyond the brain, then either we can try to say something about them by finding out what happens in the brain or, perhaps, we might consider a kind of subtraction technique. If we were to do a hypothetical brain scan study in which an individual experienced the *Yechida* level, perhaps the most interesting finding would be to see nothing change in the brain at all. In such a case, the experience would not have a neurological correlate and might actually reflect truly independent spiritual process, bolstering the claim that there is a soul entirely independent from the brain. However, for now, let us consider what brain processes we might expect when people are in the various primary knowing states.

The Neurophysiology of Primary Knowing States

Now that the primary knowing states have been described, the neurophysiological correlates of these knowing states can be explored. And the question as to how the brain helps human beings perceive the realness of reality in terms of these primary states can be addressed. It is likely that the perception of reality and the primary knowing states are associated with the integrated functioning of a number of brain structures and processes.

In the most general way, parts of the brain that help with the perception and understanding of objects play a role in primary knowing states involving multiple discrete objects. The primary sensory areas, including parts of the thalamus, visual cortex, auditory cortex, and body sensory areas, help to differentiate objects. The parietal lobe helps differentiate our own self from other objects in the world—the self-other dichotomy. Damage to areas such as the parietal lobe result in patients with severe distortions in their sense of time, space, and body. Similarly, the right parietal and occipital lobes tend to be involved with the ordering of spatial relationships. Thus, it might be argued that the functioning of these areas of the brain, which help us order our sensations and our world, provide for us a sense of discrete objects, identified by the senses. If these neural pathways are functioning properly, the result is a sense of "regular" relationships and interactions between the things perceived. However, if there is a disturbance in the functioning of these pathways, either by drugs or disease, then the result is a sense of "irregular" relationships between the things perceived.

One important point though is that there is the underlying assumption that the external world in fact has regular relationships to begin with. This is certainly the general sense of scientists who historically have searched for the beauty in the regular ways objects interact with each other. The laws of physics related to energy, motion, and entropy appear to reflect a highly ordered world. In fact, there are a number of scholars and scientists who have argued that the world is so well ordered, from the smallest atoms to DNA to animals to the earth to the universe, that such ordering could only have happened if there was an intelligent creator—God.[8] Of course, as science has progressed, we have found that relationships are not always as regular as they initially appear. The field of quantum mechanics is typically very counterintuitive and bordering on illogical. Subatomic particles apparently behave in very random ways and can interact over vast distances apparently violating the speed of light. For this reason, many great scientists, including Albert Einstein, had a great deal of difficulty perceiving the reality of quantum mechanics. Einstein famously said, "God does not play dice with the universe"; in this example, the brain of Einstein wanted there to be regular relationships even though the universe was operating in an irregular way. Thus, regular and irregular relationships depend not only on the brain processes of the individual, but also on the actual state of reality. Unfortunately, based on the discussion thus far, it is not possible to "know" what the actual relationships are, only how they are perceived. Whether the perception matches the external world in a one-to-one correspondence remains to be determined by neurotheology.

The remaining characteristic of the primary knowing states related to emotions is most likely tied into the processes of the limbic system and other areas supporting emotions. The limbic system is primarily responsible for providing our emotional responses. The limbic areas of the amygdala and hippocampus have generally been activated when perceiving emotions in faces, pictures, stories, and other emotional tasks. The limbic system helps us be aware of our emotions through its connections with structures such as the frontal, temporal, and parietal lobes, so that we understand how our emotions are directed. The limbic system is also connected to the body through the hypothalamus and autonomic nervous system so that we can feel our emotions throughout our being. In this way, the perceptions of any primary knowing state can be given a positive, neutral, or negative emotional value.

The primary knowing states involving unitary reality likely incorporate the parietal lobe's holistic processes that appear to enable us to perceive a oneness in reality and a breakdown of the self-other dichotomy. Scientific studies to date suggest that deafferentation of this area is associated with a sense of no space and no time. Interestingly, the holistic nature of this state often provides a solution to many of the moral and existential problems human beings face, such as why there is good and evil, right and wrong, and life and death. These are the issues that confront individuals in the primary knowing states of discrete being. However, there is a resolution of these issues in the unitary states, since there can be no opposites. Good and evil, right and wrong, life and death simply cannot be understood as individual concepts. They are bound into a holistic unity. As we have described above, the unitary state itself may be a holistic blend of emotions as well. However, after the experience, the ability to resolve formerly irreconcilable opposites may carry with it a rush of positive emotion and a powerful sense of the validity of the unifying perception somewhat similar to the "Eureka!" phenomenon.

What these neurophysiological correlates of the primary knowing states reflect are the conditions within each of those states. The more difficult issue to answer is why they actually *feel* real. Suffice it to say here that these states are perceived as being very real.

The Reality of Unitary Reality and Baseline Reality

The sense that is subjectively attained through the absolute operation of the holistic process yields the subjective perception of absolute and total unity of being without any time or space dimension. This experience transcends any perception of multiple discrete objects in addition to the awareness of self and other. This perceived experience of unitary reality is interpreted in most world religions as either a direct perception of God, as the *unio mystica* of the Christian tradition, or complete *devekut* (unity) with God in Jewish Hasidic and kabbalistic practices, often described as the *aspaklaria hameira* (clear glass window) describing Moses's view of God (See Yevamot 49b), which is a manifestation of God though not the revelation of God's innermost nature. Certainly, the experience does not have to be theistically labeled, and it can be understood philosophically as an ecstatic experience of the absolute, the ultimate, or the transcendent. In the Buddhist tradition the experience is interpreted as the void or nirvana. After coming out of this primary knowing state, a person may interpret it as experiencing a personal God (typically associated with positive emotions) or as a totally nonpersonal experience of total nothingness (typically associated with a neutral emotion).

Furthermore, the individual who has this subjective experience is absolutely certain of its objective reality.

In the past, many psychiatrists believed that such unitary states were actually abnormal states, caused by a significant pathological or disease process. However, analysis from both a psychological as well as an anthropological point of view indicates that these states are not generally associated with psychological disorders. Perhaps the most significant aspect of these experiences that indicates they do not represent psychoses is that the individuals who have these experiences tend to function very well in their particular society. In fact, people who experience unitary states often assume high stature in their society or culture, such as shamans or Zen masters.

In Judaism, such individuals directly experiencing the oneness of God were most often masters of the Kabbalah, both Hasidic and non-Hasidic. From the Torah, it appears that the *Avot* (forefathers) Abraham, Isaac, and Jacob may have had these states when they received prophecy, and certainly Moses achieved this level (as it states in Deuteronomy 34:10, "There had never arisen in Israel a prophet like Moses, whom God 'knew' face to face"). Many other prophets in the Torah also may have achieved such a state, but it is difficult to determine. For example, the vision of Ezekiel's chariot may have occurred after such an experience, and Job's revelation after he complains to God may also have been such a revelation. Early Jewish mystics sought to achieve such a state, and the many Hasidic rebbes we have discussed, including the Baal Shem Tov (as evidenced by his multiple stories of ascending to heaven to argue on behalf of the Jews), Rav Schneur Zalman, and Rabbi Nachman of Breslov, all may have experienced such states. Additionally, other serious kabbalists who were decidedly not Hasidic may have experienced such a state, such as Rabbi Joseph Karo (Author of the *Shulchan Aruch*) as well as Rabbi Elijah ben Solomon Zalman (also known as the Gra or Vilna Gaon), in his personal kabbalistic rituals as evidenced by his intensely kabbalistic work *Kol Tur* (Voice of the Dove) regarding the messianic era. Thus, these experiences tend to be highly adaptive for individuals in society. Furthermore, these people are able to think clearly and react to baseline reality in a manner that does not indicate any cognitive dysfunction. Conversely, as we have described previously, people with psychological disorders typically have persistent and repetitive experiences, cognitive impairment, and disordered thinking and are unable to function in everyday reality.

One might think that people who have experienced profound unitary states, in addition to day-to-day baseline reality, might have great difficulty in reconciling the two. After all, the two experiences are quite different and in some ways are incompatible with each other. Thus, for those people who have experienced both realities, the difficulty lies in developing a coherent explanation of how both of these experiences of reality can be integrated.

In an attempt to reconcile the baseline reality with the reality of unitary states, a number of models can be put forward. From the scientific perspective, baseline reality is considered to be the true reality, and the experience of unitary reality is only a mental state generated by the physical brain. Thus, the unitary reality is a manifestation of the brain's functions at best and a delusion or psychosis at worst. Another possible approach is that of the Mayavadi Hindus. For the great Mayavadi philosophers and mystics, the reality of the unitary state is so great that they deny the reality of our baseline reality. They believe that our everyday experience of reality is considered to be only a realm of illusion. Thus, all of the appearances of the external world, all of the relationships between discrete objects, all of the relationships of causality, and all of the laws of science are simply an illusion. Ultimate reality is the reality of the absolute unitary state, or what the Hindu would call Brahman.

This is similar to ideas espoused by the Hasidic leader of Izhbitza, Rabbi Mordechai Yosef. His belief is that God is real as an absolute unity and the rest of reality is essentially a "dream of

God" where there is no reality other than God. Thus, in the literal understanding of *ein od milvado*, "there is nothing beside God," our existence, as it were, is simply the internal thoughts and emanations from God but independently has no existence without God and ultimately is entirely subsumed within the Godhead.[9]

Another variation comes from Buddhist philosophers who have postulated that what is going in the mind/brain is actually *no* thing. Yet this is not "nothing" as it is understood in everyday parlance, but "no thing," simply because it cannot be conceptualized outside of the constraints of the mind. The mind has only two ways of interpreting this "no thing," namely as a unitary state or as the discrete world of baseline reality. This could potentially correlate with the kabbalistic view of the highest world of *Adam Kadmon*, the "world beyond worlds" that God exists in. This world is so removed from a human being's physical state that it is impossible to be comprehended by us. It is truthfully "no thing" but not nothing, since it is in fact everything.

In the Christian view, both baseline reality and the unitary reality are equal in terms of the certainty of their existence. On the one hand, baseline reality is definitely real, but so is the perception of the unitary state that the Christian would call God. In the Christian synthesis, the priority is given to the experience of God. For the Christian, it is as if the two realities are running parallel to each other, with the unitary reality supporting the other and causing it to be. Thus, baseline reality runs parallel to the realm of God, but God is regarded as the ultimate ground, foundation, or cause of the world of everyday baseline reality.

The typical Jewish perspective also accepts the baseline reality as well as the unitary reality. However, within different groups of Jews the perspectives taken will differ significantly. Among those more kabbalistically inclined, especially among Hasidic leaders, the unitary reality will have precedence. This is similar to the Christian perspective stated above, in which the baseline reality runs parallel to the realm of God and is constantly supported by God in order to maintain its existence.

There may also be another reconciliation in which we conclude that there is an actual reality and that reality is manifested in two different modes. This is analogous to considering reality the coin and the two modes associated with different sides of the same coin. One mode is the world we all experience, that of baseline reality, and the other mode is that of absolute unity. In this model, neither experience can be systematically reduced to the other. In fact, both experiences represent what is actually real. What is perceived to be real just depends on whether one is standing in baseline reality or unitary reality. Both are primary knowing states. When one is experiencing everyday reality, that feels real. And when one experiences absolute unity, that feels real.

This model may in fact be the closest model to that of rationalistic Jewish thought. When the Jewish rationalists state that there is an unknowable spiritual existence, they can be referring to this exact concept, namely that there is a spiritual existence that must be tapped into for the religious individual to speak to God, to form a relationship with God, and to understand why he or she must perform the various rituals of Jewish practice. However, that spiritual relationship is uniquely individualistic and unknowable from the aspect of the abstract and logical functions of the brain. As such, Judaism focuses much more on the potential integration of both rationalistic and spiritual brain processes into one's entire life. Instead of just simply thinking holistically, it could be that Judaism is seeking to think "cerebrally" with all aspects of the brain balanced in one's input and decisions.

The halacha/Jewish law process is one that is very complex and utilizes many rules and regulations. Yet, ultimately, the laws and rules only make sense within a religious framework. Mixing milk and meat and eating pork are not inherently bad things that one would shy away from. However, there is both spiritual reasoning and the practical interpretation of the biblical verses that command one not to partake of that type of food. There, once again, the practitioner who has learned about what it is that he or she is getting ready to do sees an action and must think about all of the various components and elements of fulfilling the requirement appropriately, but should also hold the spiritual components in mind. This may be why some have said that Judaism is a religion of "deed, not creed."[10] While the statement can be argued, the idea that Judaism bases its religious identity on the more rational processes of the brain may not be far off. Except for the kabbalists, Judaism in general seeks to ground its religious perspective and practice, including its theology, in the living, breathing world that is filled with individuals.

However, Judaism also sees something beyond the world we live in day-to-day, and that is precisely where the beliefs of a messianic vision and ultimate unity/Messiah may still play into the greater Jewish consciousness. The story of Moses in Rabbi Akiva's *beit midrash* (house of learning) by the Talmud in *Menachot* 29b is a great indication of this dichotomy:

> Rabbi Yehuda said in the name of Rav: When Moshe went up to the heavens, he found God sitting and fastening crowns to the letters of the Torah. He asked, "Sovereign of the universe, who is delaying you (to tie all these)?" God responded, "There will be a man who will live many generations from now whose name is Akiva son of Yoseph, and he will derive heaps of laws from every little mark." Moshe said, "Master of the universe, show him to me!" God replied, "Turn around." Moshe went and sat behind the eighth row of students [in Rabbi Akiva's house of study]. He did not understand what was being said. Moshe was upset. But when the discussion reached a certain point, Rabbi Akiva's students asked, "Rabbi, what is the source of the authority of these teachings?" Rabbi Akiva replied, "*Halacha l'Moshe miSinai* [This is law given to Moshe at Sinai]." Moshe's mind was then settled. Moshe returned to God and said, "Sovereign of the universe, you have such a man [as Akiva] and yet you give the Torah through me?" God replied, "Be silent. Thus have I decided."

Moses feels faint at not understanding the complexities of the law being taught that he himself does not know! However, when Rabbi Akiva states that the law is from Moses at Sinai, Moses is comforted. How could one be comforted if Rabbi Akiva was lying about the law being from Moses? The commentators explain that the story is actually teaching that Rabbi Akiva learned many Jewish laws from the small crowns on the Hebrew letters and that these teachings were all contained within the Jewish tradition's understanding of not leaving anything unexplained. The tradition of study and learning was what Rabbi Akiva meant when he said, "This is law given to Moses at Sinai." The spiritual experience and the individual learning and studying rationally are both intrinsically linked together in Jewish law and thought.

This perspective may also reflect how many Jewish thinkers speak of a constant tension between the desire to unify with God and a desire to withdraw from God's presence and live in the physical world. "The sons of Aaron, Nadab and Abihu, each took his pan, put fire in them, and placed incense upon it, and they brought before Adonai a strange fire, which God had not commanded them" (Leviticus 10:1). God kills them for bringing the strange fire, and the commentators debate what exactly this strange fire was. Shadal (Rabbi Samuel Luzzatto, early nineteenth-century commentator) states that they brought a fire they were not commanded to due to their haughtiness. Human beings cannot achieve a complete union with the Divine unless directed to. If people try to get too close to the fire, they will be burned, and in this case it was literal. This is an interesting point about unitary states, since, for some, the experience appears to happen as part of an experience of reality. If the unitary state is perceived as impersonal, we are not asked or directed toward it, at least not by any kind of being. But in the theistic tradition of Judaism, it seems reasonable to suggest that God directs a person toward that experience. It would seem very unlikely for a person's brain to "accidentally" become one with God.

The fact is that Judaism has many diverse perspectives about how to approach God, but the constant need to reassess one's own relationship to the creator is a common and widely utilized concept. Rabbi Joseph B. Soloveitchik (in his work *Lonely Man of Faith*) popularized this concept in his description of the two "stories" of Adam, with Adam I of the "natural work community" being the scientist world conqueror, and Adam II of the "covenantal faith community" being spiritually connected and desiring something beyond this world. According to this position, the "lonely man of faith" is lonely because he cannot reconcile these two positions and must continue to walk the tightrope of balancing the current experience of discrete reality with the possibility of there being a larger unitary reality as well. One of the great challenges of neurotheology will be to continue to deal with this issue of how various primary knowing states are experienced and expressed and how differences between them can be reconciled. Such a reconciliation lies at the heart of the epistemological question regarding the fundamental nature of reality.

What Is Really Real?

The question of what is really real becomes much more complicated than it already seems when we invoke a neurophysiological explanation of human experience. If we agree that the mind/brain is what we use to perceive reality, then we are left with the difficult (or impossible) task of somehow getting outside of our mind or brain in order to prove what we perceive to be reality. Many of the meditative philosophies use this concept to explain why their approach is important for reaching a greater experience of reality. They argue that baseline reality is an illusion or less real and that the true reality requires an excursion of the mind into a realm beyond normal experience, not unlike the cave allegory by Plato. We have to turn our mind around or even leave the cave of our brain if we are going to truly know reality. This is what we have described previously as the absolute unitary state. We have already described in this chapter that the unitary state is a primary knowing state, but these are all states of the mind/brain. Further, the aspect of such states that makes them so real is that they are experienced by the mind/brain as being intensely real. Although a neurotheological approach might suggest that such unitary states have their basis in neuroanatomy, neurophysiology, and the flux of neurotransmitters, it is equally true that the experience of baseline reality, which both the average person and the average scientist construe to be really real, is based on exactly the same parameters. Thus, one can never get at what is "really out there" without its being processed, one way or another, through the brain.

Many may find it deeply troubling that the experience of God, the sense of the absolute, the sense of mystery and beauty in the universe, the most profoundly moving experiences of which humans are capable, might be reducible to neurophysiological processes. However, such an interpretation misses a few rather important points. First of all, our experience of baseline reality (e.g., of chairs, tables, love, hate), indeed of our whole physical and psychological environment, can also be reduced to brain chemistry. So what criteria can we use to evaluate whether God, other unitary experiences, or our everyday world is more "real"? Can we use our subjective sense of the absolute certainty of the objective reality of our everyday world to establish that that world is "really real"?

To simplify the issue somewhat, let us for the moment contrast the most extreme unitary state with baseline reality. In such an exercise one can see that there is no question that the absolute unitary state wins out as being "more real." People who have experienced an absolute unitary state regard it as being more fundamentally real than baseline reality. In our survey of people's most intense spiritual experiences, the data supports this very contention. Over 90 percent of the survey respondents reported that their intense spiritual experience felt as real or more real than everyday reality. Even the memory of it carries the sense of greater fundamental reality than that generated by their experiences of day-to-day living. From our Survey of Rabbis, many of the respondents who reported a spiritual experience related that it had a lasting impact on them and that they think back to it regularly. If we use the criterion, therefore, of the sense of certainty of the reality of that state, the absolute unitary state would appear to win out as being "more real" and hence more representative of the "true" reality.

To further clarify this point, let us compare four characteristics of the experience of baseline reality with that of unitary reality. The experience of baseline reality demonstrates the following four fundamental properties:[11]

1. A strong sense of the reality of what is experienced.
2. Endurance of that reality through very long periods of time, usually only interrupted by sleeping.
3. The sense that when elements in baseline reality disappear, they have in fact ceased to be.
4. High cross-subjective validation both for details of perception and core meaning. In other words, other people corroborate our perceptions of the world, that is, reality is a collective hunch.

The experience of unitary reality has the following criteria:

1. An extremely strong sense of reality, to the point of its being absolutely compelling under almost all circumstances.
2. Endurance for short periods of time relative to the sense of time of baseline reality.
3. A sense of its underlying persistence and continued existence even when the perception of the overall state has ended.
4. High cross-subjective validation for the core perceptions, but moderate to low cross-subjective validation for perceptual detail in those unitary states.

At the moment, neurotheology offers no way to determine whether the unitary state or baseline reality state reflects what is truly more real. Clearly, baseline reality has some significant claim to being ultimate reality. However, the unitary states can be so compelling that it is difficult to overlook their description of reality. This being the case, it seems difficult to maintain that because unitary states can be understood in terms of neuropsychological processes, such unitary states are therefore derivative from baseline reality. Indeed, the reverse argument could be made just as well. Neuropsychology can give no answer as to which state is more truly reflective of some actual reality. We are reduced to saying that each is real in its own way and for its own adaptive ends. Thus, none of the primary knowing states mentioned exist beyond any of the other states insofar as ultimate reality goes, even though when a person is in one of these states, he or she has the sense that other states are less real. This may be the only thing that we can use to help determine what is really real until someone determines a method for going beyond the mind/brain's perception of ultimate reality. This may not be very epistemologically satisfying, but up to now there are no other alternatives.

Therefore, we must conceive of the brain as a machine that operates upon whatever it is that fundamental reality may be and produces at the very least two basic versions. One version is what human beings refer to as baseline reality, and the other version is that of absolute unitary states. Both perceptions are accompanied by a profound subjective certainty of their objective reality. Whatever is prior to the experience of absolute unity and the baseline reality of everyday life is most likely unknowable, since that which is in any way known must be translated, and in this sense transformed, by the brain. While as a religious Jew and rabbi, David is able to fully believe in God and the Torah, as well as the entire Jewish tradition and messages, and Andrew fully believes in science and its principles as key requirements in the pursuit of truth, from the perspective of neurotheology, neither can claim his view of ultimate reality is 100 percent supported. In fact, both of us believe in the benefits of science and religion working together toward greater knowledge. Even though for David this may be with the understanding that this greater knowledge will only aid his religious beliefs, and for Andrew this may not be the case, both of us are excited about the possibilities and opportunities neurotheology has to offer when analyzing Judaism. Only by engaging our brains can we thoroughly participate in both Judaism and scientific research, allowing the experiences of reality, whichever they may be, rationalistic or spiritual, to be fully explored through the lens of Jewish neurotheology. Through such a process, we would hope that everyone can explore the infinite universe around them and find the fascination and joy in such an exploration.

What is divine wisdom?
Divine wisdom is the inner delight of the Infinite,
condensed and crystallized until fit for human consumption.
What is a mitzvah?
A mitzvah is divine wisdom condensed and crystallized
until it can be performed as a physical action.
That is why in the study of Torah there is infinite delight.
That is why in the act of a mitzvah there is unlimited joy.

—From Chabad.org Daily Thought 6/16/2017,
citing *Maamar Arbaah Rashei Shanim Heim*, 5731;
see also on *Tanya*, *Igeret HaKodesh*, 29

Notes

Chapter 1

1. Skorecki K, Selig S, Blazer S, Bradman R, Bradman N, Waburton PJ, Ismajlowicz M, Hammer MF. Y chromosomes of Jewish priests. *Nature.* 1997;385(6611):32
2. Meshberger FL. 1990. An interpretation of Michelangelo's Creation of Adam based on neuroanatomy. *Journal of the American Medical Association* 264:1837–41.
3. Schroeder GL. 2009. *The Science of God: The Convergence of Scientific and Biblical Wisdom.* New York: Free Press.
4. Spinoza B. 1989. *Tractatus Theologico-Politicus*: Gebhardt Edition, Netherlands: Brill.
5. Spinoza B. 1989. *Tractatus Theologico-Politicus*: Gebhardt Edition, Netherlands: Brill.
6. Gould SJ. 1999. *Rocks of Ages.* New York: Ballantine.
7. Soloveitchik, J. D. (1992). *The Lonely Man of Faith.* New York: Doubleday.
8. Hirsch, S. (1984). *The collected writings*: Volume IX: Timeless Hashkafah: Two Letters – on the Aggadah. Pg. 209 New York: P. Feldheim
9. Newberg AB. 2010. *Principles of Neurotheology.* Surrey, UK: Ashgate Publishing Limited.

Chapter 2

1. Newberg AB. 2010. *Principles of Neurotheology.* Surrey, UK: Ashgate Publishing Limited.
2. For Maimonides position see *Mishneh Torah* Laws of Foundations of the Torah 4:8 and Guide for the Perplexed 3:54. For Rav Saddiah Gaon, see *Saadia Gaon, The Book of Beliefs and Opinion* . New Haven: Yale University Press, 1976 (243–244)
3. The *Zohar* explicitly states the components of *Nefesh, Ruach,* and *Neshama*, and the Ari in the system of Lurianic Kabbalah adds the two higher components of *Chaya* and *Yechida*, though according to the Lurianic system, these two higher levels were also hinted at and a piece of the Zoharic Kabbalistic system as well.

4. d'Aquili EG, Newberg AB. 1999. *The Mystical Mind: Probing the Biology of Religious Experience*. Minneapolis, MN: Fortress Press.
5. Wilson EO. 1999. *Concilience*. New York: Vintage Press.
6. Newberg AB, Waldman MR. 2009. *How God Changes Your Brain*. New York: Ballantine.
7. James W. [1890] 1963. *Varieties of Religious Experience*. New York: University Books.
8. Gerrish BA, MacKintosh HR, Stewart JS. 1999. *The Christian Faith by Friedrich Schleiermacher*. Edinburgh: T. & T. Clark Publishers.
9. Otto R. 1958. *Idea of the Holy*. Oxford: Oxford University Press.
10. Jung CG. 1975. The Collected Works: Volume 11. Psychology and Religion: West and East. Princeton, NJ: Princeton University Press.
11. Durkheim E., cited in Morris B. 1987. *Anthropological Studies of Religion: An Introductory Text*. New York: Cambridge University Press.
12. Geertz C. 1985. Religion as a cultural system. In Bantom M (ed.), *Anthropological Approaches to the Study of Religion*. London: Tavistock.
13. Atran S. 2002. *In Gods We Trust: The Evolutionary Landscape of Religion*. New York: Oxford University Press. Boyer P. 2002. *Religion Explained*. New York: Basic Books.
14. See as well Schachter, Rabbi Hershel. (1988). Land for Peace: A Halachic Perspective. *RJJ Journal of Halacha and Contemporary Society*, 16, 72–95.
15. *Merriam Webster Online Dictionary.*
16. See *Tzipita L'Yeshuah* Section 1
17. See *Michtav M'Eliyahu* Vol 3 pages 178–179
18. See *Sichot Ha'Ran* (Lectures of Rabbi Nachman) #33
19. See *Mishnat Chachamim* Brooklyn 5624 Section 23
20. Dawkins R. 1997. Is science a religion? *The Humanist*, January/February issue.
21. Newberg AB, Waldman MR. 2006. *Why We Believe What We Believe: Uncovering Our Biological Need for Meaning, Spirituality, and Truth*. New York: Free Press.
22. This perspective has come under scrutiny from much of the Kabbalistic academic scholarship, many of whom believe the original ten *sefirot* were actual parts of God split in some way. However, in traditional Jewish thought, this perspective was classicaly viewed as heresy and the texts have not been interpreted to mean a literal splitting of the Godhead.
23. Stark R. 2008. *What Americans Really Believe: New Findings from the Baylor Surveys of Religion*. Waco, TX: Baylor University Press.
24. David Goldstein quoted in *Science News*, October 3, 1998.

Chapter 3

1. Lamm N. 2010. *Torah Umadda*. Jerusalem: Koren Publishers Jerusalem.
2. Yang J, Monti DA. 2017. *Clinical Acupuncture and Ancient Chinese Medicine.* New York: Oxford University Press.
3. Freud S; Strachey J, Gay P (eds.). 1989. *The Future of an Illusion*. New York: W. W. Norton & Company.
4. Swinburne R. 1997. *Simplicity as Evidence for Truth*. Milwaukee, WI: Marquette University Press.
5. *Ibid*
6. Liu TT, Brown GG. 2007. Measurement of cerebral perfusion with arterial spin labeling: Part 1. Methods. *Journal of the International Neuropsychological Society* 13(3):517–25.
7. Newberg AB, Wintering N, Yaden DB, Zhong L, Bowen B, Averick N, Monti DA. 2017. Effect of a one-week spiritual retreat on dopamine and serotonin transporter binding: a preliminary study. *Religion, Brain & Behavior*. doi:10.1080/2153599X.2016.1267035.
8. Moghbel M, Newberg A, Alavi A. 2016. Positron emission tomography: ligand imaging. In Masdeu JC, González RG. *Handbook of Clinical Neurology*, pp. 229–40.
9. Yaden DB, Iwry J, Newberg AB. 2017. Neuroscience and religion: surveying the field. In Clements NK. *Religion: Mental Religion*. Farmington Hills, MI: Macmillan Reference USA, pp. 277–99.
10. Hill PC, Hood RW. 1999. *Measures of Religiosity*. Birmingham, AL: Religious Education Press.

Chapter 4

1. Helgason CM. 1987. Commentary on the significance for modern neurology of the 17th century B.C. Surgical Papyrus. *Canadian Journal of Neurological Sciences*. 14(4):560–63.
2. Simpson D. 2005. Phrenology and the neurosciences: contributions of F. J. Gall and J. G. Spurzheim. *ANZ Journal of Surgery* 75(6):475.
3. Broca P. 2015. Comparative anatomy of the cerebral convolutions: the great limbic lobe and the limbic fissure in the mammalian series. *Journal of Comparative Neurology* 523(17):2501–54.
4. Freud S; Strachey J (trans.). 2010. *The Interpretation of Dreams*. New York: Basic Books.
5. Freud S; Strachey J (trans.). 2010. *The Interpretation of Dreams*. New York: Basic Books.
6. Kandel ER, Schwartz JH, Jessell TM, Siegelbaum SA, Hudspeth AJ

(eds.). 2013. *Principles of Neural Science.* 5th ed. New York: McGraw-Hill Companies.

7. *Ibid*
8. Hugdahl K. 1996. Cognitive influences on human autonomic nervous system function. *Current Opinion in Neurobiology* 6:252–58. Newberg AB. 2010. *Principles of Neurotheology*. Surrey, UK: Ashgate Publishing Limited.
9. Kandel ER, Schwartz JH, Jessell TM, Siegelbaum SA, Hudspeth AJ (eds.). 2013. *Principles of Neural Science.* 5th ed. New York: McGraw-Hill Companies.
10. Gazzaniga MS, LeDoux JE. 1978. *The Integrated Mind.* New York: Plenum Press.
11. Gazzaniga MS, LeDoux JE. 1978. *The Integrated Mind.* New York: Plenum Press. Kandel ER, Schwartz JH, Jessell TM, Siegelbaum SA, Hudspeth AJ (eds.). 2013. *Principles of Neural Science.* 5th ed. New York: McGraw-Hill Companies.
12. Bogen, JE. 1969. The other side of the brain, II: an appositional mind. *Bulletin of Los Angeles Neurological Society* 34:135–62.
13. Eskenazi TC, Weiss AL (Editors). 2008. *The Torah: A Women's Commentary.* New York: CCAR Press.
14. Kandel ER, Schwartz JH, Jessell TM, Siegelbaum SA, Hudspeth AJ (eds.). 2013. *Principles of Neural Science.* 5th ed. New York: McGraw-Hill Companies.
15. Leinonen L, Hyvarinen J, Nyman J, Linnakoski D. 1979. Functional properties of neurons in the lateral part of associative area 7 of awake monkeys. *Experimental Brain Research* 34:299–320.
16. Lynch JC. 1980. The functional organization of posterior parietal association cortex. *Behavioral Brain Sciences* 3:485–99. di Pellegrino G, Làdavas E. 2015. Peripersonal space in the brain. *Neuropsychologia* 66:126–33.
17. Benton A. 1979. Visuoperceptive, visuospatial and visuoconstructive disorders. In Heilman KM, Valenstein E (eds.). *Clinical Neuropsychology.* Oxford: Oxford University Press, pp. 186–232. Ratcliff DA, Davies-Jones GAB. 1972. Defective visual localization in focal brain wounds. *Brain* 95:49–60.
18. Cléry J, Guipponi O, Wardak C, Ben Hamed S. 2015. Neuronal bases of peripersonal and extrapersonal spaces, their plasticity and their dynamics: knowns and unknowns. *Neuropsychologia* 70:313–26.
19. Stuss DT, Benson DF. 1986. *The Frontal Lobes.* New York: Raven Press. Miller BL, Cummings JL. 2017. *The Human Frontal Lobes: Functions and Disorders.* 3rd ed. New York: Guilford Press.

20. Adrianov OS. 1978. Projection and association levels of cortical integration. In Brazier MAB, Petsche H (eds.). *Architectonics of the Cerebral Cortex*. New York: Raven Press. Geschwind N. 1965. Disconnexion syndromes in animals and man. *Brain* 88:585–644.
21. Miller BL, Cummings JL. 2017. *The Human Frontal Lobes: Functions and Disorders*. 3rd ed. New York: Guilford Press.
22. Pohl W. 1973. Dissociation of spatial discrimination deficits following frontal and parietal lesions in monkeys. *Journal of Comparative Physiological Psychology* 82:227–39. Mishkin M, et al. 1977. Kinesthetic discrimination after prefrontal lesions in monkeys. *Brain Research* 130:163–68.
23. Newberg A, Pourdehnad M, Alavi A, d'Aquili E. 2003. Cerebral blood flow during meditative prayer: preliminary findings and methodological issues. *Perceptual and Motor Skills* 97:625–30.
24. Pribram KH, McGuinness D. 1975. Arousal, activation, and effort in the control of attention. *Psychological Review* 82:116–49. Pribram, KH. 1981. Emotions. In Filskov SK, Boll TJ (eds.). *Handbook of Clinical Neuropsychology*. New York: Wiley.
25. Fuster, JM. 1980. *The Prefrontal Cortex: Anatomy, Physiology, and Neuropsychology of the Frontal Lobe*. New York: Raven Press.
26. Miller BL, Cummings JL. 2017. *The Human Frontal Lobes: Functions and Disorders*. 3rd ed. New York: Guilford Press.
27. Bruce CJ, Desimone R, Gross CG. 1986. Both striate and superior colliculus contribute to visual properties of neurons in superior temporal polysensory area of Macaque monkey. *Journal of Neurophysiology* 58:1057–76. Burton H, Jones EG. 1976. The posterior thalamic region and its cortical projections in new world and old world monkeys. *Journal Comparative Neurology* 168:249–302. Seltzer B, Pandya DN. 1978. Afferent cortical connections and architectonics of the superior temporal sulcus and surround cortex in the rhesus monkey. *Brain Research* 149:1–2.
28. Kandel ER, Schwartz JH, Jessell TM, Siegelbaum SA, Hudspeth AJ (eds.). 2013. *Principles of Neural Science*. 5th ed. New York: McGraw-Hill Companies.
29. *Ibid*
30. Davis M. 1992. The role of the amygdala in fear and anxiety. *Annual Review of Neuroscience* 15:353–75.
31. Anderson P, Morris R, Amaral D, Bliss T, O'Keefe J. 2007. The hippocampal formation. In *The Hippocampus Book*. New York: Oxford University Press.

32. Jones EG. 2007. *The Thalamus*. Cambridge: Cambridge University Press.
33. *Ibid*
34. Newman J. 1995. Thalamic contributions to attention and consciousness. *Consciousness and Cognition*. 4(2):172–93.
35. Wager T. 2002. Functional neuroanatomy of emotion: a meta-analysis of emotion activation studies in PET and fMRI. *NeuroImage* 16(2):331–48. Craig AD. 2009. How do you feel—now? The anterior insula and human awareness. *Nature Reviews in Neuroscience* 10(1):59–70.
36. Kandel ER, Schwartz JH, Jessell TM, Siegelbaum SA, Hudspeth AJ (eds.). 2013. *Principles of Neural Science.* 5th ed. New York: McGraw-Hill Companies.
37. Strata P, Scelfo B, Sacchetti B. 2011. Involvement of cerebellum in emotional behavior. *Physiology Research* 60(Suppl 1):S39–48.
38. Hoppe KD. 1977. Spilt brains and psychoanalysis. *Psychoanalytic Quarterly* 46:220–44.
39. Green JD, Adey WR. 1956. Electrophysiological studies of hippocampal connections and excitability. *Electroencephalography and Clinical Neurophysiology* 8:245–62. Nauta WJH. 1958. Hippocampal projections and related neural pathways to the midbrain in cat. *Brain* 81:319–40. Joseph R, Forrest N, Fiducis D, Como P, Siegal J. 1981. Behavioral and electrophysiological correlates of arousal. *Physiological Psychology* 9:90–95.
40. Joseph, R. 1990. *Neuropsychology, Neuropsychiatry, and Behavioral Neurology*. New York: Plenum, pp. 116–20.
41. Elias AN, Guich S, Wilson AF. 2000. Ketosis with enhanced GABAergic tone promotes physiological changes in transcendental meditation. *Medical Hypothesis* 54(4):660–62. Elias AN, Wilson AF. 1995. Serum hormonal concentrations following transcendental meditation—potential role of gamma aminobutyric acid. *Medical Hypotheses* 44(4):287–91.
42. Foote S. 1987. Extrathalamic modulation of cortical function. *Annual Review of Neuroscience* 10:67–95.
43. Aghajanian G, Sprouse J, Rasmussen K. 1987. Physiology of the midbrain serotonin system. In Meltzer H (ed.). *Psychopharmacology: The Third Generation of Progress*. New York: Raven Press, pp. 141–49.
44. Van Praag H, De Haan S. 1980. Depression vulnerability and 5-hydroxytryptophan prophylaxis. *Psychiatric Research* 3:75–83.
45. Aghajanian GK, Marek GJ. 1999. Serotonin and hallucinogens. *Neuropsychopharmacology*21 (2 Suppl):16S–23S. Bujatti M, Riederer P. 1976. Serotonin, noradrenaline, dopamine metabolites in transcendental

meditation-technique. *Journal of Neural Transmission* 39(3):257–67.

46. Walton KG, Pugh ND, Gelderloos P, Macrae P. 1995. Stress reduction and preventing hypertension: preliminary support for a psychoneuroendocrine mechanism. *Journal of Alternative and Complementary Medicine* 1(3):263–83.
47. Foote S. 1987. Extrathalamic modulation of cortical function. *Annual Review of Neuroscience* 10:67–95.
48. Bujatti M, Riederer P. 1976. Serotonin, noradrenaline, dopamine metabolites in transcendental meditation-technique. *Journal of Neural Transmission* 39(3):257–67. Walton KG, Pugh ND, Gelderloos P, Macrae P. 1995. Stress reduction and preventing hypertension: preliminary support for a psychoneuroendocrine mechanism. *Journal of Alternative and Complementary Medicine* 1(3):263–83. Infante JR, Torres-Avisbal M, Pinel P, Vallejo JA, Peran F, Gonzalez F, Contreras P, Pacheco C, Roldan A, Latre JM. 2001. Catecholamine levels in practitioners of the transcendental meditation technique. *Physiology and Behavior* 72(1–2):141–46.
49. Walton KG, Pugh ND, Gelderloos P, Macrae P. 1995. Stress reduction and preventing hypertension: preliminary support for a psychoneuroendocrine mechanism. *Journal of Alternative and Complementary Medicine* 1(3):263–83. Sudsuang R, Chentanez V, Veluvan K. 1991. Effects of Buddhist meditation on serum cortisol and total protein levels, blood pressure, pulse rate, lung volume and reaction time. *Physiology and Behavior* 50:543–48. Jevning R, Wilson AF, Davidson JM. 1978. Adrenocortical activity during meditation. *Hormones and Behavior* 10(1):54–60.

Chapter 5

1. Gianotti LR, Mohr C, Pizzagalli D, Lehmann D, Brugger P. 2001. Associative processing and paranormal belief. *Psychiatry and Clinical Neurosciences* 55(6):595–603.
2. Hamer D. 2004. *The God Gene: How Faith Is Hardwired into Our Genes.* New York: Random House.
3. Kopsida E, Stergiakouli E, Lynn PM, Wilkinson LS, Davies W. 2009. The role of the Y chromosome in brain function. *Open Neuroendocrinology Journal* 2:20–30.
4. Dessler, E. E., & Carmell, A. (1988). Strive for Truth! Part Two Michtav Me'Eliyahu (Vol. 2: pp 52-57). Jerusalem: Feldheim.
5. See Ari's *Sha'ar Ha'hakdamot—Gate of Introductions First Drasha/Lecture "Be'olam Ha'nikudam"*—The World of Points
6. Kellner M. *Moment Magazine*, July 31, 2014.

Chapter 6

1. Jankowski KF, Takahashi H. 2014. Cognitive neuroscience of social emotions and implications for psychopathology: examining embarrassment, guilt, envy, and schadenfreude. *Psychiatry and Clinical Neuroscience* 68(5):319–36.
2. d'Aquili EG. 1993. The myth-ritual complex: a biogenetic structural analysis. In Ashbrook JB (ed.). *Brain, Culture, and the Human Spirit*. New York: Lanham Press.
3. Jung CG. 1958. *Psyche and Symbol*. New York: Doubleday Anchor Books. Levi-Strauss C. 1963. *Structural Anthropology*. New York: Anchor Books.
4. Kushner H. 2004. *When Bad Things Happen to Good People*. New York: Anchor Reprint.
5. Frankl VE. 2006. *Man's Search for Meaning*. Boston: Beacon Press.
6. Tauber, Y, Schneerson M. 2016. *The inside story : a Chassidic perspective on biblical events, laws and personalities: To be or Not to be (2-13)*. Brooklyn, NY: Meaningful Life Center.
7. M. Breuer, *Pirkei Moadot*,Vol. 1, Jerusalem 1986, pp. 14–16.
8. Soloveitchik JB. 2006. *Lonely Man of Faith*. New York: Doubleday.
9. Tauber, Y, Schneerson M. 2016. *The inside story : a Chassidic perspective on biblical events, laws and personalities: To be or Not to be (2-13)*. Brooklyn, NY: Meaningful Life Center.
10. Soloveitchik, J, Shatz D, Wolowelsky J, Ziegler, R. (2008). *Abraham›s journey : reflections on the life of the founding patriarch*. Jersey City, NJ: Published for Toras HoRav Foundation by KTAV Pub. House.

Chapter 7

1. d'Aquili EG, Newberg AB. 1999. *The Mystical Mind: Probing the Biology of Religious Experience*. Minneapolis, MN: Fortress Press. Newberg AB, d'Aquili EG, Rause V. 2001. *Why God Won't Go Away: Brain Science and the Biology of Belief*. New York: Ballantine.
2. Schein MW, Hale EB. 1965. Stimuli Eliciting Sexual Behavior. In Beach FA (ed.). *Sex and Behavior*. New York: John Wiley & Sons. Tinbergen N. 1951. *The Study of Instinct*. London: Oxford University Press. Rosenblatt JS. 1965. Effects of experience on sexual behavior in male cats. In Beach FA (ed.). *Sex and Behavior*. New York: John Wiley & Sons.
3. d'Aquili EG, Newberg AB. 1999. *The Mystical Mind: Probing the Biology of Religious Experience*. Minneapolis, MN: Fortress Press.
4. Damasio A. 1999. *The Feeling of What Happens: Body and Emotion in the Making of Consciousness*. New York: Harcourt Brace.

5. For further information about the details of the larger debates between the Hasidim and the Mitnagdim of the time, particularly focused on Rabbi Elijah of Vilna, also known as "The Vilna Gaon," see Stern, Eliyahu. (2013). The Gaon versus Hasidism. In *The Genius: Elijah of Vilna and the Making of Modern Judaism* (pp. 83–114). Yale University Press.
6. d'Aquili EG, Newberg AB. 1999. *The Mystical Mind: Probing the Biology of Religious Experience*. Minneapolis, MN: Fortress Press.
7. d'Aquili EG, Newberg AB. 1993. Liminality, trance and unitary states in ritual and meditation. *Studia Liturgica* 23:2–34.
8. Bradley MM, Lang PJ, Cuthbert BN. 1993. Emotion, novelty, and the startle reflex: habituation in humans. *Behavioral Neuroscience* 107(6):970–80.
9. Johnson KD, Rao H, Wintering N, Dhillon N, Hu S, Zhu S, Korczykowski M, Johnson K, Newberg AB. 2014. Pilot study of the effect of religious symbols on brain function: association with measures of religiosity. *Spirituality in Clinical Practice* 1(2):82–98.
10. Greening S, Norton L, Virani K, Ty A, Mitchell D, Finger E. 2014. Individual differences in the anterior insula are associated with the likelihood of financially helping versus harming others. *Cognitive, Affective, & Behavioral Neuroscience* 14(1):266–77.
11. van Izendoorn MH, Bakermans-Kranenburg MJ, Pannebakker F, Out D. 2010. In defense of situational morality: genetic, dispositional and situational determinants of children's donating to charity. *Journal of Moral Education* 39:1–20.
12. Halevi, Rabbi Hayyim David Halevi's "*Mekor Hayim haShalem.*"
13. See *Yoma* 39a and Rashi ad loc. See also Dubov, N. D. (n.d.). Kelipot and Sitra Achra. Retrieved from https://www.chabad.org/library/article_cdo/aid/361900/jewish/Kelipot-and-Sitra-Achra.htm which requests we add the following comment: The content in this page is produced by Chabad.org, and is copyrighted by the author and/or Chabad.org.
14. Ding X, Tang YY, Cao C, Deng Y, Wang Y, Xin X, Posner MI. 2014. Short-term meditation modulates brain activity of insight evoked with solution cue. *Social Cognitive and Affective Neuroscience* 10(1):43–49.
15. See Benyosef S. H. (2006). The Additional Soul. In *Living the Kabbalah: A Guide to the Sabbath and Festivals in the Teachings of Rabbi Rafael Moshe Luria* (pp 20–36). Jerusalem, Israel: Feldheim.
16. Newberg AB, Wintering N, Yaden DB, Zhong L, Bowen B, Averick N, Monti DA. 2017. Effect of a one-week spiritual retreat on dopamine and serotonin transporter binding: a preliminary study. *Religion, Brain & Behavior*. doi: 10.1080/2153599X.2016.1267035.
17. Schjoedt U, Stødkilde-Jørgensen H, Geertz AW, Roepstorff A. 2009.

Highly religious participants recruit areas of social cognition in personal prayer. *Social, Cognitive, and Affective Neuroscience* 4(2):199–207.

18. Buber, M. (2010). *I and Thou.* Mansfield Centre, CT: Martino Publishing. Reprint from original 1937 translation of his initial publication in 1923.
19. Heschel, A. J. (1955). *God in Search of Man: A Philosophy of Judaism.* New York, NY: Farrar, Straus, and Cudahy.
20. Recently, there has been a significant increase in multiple works collecting the halachik responsa to new techniologic and societal innovations and exploring their impact on modern Jewish practice. For a popular recent Orthodox publication, one is recommended to read Lichtenstein, D. (2014). *Headlines: Halachic Debates of Current Events.* As well as Lichtenstein, D. (2017). *Headlines 2: Halachic Debates of Current Events.* For more in-depth analysis, one is recommended to look at Rabbi J. David Bleich's Contemporary Halakhic Problems series (Currently up to Volume 6)
21. Heschel AJ. 1955. *God in Search of Man : A Philosophy of Judaism.* New York: Farrar, Straus and Giroux

Chapter 8

1. Oman D, Kurata JH, Strawbridge WJ, Cohen RD. 2002. Religious attendance and cause of death over 31 years. *International Journal of Psychiatry and Medicine* 32(1):69–89. McCullough ME, Hoyt WT, Larson DB, Koenig HG, Thoresen C. 2000. Religious involvement and mortality: a meta-analytic review. *Health Psychology* 19(3):211–22.
2. Strawbridge WJ, Cohen RD, Shema SJ, Kaplan GA. 1997. Frequent attendance at religious services and mortality over 28 years. *American Journal of Public Health.* 87(6):957–61. Hummer RA, Rogers RG, Nam CB, Ellison CG. 1999. Religious involvement and U.S. adult mortality. *Demography* 36(2):273–85. Oman D, Reed D. 1998. Religion and mortality among the community-dwelling elderly. *American Journal of Public Health* 88(10):1469–75. Helm HM, Hays JC, Flint EP, Koenig HG, Blazer DG. 2000. Does private religious activity prolong survival? A six-year follow-up study of 3,851 older adults. *Journal of Gerontology: Series A, Biological Science and Medical Science* 55(7):M400–5. Koenig HG, Hays JC, Larson DB, et al. 1999. Does religious attendance prolong survival? A six-year follow-up study of 3,968 older adults. *Journal of Gerontology: Series A, Biological Science and Medical Science* 54(7):M370–76.
3. Kark JD, Shemi G, Friedlander Y, Martin O, Manor O, Blondheim SH. 1996. Does religious observance promote health? Mortality in secular

vs religious kibbutzim in Israel. *American Journal of Public Health* 86(3):341–46.

4. de Gouw HW, Westendorp RG, Kunst AE, Mackenbach JP, Vandenbroucke JP. 1995. Decreased mortality among contemplative monks in the Netherlands. *American Journal of Epidemiology* 141(8):771–75.
5. Abbotts J, Williams R, Ford G, Hunt K, West P. 1997. Morbidity and Irish Catholic descent in Britain: an ethnic and religious minority 150 years on. *Social Science and Medicine* 45(1):3–14.
6. Comstock GW, Partridge KB. 1972. Church attendance and health. *Journal of Chronic Disease* 25(12):665–72.
7. Koenig HG, George LK, Hays JC, Larson DB, Cohen HJ, Blazer DG. 1998. The relationship between religious activities and blood pressure in older adults. *International Journal of Psychiatry and Medicine.* 28(2):189–213. Hixson KA, Gruchow HW, Morgan DW. 1998. The relation between religiosity, selected health behaviors, and blood pressure among adult females. *Preventive Medicine* 27(4):545–52.
8. Oxman TE, Freeman DH Jr., Manheimer ED. 1995. Lack of social participation or religious strength and comfort as risk factors for death after cardiac surgery in the elderly. *Psychosomatic Medicine* 57(1):5–15.
9. Pressman P, Lyons JS, Larson DB, Strain JJ. 1990. Religious belief, depression, and ambulation status in elderly women with broken hips. *American Journal of Psychiatry* 147(6):758–60.
10. Van Ness PH, Kasl SV, Jones BA. 2003. Religion, race, and breast cancer survival. *International Journal of Psychiatry and Medicine* 33(4):357–75.
11. Hodges SD, Humphreys SC, Eck JC. 2002. Effect of spirituality on successful recovery from spinal surgery. *Southern Medical Journal* 95(12):1381–84.
12. Yates JW, Chalmer BJ, St James P, Follansbee M, McKegney FP. 1981. Religion in patients with advanced cancer. *Medical Pediatric Oncology* 9(2):121–28. Kune GA, Kune S, Watson LF. 1992. The effect of family history of cancer, religion, parity and migrant status on survival in colorectal cancer: the Melbourne Colorectal Cancer Study. *European Journal of Cancer* 28A(8–9):1484–87.
13. Hirsch SR, (2002). *Horeb: A Philosophy of Jewish Laws and Observance* (7th ed.). London, UK: Soncino Pr Ltd, pg. 317
14. Monti DA, Bazzan AJ. 2008. *The Great Life Makeover* New York: HarperCollins.
15. Friedlander Y, Kark JD, Kaufmann NA, Stein Y. 1985. Coronary heart disease risk factors among religious groupings in a Jewish population sample in Jerusalem. *American Journal of Clinical Nutrition* 42(3):511–21.

16. Miller L, Gur M. 2002. Religiousness and sexual responsibility in adolescent girls. *Journal of Adolescent Health* 31(5):401–6.
17. Blomgren J, Martikainen P, Grundy E, Koskinen S. 2010. Marital history 1971–91 and mortality 1991–2004 in England & Wales and Finland. *Journal of Epidemiology and Community Health* 66(1):30–36.
18. Samitz G, Egger M, Zwahlen M. 2011. Domains of physical activity and all-cause mortality: systematic review and dose-response meta-analysis of cohort studies. *International Journal of Epidemiology* 40(5):1382–400.
19. McMorris T, Collard K, Corbett J, Dicks M, Swain JP. 2008. A test of the catecholamines hypothesis for an acute exercise-cognition interaction. *Pharmacology and Biochemistry in Behavior* 89(1):106–15. Hillman CH, Erickson KI, Kramer AF. 2008. Be smart, exercise your heart: exercise effects on brain and cognition. *Nature Reviews in Neuroscience* 9(1):58–65. Tomporowski PD. 2003. Effects of acute bouts of exercise on cognition. *Acta Psychologica* 112(3): 297–324.
20. Aizer AA, Chen MH, McCarthy EP, et al. 2013. Marital status and survival in patients with cancer. *Journal of Clinical Oncology* 31(31):3869–76.
21. Kaffman M, Shoham S, Palgi M, Rosner M. 1986. Divorce in the kibbutz: past and present. *Contemporary Family Therapy* 8:301–15.
22. Grundmann E. 1992. Cancer morbidity and mortality in USA Mormons and Seventh-day Adventists. *Archives d'Anatomie et de Cytologie Patholiques* 40(2–3):73–78.
23. Miller WR. 1998. Researching the spiritual dimensions of alcohol and other drug problems. *Addiction* 93(7):979–90.
24. Bai M, Lazenby M. 2015. A systematic review of associations between spiritual well-being and quality of life at the scale and factor levels in studies among patients with cancer. *Journal of Palliative Medicine* 18(3):286–98. Lucette A, Ironson G, Pargament KI, Krause N. 2016. Spirituality and religiousness are associated with fewer depressive symptoms in individuals with medical conditions. *Psychosomatics* 57(5):505–13.
25. Amihai I, Kozhevnikov M. 2015. The influence of Buddhist meditation traditions on the autonomic system and attention. *Biomedical Research International* 2015:731579. Newberg AB, Iversen J. 2003. The neural basis of the complex mental task of meditation: neurotransmitter and neurochemical considerations. *Medical Hypothesis* 61(2):282–91.
26. Buttle H. 2015. Measuring a journey without goal: meditation, spirituality, and physiology. *Biomedical Research International.* 2015:891671. doi: 10.1155/2015/891671.
27. Davidson RJ, Kabat-Zinn J, Schumacher J, et al. 2003. Alterations in

brain and immune function produced by mindfulness meditation. *Psychosomatic Medicine* 65(4):564–70.

28. Pargament KI, Koenig HG, Tarakeshwar N, Hahn J. 2001. Religious struggle as a predictor of mortality among medically ill elderly patients: a 2-year longitudinal study. *Archives of Internal Medicine* 161(15):1881–85.
29. Quoted from Health Fitness Revolution Online, March 31, 2016.
30. Brummett BH, Helms MJ, Dahlstrom WG, Siegler IC. 2006. Prediction of all-cause mortality by the Minnesota Multiphasic Personality Inventory Optimism-Pessimism Scale scores: study of a college sample during a 40-year follow-up period. *Mayo Clinic Proceedings* 81(12):1541–44.
31. Green A. 1992. *Tormented Master: The Life and Spiritual Quest of Rabbi Nahman of Bratslav.* Woodstock, VT: Jewish Lights Publishing, pp. 41–42.
32. Anson J, Anson O. 2001. Death rests a while: holy day and Sabbath effects on Jewish mortality in Israel. *Social Science and Medicine* 52(1):83–97.
33. Phillips DP, Feldman KA. 1973. A dip in death before ceremonial occasions: some new relationships between social integration and mortality. *American Sociological Review* 38:678–96.
34. Phillips DP, King EW. 1988. Death takes a holiday: mortality surrounding major social occasion. *Lancet* 24:728–32.
35. Anson J, Anson O. 2001. Death rests a while: holy day and Sabbath effects on Jewish mortality in Israel. *Social Science and Medicine* 52(1):83–97.
36. Kutz I. 2002. Samson, the Bible, and the DSM. *Archives of General Psychiatry* 59(6):565; author reply 565–66.
37. La Pierre LL. 2003. JCAHO safeguards spiritual care. *Holistic Nursing Practice* 17(4):219.
38. Stefanek M, McDonald PG, Hess SA. 2005. Religion, spirituality and cancer: current status and methodological challenges. *Psycho-Oncology* 14(6):450–63.
39. Lo B, Ruston D, Kates LW, et al. 2002. Discussing religious and spiritual issues at the end of life: a practical guide for physicians. *Journal of the American Medical Association* 287(6):749–54.
40. Daaleman TP, Nease DE, Jr. 1994. Patient attitudes regarding physician inquiry into spiritual and religious issues. *Journal of Family Practice* 39(6):564–68. King DE, Bushwick B. 1994. Beliefs and attitudes of hospital inpatients about faith healing and prayer. *Journal of Family Practice* 39(4):349–52.
41. Ellis MR, Vinson DC, Ewigman B. 1999. Addressing spiritual concerns of patients: family physicians' attitudes and practices. *Journal of Family Practice* 48(2):105–9. Armbruster CA, Chibnall JT, Legett S.

2003. Pediatrician beliefs about spirituality and religion in medicine: associations with clinical practice. *Pediatrics* 111(3):e227–35.

Chapter 9

1. Kirkpatrick LA. 1997. An attachment-theory approach to the psychology of religion. In Spilka B, McIntosh DN (eds.). *The Psychology of Religion: Theoretical Approaches*. Boulder, CO: Westview Press.
2. McIntosh DN. 1997. Religion-as-schema, with implications for the relation between religion and coping. In Spilka B, McIntosh DN (eds.). *The Psychology of Religion: Theoretical Approaches*. Boulder, CO: Westview Press.
3. Weinberger DR, Harrison P. 2011. *Schizophrenia.* 3rd ed. Oxford: Blackwell Publishing.
4. Dawkins R. 2006. *The God Delusion*. New York: Houghton Mifflin Company.
5. See Rabbi Kooks position in *The Eulogy in Yerushalayim* in *Maamarei Ha'Reiyah* pg.98 on Theodore Herzl as a possible spark of the "Messiah son of Joseph." This is based on the position (about to be discussed in the chapter) often cited in the Tradition that there will be two Messiahs, one "son of Joseph" who will start the redemption, and one "son of David" who will complete the process. In the speech, it appears clear that there is a concept for Rav Kook of a redemptive period without a specific Messiah, but that is still part of the messianic process. See also the Vilna Gaon's work "*Kol Ha-Tor* Chapter 2: Section 1:156 where he emphasizes the utility of human initiative in the messianic process.
6. For a further analysis of this topic, see Weiderblank N. (2018). 26.10 Appendix A: Kol Ha-Tor, R. Kook, and the Role of Secular Zionism in the Process of Redemption. In *Illuminating Jewish Thought: Explorations of Free Will, the Afterlife, and the Messianic Era.* New Milford, CT: Maggid Books.
7. See *Igrot Ha'Rav Kook* (Letters of Rabbi Kook) Part 3 pg. 155
8. MacKenzie MB, Kocovski NL. 2016. Mindfulness-based cognitive therapy for depression: trends and developments. *Psychological Research and Behavioral Management* 19(9):125–32. Koenig HG, Pearce MJ, Nelson B, Erkanli A. 2016. Effects on daily spiritual experiences of religious versus conventional cognitive behavioral therapy for depression. *Journal of Religion and Health* 55(5):1763–77.
9. Kelly JF, Stout RL, Magill M, Tonigan JS, Pagano ME. 2011. Spirituality in recovery: a lagged mediational analysis of Alcoholics Anonymous'

principal theoretical mechanism of behavior change. *Alcoholism, Clinical and Experimental Research* 35(3):454–63.

10. Friedman M. 2002–2003. Psychotherapy and teshuvah: Parallel and overlapping systems for change. *Torah U-Madda Journal* 11:238–53.
11. Devinsky O, Lai G. 2008. Spirituality and religion in epilepsy. *Epilepsy and Behavior* 12(4):636–43.
12. Jasper H, Penfield W. 1954. *Epilepsy and the Functional Anatomy of the Human Brain.* 2nd ed. London: Little, Brown and Co.
13. Booth JN, Koren SA, Persinger MA. 2005. Increased feelings of the sensed presence and increased geomagnetic activity at the time of the experience during exposures to transcerebral weak complex magnetic fields. *International Journal of Neuroscience* 115(7):1053–79.
14. Granqvist P, Fredrikson M, Unge P, Hagenfeldt A, Valind S, Larhammar D, Larsson M. 2005. Sensed presence and mystical experiences are predicted by suggestibility, not by the application of transcranial weak complex magnetic fields. *Neuroscience Letters* 379(1):1–6.
15. Saver JL, Rabin J. 1997. The neural substrates of religious experience. *Journal of Neuropsychiatry and Clinical Neurosciences* 9(3):498–510.
16. Damasio A. 1994. *Descartes Error: Emotion, Reason, and the Human Brain.* New York: Avon Books.
17. Bullock WA, Gilliland K. 1993. Eysenck's arousal theory of introversion-extraversion: A converging measures investigation. *Journal of Personality and Social Psychology* 64(1):113–23.
18. Kumari V, ffytche DH, Williams SC, Gray JA. 2004. Personality predicts brain responses to cognitive demands. *Journal of Neuroscience* 24(47):10636–41.
19. Forsman LJ, de Manzano O, Karabanov A, Madison G, Ullén F. 2012. Differences in regional brain volume related to the extraversion-introversion dimension—a voxel based morphometry study. *Neuroscience Research* 72(1):59–67.
20. Leutgeb V, Leitner M, Wabnegger A, Klug D, Scharmüller W, Zussner T, Schienle A. 2015. Brain abnormalities in high-risk violent offenders and their association with psychopathic traits and criminal recidivism. *Neuroscience* 308:194–201.
21. Kuhn CM, Schanberg, SM. 1998. Responses to maternal separation: mechanisms and mediators. *International Journal of Developmental Neuroscience* 16:261–70. Black JE. 1998. How a child builds its brain: some lessons from animal studies of neural plasticity. *Preventive Medicine* 27:168–71.
22. Huttenlocher PR, deCourten C. 1987. The development of the striate cortex in man. *Human Neurobiology* 6:1–9. Huttenlocher PR. 1979.

Synaptic density in human frontal cortex: developmental changes and effects of aging. *Brain Research* 163:195–205.

23. Chugani HT, Phelps ME, Mazziotta JC. 1987. Positron emission tomography study of human brain functional development. *Annals of Neurology* 22:487–97.
24. Chugani HT, Phelps ME, Mazziotta JC. 1989. Metabolic assessment of functional maturation and neuronal plasticity in the human brain. In von Euler C, Forssberg H, Lagercrantz H (eds.). *Neurobiology of Early Infant Behavior*. New York: Stockton Press.
25. d'Aquili EG, Newberg AB. 1999. *The Mystical Mind: Probing the Biology of Religious Experience*. Minneapolis, MN: Fortress Press.
26. See Maimonides *Mishneh Torah* Laws of Repentance 7:7 as well as Soloveitchik JD, Peli PH (1974). *Al Hateshuva*. Jerusalem: World Zionist Organization Dept. of Torah Education and Culture in the Diaspora (pp. 170). See also Bereishit Rabbah 22:17, Talmud *Yoma* 86b, and Rav Kook *Orot Ha'teshuva* 7:1 for further elucidation of this topic and perspective on repentance.
27. Soloveitchik J.D., Wolowelsky J.B., & Ziegler R. (2017). *Halakhic Morality: Essays on Ethics and Masorah*. New Milford, CT, USA: Maggid Books. p.50.
28. Cited in Soloveitchik JB. *Halakhic Morality: Essays on Ethics and Mesorah.* p. 51.
29. *Ibid* pp.51-51
30. Soloveitchik J.D., Wolowelsky J.B., & Ziegler R. (2017). Halakhic Morality: Essays on Ethics and Masorah. New Milford, CT, USA: Maggid Books. p. 103
31. Soloveitchik JB. *Halakhic Morality: Essays on Ethics and Mesorah*. p. 78.
32. *Ibid* p. 103

Chapter 10

1. See Weiderblank N. (2018). *Illuminating Jewish Thought: Explorations of Free Will, the Afterlife, and the Messianic Era.* New Milford, CT: Maggid Books for an in-depth analysis of traditional Jewish sources on the nature of the Jewish soul.
2. Kaplan JT, Freedman J, Iacoboni M. 2007. Us versus them: political attitudes and party affiliation influence neural response to faces of presidential candidates. *Neuropsychologia* 45(1):55–64.
3. See Soloveitchik, J. B. (1966). Sacred and Profane, Kodesh and Chol in World Perspective. *Gesher*,3(1), 5-29.
4. Radua J, Del Pozo NO, Gómez J, Guillen-Grima F, Ortuño F. 2014. Meta-analysis of functional neuroimaging studies indicates that an increase

of cognitive difficulty during executive tasks engages brain regions associated with time perception. *Neuropsychologia* 58:14–22.

5. Mahy CE, Moses LJ, Pfeifer JH. 2014. How and where: theory-of-mind in the brain. *Developmental Cognitive Neuroscience* 9:68–81.
6. James W. [1890] 1963. *Varieties of Religious Experience*. New York: University Books.
7. De Chardin P; Wall B (trans.). 1975. *The Phenomenon of Man*. New York: HarperCollins.
8. See Rav Kook's Chazon Hatzimchanut V'Hashalem (Vision of Vegetarianism and Peace)
9. Brown WS, Murphy N, Malony HN. 1998. *Whatever Happened to the Soul?* Minneapolis: Fortress Press.
10. Soloveitchik, J, Wolowelsky J, Ziegler R. 2017. *Halakhic Morality: Essays on Ethics and Masorah*. New Milford, CT: Maggid Books.p.197.
11. *Ibid* pp.197-198
12. Newberg A, Wintering NA, Morgan D, Waldman MR. 2006. The measurement of regional cerebral blood flow during glossolalia: a preliminary SPECT study. *Psychiatry Research: Neuroimaging* 148(1):67–71. Newberg AB, Wintering NA, Yaden DB, Waldman MR, Reddin J, Alavi A. 2015. A case series study of the neurophysiological effects of altered states of mind during intense Islamic prayer. *Journal of Physiology: Paris*. pii: S0928–4257(15)00018–2. doi: 10.1016/j.jphysparis.2015.08.001.
13. Newberg AB, Wintering NA, Yaden DB, Waldman MR, Reddin J, Alavi A. 2015. A case series study of the neurophysiological effects of altered states of mind during intense Islamic prayer. *Journal of Physiology: Paris*. 109; 214–220.
14. Piyadassi T. 1964. *The Buddha's Ancient Path*. London: Rider, p. 12.
15. Obeyesekere G. 1968. Theodicy, sin, and salvation in a sociology of Buddhism. In Leach ER (ed.). *Dialectic in Practical Religion*. Cambridge: Cambridge University Press, pp. 7–40.
16. Tatz A, Gottlieb D. 2005. *Letters to a Buddhist Jew*. Southfield, MI: Targum Press, pp. 274–75.

Chapter 11

1. Armstrong K. 1993. *A History of God*. New York: Ballantine.
2. Campbell J. 1972. *Myths to Live By*. New York: Viking Press.
3. Dunning HR. 1988. *Grace, Faith & Holiness: A Wesleyan Systematic Theology*. Kansas City: Beacon Hill Press.
4. Pope WB. 2016. *A Compendium of Christian Theology*. Sydney: Wentworth Press.
5. Hodge C. 1999. *Systematic Theology*. Peabody, MA: Hendrickson

Publishers.
6. Hopkins EH. 2009. *Systematic Theology*. Valley Forge, PA: Judson Press.
7. Barth K. 1991. *The Gottingen Dogmatics*. Grand Rapids, MI: Wm. B. Eerdmans Publishing Co.
8. Deane SN. 2001. *Anselm Basic Writings*. Peru, IL: Open Court Publishing Co.
9. Soloveitchik J.D., Wolowelsky J.B., & Ziegler R. (2017). Halakhic Morality: Essays on Ethics and Masorah. New Milford, CT, USA: Maggid Books. P.179
10. Lichtenstein A. 2004. *Leaves of Faith: The World of Jewish Learning.* Vol. 2. Jersey City, NJ: Ktav, pp. 50–51.
11. Amital Y. *Jewish Values in a Changing World.* Jersey City, NJ: Ktav, 2005, p. 23.
12. K Kohler, K. (1906). Revelation. In The Jewish Encyclopedia: A Descriptive Record of the History, Religion, Literature, and Customs of the Jewish People from the Earliest Times to the Present Day (Vol. 10, pp. 396–397). New York, NY: Funk & Wagnalls Co. http://www.jewishencyclopedia.com/articles/12713-revelation
13. *ibid*
14. *ibid*
15. *ibid*
16. See *Tanya: Shaar HaYichud* (Gate of Unity) Chapter 10
17. Spinoza B; Shirley S (trans.). 1981. *Tractatus Theologico-Politicus*. New York: Leiden.
18. From Sefaria.org version
19. Soloveitchik JB. 2006. *Lonely Man of Faith*. New York: Doubleday, p. 82.
20. Buber M; Kaufman W, (trans.). 1996. *I and Thou*. New York: Touchstone.
21. Schopenhauer, A, Haldane RB, Kemp J (trans.). 1961. *The World as Will and Idea*. Garden City, NY: Doubleday.

Chapter 12

1. Kane R (ed.). 2002. *Oxford Handbook of Free Will*. New York: Oxford University Press.
2. Libet B, Gleason CA, Wright EW, Pearl DK. 1983. Time of conscious intention to act in relation to onset of cerebral activity (Readiness-Potential). *Brain* 106(3):623–42.
3. Mele AR. 2008. Psychology and free will: a commentary. In Baer J, Kaufman JC, Baumeister RF (eds.). *Are We Free? Psychology and Free Will.* New York: Oxford University Press, pp. 325–46.

4. See Quote from Rav Tzadok HaCohen of Lublin in *Pri Tzaddik:* Bamidbar (Numbers) From Rosh Chodesh Menachem Av 4.236-237. For further analysis see Weiderblank N. 2018. 14.5 Kabbalistic Perspectives: R. Tzadok and R. Kook. In *Illuminating Jewish Thought: Explorations of Free Will, the Afterlife, and the Messianic Era.* New Milford, CT: Maggid Books.
5. Harris D., *The Moral Landscape*. 2010. New York: Free Press.
6. See again Weiderblank N. 2018. *Illuminating Jewish Thought: Explorations of Free Will, the Afterlife, and the Messianic Era.* New Milford, CT: Maggid Books for a rather complete discussion on the varying traditional perspectives in Jewish philosophy on free will.
7. Soloveitchik J.B., and Berger M.S. The Emergence of Ethical Man. Ktav Pub. House, 2005.
8. Carmy S. 2006. Use it or lose it: on the moral imagination of free will. In *Judaism, Science, and Moral Responsibility*. The Orthodox Forum. New York: Rowman & Littlefield, pp. 104–51.
9. Greene JD, Nystrom LE, Engell AD, Darley JM, Cohen JD. 2004. The neural bases of cognitive conflict and control in moral judgment. *Neuron* 44(2):389–400. Xue SW, Wang Y, Tang YY. 2013. Personal and impersonal stimuli differentially engage brain networks during moral reasoning. *Brain and Cognition* 81(1):24–28.
10. Lim J, Kurnianingsih YA, Ong HH, Mullette-Gillman OA. 2017. Moral judgment modulation by disgust priming via altered fronto-temporal functional connectivity. *Science Reports* 7(1):10887. doi: 10.1038/s41598-017-11147-7. Decety J, Michalska KJ, Kinzler KD. 2012. The contribution of emotion and cognition to moral sensitivity: a neurodevelopmental study. *Cerebral Cortex* 22(1):209–20.
11. Newberg AB, d'Aquili EG, Newberg SK, DeMarici V. 2000. The neuropsychological basis of forgiveness. In McCullough ME, Pargament KI, Thoresen CE (eds.). *Forgiveness: Theory, Practice, and Research.* New York: Guilford Press, pp. 91–110.
12. Eisenberger NI. 2012. The neural bases of social pain: evidence for shared representations with physical pain. *Psychosomatic Medicine* 74(2):126–35.
13. Palagi E, Norscia I. 2015. The season for peace: reconciliation in a despotic species (Lemur catta). *PLoS One* 10(11):e0142150. doi: 10.1371/journal.pone.0142150. Holobinko A, Waring GH. 2010. Conflict and reconciliation behavior trends of the bottlenose dolphin (Tursiops truncatus). *Zoo Biology* 29(5):567–85.
14. Gazzaniga MS. 2005. *The Ethical Brain.* New York: Dana Press.

15. Lichtenstein A. *Leaves of Faith*. Vol. 2. Jersey City, NJ: Ktav, 2004, pp. 34, 48.

Chapter 13

1. Kook A., Na'or B. (2000). *In the desert--a vision : (Midbar Shur) : thoughts on the portion of the week and the holidays delivered by Rabbi Abraham Isaac Hakohen Kook in Zoimel, 5654–5656 (1894–1896): Toldot (37–39)*. Spring Valley, NY: Orot. Schneersohn, S. D. (2012). Overcoming Folly - Kuntres Umaayon (Chasidic Heritage Series). Pgs 64–67. Brooklyn, NY: Kehot Publication Society.
2. d'Aquili EG, Newberg AB. 1999. *The Mystical Mind: Probing the Biology of Religious Experience*. Minneapolis, MN: Fortress Press.
3. Bucke RM. 1961. *Cosmic Consciousness*. Secaucus, NJ: Citadel Press.
4. MacLean KA, Johnson MW, Griffiths RR. 2011. Mystical experiences occasioned by the hallucinogen psilocybin lead to increases in the personality domain of openness. *Journal of Psychopharmacology* 25(11):1453–61.
5. Yaden DB, Nguyen KDL, Kern ML, Belser AB, Eichstaedt JC, Iwry J, Wintering N, Hood RW, Newberg AB. 2016. Of roots and fruits: a comparison of psychedelic and non-psychedelic mystical experiences. *Journal of Humanistic Psychology* 57(4):338–53.
6. Dalai Lama, Jinpa T. 2005. *The Essence of the Heart Sutra: The Dalai Lama's Heart of Wisdom Teachings*. Somerville, MA: Wisdom Publications.
7. Leet L. 2003. *The Kabbalah of the Soul*. Rochester, VT: Inner Traditions.
8. Collins FS. 2006. *The Language of God.* New York: Free Press.
9. See *Mei Shiloach* by Rabbi Mordechai Izbizher on Exodus: Volume 2—Exodus: Yitro 20:20 #16 (on verse 20).
10. See discussion by Rabbi Jonathan Sacks in his "Covenant & Conversation Exodus: The Book of Redemption" Section on Parshat Yitro - 17 January 2011 - 17 Shevat 5771 "Deed and Creed"
11. d'Aquili EG, Newberg AB. 1999. *The Mystical Mind: Probing the Biology of Religious Experience*. Minneapolis, MN: Fortress Press.

Index

C

D

E

I

J

K

L

M

N

S

ANDREW B. NEWBERG M.D. is Director of Research at the Marcus Institute of Integrative Health at Thomas Jefferson University Hospital and Medical College and has published over two hundred articles, essays, and book chapters. He is considered a pioneer in the field of neurotheology which seeks to link neuroscience with religious and spiritual experience. Dr. Newberg has published over two hundred articles, essays and book chapters. He is the author and co-author of eight books including the bestselling, *How God Changes Your Brain* (Ballantine, 2009), and *Why God Won't Go Away* (Ballantine, 2001.)

RABBI DAVID HALPERN, M.D. is a resident in the Department of Psychiatry and Human Behavior at the Thomas Jefferson University Hospital and Medical College. Dr. Halpern received Rabbinic Ordination from Rabbi Isaac Elchanan Theological Seminary of Yeshiva University, where he studied Jewish law, philosophy, and their intersection with the field of medicine. He collaborated with Dr. Newberg at the Marcus Institute of Integrative Health on the topic of Jewish Neurotheology to create this book *The Rabbi's Brain.*